Springer Tracts in Mechanical Engineering

About this Series

Springer Tracts in Mechanical Engineering (STME) publishes the latest developments in Mechanical Engineering - quickly, informally and with high quality. The intent is to cover all the main branches of mechanical engineering, both theoretical and applied, including:

- Engineering Design
- Machinery and Machine Elements
- Mechanical structures and stress analysis
- Automotive Engineering
- Engine Technology
- Aerospace Technology and Astronautics
- Nanotechnology and Microengineering
- Control, Robotics, Mechatronics
- MEMS
- Theoretical and Applied Mechanics
- Dynamical Systems, Control
- Fluids mechanics
- Engineering Thermodynamics, Heat and Mass Transfer
- Manufacturing
- Precision engineering, Instrumentation, Measurement
- Materials Engineering
- Tribology and surface technology

Within the scopes of the series are monographs, professional books or graduate textbooks, edited volumes as well as outstanding PhD theses and books purposely devoted to support education in mechanical engineering at graduate and post-graduate levels.

More information about this series at http://www.springer.com/series/11693

Qingjie Cao
Alain Léger

A Smooth and Discontinuous Oscillator

Theory, Methodology and Applications

Springer

Qingjie Cao
School of Astronautics
Harbin Institute of Technology
Harbin
China

Alain Léger
Laboratoire de Mécanique et d'Acoustique, CNRS
Marseille Cedex 13
France

ISSN 2195-9862 ISSN 2195-9870 (electronic)
Springer Tracts in Mechanical Engineering
ISBN 978-3-662-53092-4 ISBN 978-3-662-53094-8 (eBook)
DOI 10.1007/978-3-662-53094-8

Library of Congress Control Number: 2016947202

Printed on acid-free paper

This Springer imprint is published by Springer Nature
The registered company is Springer-Verlag GmbH Berlin Heidelberg

*The original version of the book was revised:
For detailed information please see erratum.
The erratum to this book is available
at DOI 10.1007/978-3-662-53094-8_17*

Foreword

I am proud to write the Foreword for this volume by Prof. Qingjie Cao and his co-author. I came to know Prof. Qingjie Cao from the last decade beginning of this century at Aberdeen, UK, Kyoto, Japan, St. Petersburg, Russian, and Harbin China. I observed his sincere attitude for nonlinear dynamics, especially the creation of the smooth and discontinuous (SD) oscillator. This work opened my eyes to the phenomena in both smooth and discontinuous dynamics. As for my knowledge of nonlinear dynamics, I felt something fresh in the creation of a discontinuous oscillator and a generalized Hooke's law due to hyperbolicity transition. So, perhaps a personal word is in order here on the creation of the SD oscillator, the theory and applications of which will give a better knowledge of the natural world.

Many works have been published, many books and many papers as well, which deal with the qualitative dynamics of systems arising from any field of natural science, either in the case of simple models, or in the case of large-scale systems, even in continuous media such as those dealt with in fluid mechanics.

This book comes back to a very simple mass spring oscillator, followed to wisdom "To know something new to visit the old." A case at smaller strains is so simple that it might be included in any undergraduate class on vibrations, oscillations, pre-buckling, etc. A case at larger strains is that nonlinear terms are usually introduced which could deserve to represent any model of nonlinearity. This allows to extend the analysis of nonlinear vibrations or post-buckling theories. In the present case, the nonlinearity arises from the changes of the geometry. What was new and really impressive in the so-called SD oscillator was the simple idea that taking the exact nonlinearity without any approximation might lead not only to a better approximation but also to new phenomena. At first, no restriction is required for the analysis to give the exact trajectory. Then the authors observe that changing a parameter from a given nonzero value to zero, which amounts to modifying the geometry and is associated with the name "Smooth to Discontinuous" changes the qualitative properties.

This oscillator seems to have really new properties and gives a new step in the understanding of nonlinear phenomena. During the last decade, the authors accumulate a huge quantity of facts and statistics collected together for reference or analysis due to a large number of theoretical and experimental studies. The book they wrote is aimed to share some of these findings.

It starts with the basic description of the mechanical system with a survey of its new properties. Then it investigates in detail each of the announced properties, backs up the results by several different ways for numerical calculations, and closes by experiments and engineering applications.

The book uses either simple tools as long as they are sufficient to give perfectly rigorous results, or difficult and recent mathematical concepts when necessary. It will certainly be very useful. Looking back, fifty years after the beginning of chaos in a nonlinear circuit determined by a special Duffing system, I am extremely pleased to see that progresses continue to being made, even through the behavior of very simple mechanical systems.

As a last comment, the classical chapter on challenges and future works, which is often conventional and formal, includes here very interesting suggestions, some of them of a great theoretical importance.

Kyoto
November 2016

Yoshisuke Ueda

Preface

The most interesting phenomenon in nonlinear dynamics and certainly one of the most important changes in the understanding of the world in the last five or six decades, is the discovery of chaos. In 1961, Yoshisuke Ueda discovered the so-called Ueda Attractor as a young researcher using an analog computer when he was studying the responses of a nonlinear circuit described by a special type of Duffing oscillator. This discovery undoubtedly changed the way of looking at the world, which until then was perceived either deterministic or random. It is now in the common knowledge that it has had fundamental effects in many theories including oscillations, wave propagation and control, in any field of physics, in fluid mechanics, in meteorology, in astronomy, in biology, in economics, in population dynamics and others. In essence, the world is really concerned with the discovery of chaos as a milestone in current knowledge and research perspectives.

In 1963, Edward Lorenz proposed a new model of the weather system, which became known as the Lorenz attractor by trying to extract the main properties of the atmospheric turbulence through a truncation of Navier–Stokes equations. This model of turbulence followed its previous observations of the extreme sensitivity to initial conditions. It is now associated with the so-called butterfly effect, which again brought a new outlook on what was previously taken as randomness. A few years later, a simplified model having the same objective of understanding turbulence has been proposed using iterates of maps and gave the Henon attractor. Lots of works then followed in the fields of mathematics and theoretical physics to improve our understanding of the mechanisms which lead to chaos, period doubling bifurcations, transitions to torus of higher dimensions, intermittency, and others nonlinear phenomena.

An archetypal mechanical oscillator was firstly proposed in 2006, which has the fundamental property of being either continuous and smooth or discontinuous depending on the value of a certain parameter. This system is now defined as the SD oscillator to refer to this property with the generalized Hooke's law where the stiffness can be positive, negative or zero (quasi-zero) depending on the geometrical configuration of the system. It is a simple mass-spring system in which the only difference with what has been done before, for example in vibration or buckling models, is that the change of the geometry during the motion was taken into account without any approximation whatever in the displacement or on the frequency, exactly in the same way as if the equation of a simple pendulum keeps $\sin x$ in the right-hand side instead of any polynomial expansion.

The SD oscillator can be seen as an archetypal system, by which we mean that it is a prototype of a dynamical system where the nonlinearity is irrational. This firstly follows from the fact that there are no approximations in the large displacements, and secondly that the nonlinearity is either smooth, as long as a geometrical parameter related to the relative positions of the spring and the mass is different from zero, or discontinuous when the geometrical parameter reaches zero. In the smooth regime, it bears significant similarities with classical types of nonlinear oscillators, but at the nonsmooth limit it involves substantial departures from the standard ones. The ability of the SD oscillator to provide a new outlook on nonlinear phenomena relies upon any feature of qualitative analyses: sets and properties of equilibria, periodic solutions, co-dimension bifurcations, and chaotic attractors, etc., all these features involving the transition when the nonlinearity changes from smooth to discontinuous. Another feature of this oscillator is the transition from a single stability to a bistability system as the geometrical parameter crosses the critical value, where the stiffness at the origin can be positive, or negative depending on the value of the geometrical parameter. This is always known as the generalized Hooke's law.

The main purpose of this book is to provide an unconventional way to understand qualitative phenomena of the natural world, through the dynamics of a simple mechanical system which may have either a smooth or a nonsmooth nonlinearity. In other words, following the steps of the discovery of chaos, it aims at showing that a deep analysis of an appropriate simple model may have important consequences in the understanding of the world. This book is also strongly motivated by an invitation to professionals in science and engineering to pay special attention to the modeling, the analysis, and the applications of nonlinear phenomena. The authors would like to give a real encouragement to students and researchers who are interested in developing a cross-discipline of nonlinear science at the intersection of mathematics, theoretical physics, and engineering sciences.

In addition to the collaborations indicated at the beginning of each chapter, the authors would like to express their utmost thanks to the following Doctors and Professors, whatever the specificity of their contribution:

Enli Chen
Yushu Chen
Ming Feng
Shengliang Fu
Celso Grebogi
Kai Guo
Ning Han
Zhifeng Hao

Ling Hong
Zhixin Li
Mei Lin
Guanwei Luo
Cunzhi Pan
Ekaterina Pavlovskaia
Elaine Pratt
J. Michael T. Thompson
Ruilan Tian
Dan Wang
Jianhua Wang
Marian Wiercigroch
Jingtang Xing
Yeping Xiong
Shaopu Yang
Xinwei Yang
Xiaoyan Zhang
Yongxiang Zhang

Harbin, China
Marseille, France
October 2016

Qingjie Cao
Alain Léger

Acknowledgements

The first author would like to acknowledge the financial support from Natural Science Foundation of China (Grant No. 10872136, 11072065, 11372082 and 11572096) and the National Basic Research Program of China (Grant No. 2015CB057405).

Contents

Chapter 1
Introduction

This introduction involves three parts. The first part presents the brief historical points about the development of oscillators which are the critical important impact on the history of nonlinear dynamics. The second part shows several examples which might help readers to understand the limitation of the classical theory of nonlinear dynamics, which stresses the reader's attention that it is necessary to be careful when truncating a nonlinear system into a polynomial one, and we give the outline of this book.

1.1 Historical Points

Classical oscillators have been widely used in modern sciences and technologies. They are systems that work on the principle of oscillations with a periodic fluctuation between different states. This is based on energy exchange and plays an unique role in the theory of nonlinear dynamics and its practical applications.

The simplest oscillator is the harmonic one which is made of a linear spring and a lumped mass vibrating around its equilibrium position according to Hooke's law. This gives the linear relationship between actions (the applied external force) and reactions (the mass displacement), that is the deformation of the spring is proportional to the external applied force. From a theoretical point of view, the concept of stiffness has been firstly proposed by Robert Hooke [1], whose name is associated with the linear relationship between the deformation of the spring and the force applied to it. At the time when Hooke's law and Newton's three laws of motion were established, the mechanism of harmonic vibration was not fully understood. Brothers James and John Bernoulli created the fundamental framework to study mechanics. James Bernoulli [2] developed the force balance technique to formulate the equation of motion of the spring-mass system using the Hooke's law and the Newton's second law of motion [3]. John Bernoulli determined the natural frequency as the square root of the ratio

Q. Cao and A. Léger, *A Smooth and Discontinuous Oscillator*, Springer Tracts in Mechanical Engineering, DOI 10.1007/978-3-662-53094-8_1

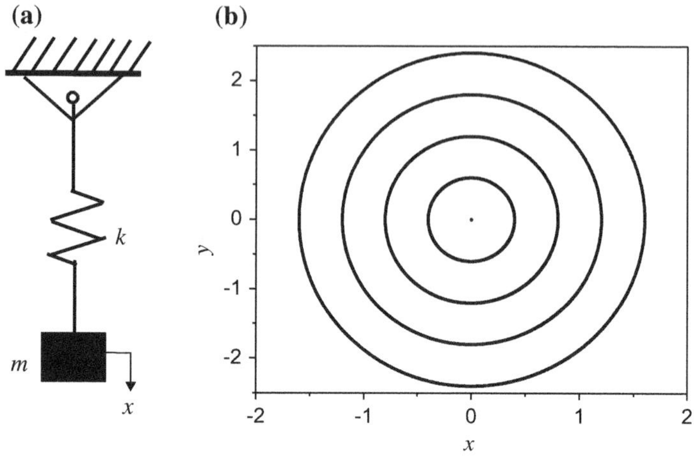

Fig. 1.1 **a** Harmonic oscillator and **b** its phase portraits comprised of a family of periodic trajectories surrounding the equilibrium point (0, 0)

between the stiffness and the mass, details seen in [4], which is apparently the first theoretical formulation of the harmonic oscillator.

The abstract harmonic oscillator with a lumped mass m and a linear spring of stiffness k is shown in Fig. 1.1a. Its equation of motion is given as

$$\ddot{x} + \omega_n^2 x = 0, \tag{1.1}$$

where x is the displacement and $\omega_n = \sqrt{\frac{k}{m}}$ is the natural frequency. The Hamiltonian of this simple system reads

$$H(x, y) = \frac{1}{2}y^2 + \omega_n^2 \frac{1}{2}x^2 = E,\ y = \dot{x}, \tag{1.2}$$

which leads to a family of harmonic solutions of frequency ω_n. The structure of the phase portrait in the $(x, \dot{x})$ plane is shown in Fig. 1.1b.

Leonhard Euler [5] wrote down the equation of motion for the case where the oscillator is harmonically forced with a driving frequency ω and without damping, which is expressed by the following equation

$$m\ddot{x} + kx = F \cos \omega t, \tag{1.3}$$

in which the resonant condition then appeared through the ratio $\Omega = \frac{\omega}{\omega_n} = 1, \omega_n^2 = \frac{k}{m}$. Hermann Von Helmholtz [6] introduced a displacement involving quadratic term into the harmonically excited oscillator, which reads

$$m\ddot{x} + k_1 x + k_2 x^2 = F \cos \omega t, \tag{1.4}$$

which is now referred to as the Helmholtz equation. The works done after Hermann Von Helmholtz, by John William Strutt, Third Baron Rayleigh, investigated the free vibration of the oscillator with a cubic nonlinear term, that is

$$m\ddot{x} + k_1 x + k_2 x^3 = 0, \tag{1.5}$$

which is identical to the Duffing system without damping and forcing. Details can be seen in [7].

The most important work on nonlinear oscillators is due to Georg Duffing [8] who introduced a harmonically excited nonlinear oscillator with both a quadratic and a cubic stiffness into the initial linear oscillator in order to understand some effects observed in many problems of solid mechanics. He is the first person who introduced the harmonic excitation to investigate the pseudo-harmonic oscillations and who obtained the frequency responses with the nonlinear phenomena such as jumps and hysteresis. Georg Duffing also constructed an experimental rig to verify his analytical results. He used a heavy pendulum driven by a harmonic force of which the equation of motion has a restoring force consisting of a linear and cubic terms with respect to the angular displacement for the symmetrical case, or an additional quadratic term for the asymmetric case. The equation is written as

$$\ddot{x} + \alpha x - \beta x^2 - \gamma x^3 = k \sin \omega t, \tag{1.6}$$

which is currently known as the harmonically excited undamped Duffing oscillator. In general there are several different forms of Duffing equation, but the standard one [9] frequently used in the literature is the symmetrical one which reads

$$\ddot{x} + 2\xi \dot{x} + \alpha x + \gamma x^3 = f_0 \cos \omega t, \tag{1.7}$$

This oscillator exhibits several new features that have been demonstrated in literature such as quasi-zero stiffness when $\alpha = 0$, single stability when $\alpha > 0, \gamma > 0$, bistability when $\alpha < 0, \gamma > 0$, single stability with a pair of saddle points or a pendulum type dynamics when $\alpha > 0, \gamma < 0$ and instability with hyperbolicity when $\alpha < 0, \gamma \leq 0$, as $\xi = 0$ and $f_0 = 0$ is taken..

It is now believed that Duffing equation has been one of the most important contribution to the fundamental research and practical applications [10]. This was a milestone in the foundation of the theory of nonlinear dynamics. This equation has been extensively studied and played a very important role in the area of nonlinear dynamics. Yoshisuke Ueda [11, 12] first discovered chaos by constructing a chaotic attractor which was named as the Ueda attractor, on 27 November 1961 at Kyoto University. At that time Ueda was a postgraduate student working on the forced Duffing oscillator with only a cubic nonlinearity. Nayfeh [13] gave the bifurcations in a forced Duffing oscillator [14, 15] with a softening stiffness. Guckenheimer and Holmes [16] studied a double-well Duffing oscillator and provided the simplest model for the forced vibration of a cantilever beam in the nonuniform field of two permanent magnets.

There are other oscillators which have contributed significantly to the development of nonlinear dynamics. Specifically, Van der Pol [17] when studying electric circuits developed a new dynamical system called the Van der Pol oscillator, involving a nonlinear damping arising from the modeling of a triode valve. Lorenz [18] presented an analysis of a coupled set of three quadratic ordinary differential equations, which is known as the Lorenz system, representing three modes of the Oberbeck-Boussinesq equations for fluid convection in a two-dimensional layer. Guanrong Chen [19] and Jinhu Lü [20] proposed a generalized Lorenz family, whose strange attractors were observed from a physical electronic circuit with a single nonlinearity. These works not only enriched the fundamental aspects of nonlinear dynamics but also stimulated applications in a wide range in engineering disciplines.

1.2 Applications and Questions

To our knowledge one of the earliest studies of nonlinear phenomena is the motion of a simple pendulum and which has greatly influenced the history of physics. Galileo discovered the law of isochronism [21], which marked the beginning of scientific research on pendulums. Later on, in 1674 Huygens discovered that large pendulum oscillations deviate from the law of isochronism which is the remarkable nonlinear phenomena [22]. The simplest pendulum is the mathematical pendulum shown in Fig. 1.2a and its equation of motion reads:

$$\ddot{x} + \frac{g}{\ell} \sin x = 0, \tag{1.8}$$

where constants g and ℓ are the gravitational acceleration and the length of the pendulum.

The structure of the phase portraits of the pendulum can be determined by solving Eq. (1.8) directly, which results in the Hamiltonian $H(x, y) = \frac{1}{2}y^2 + \frac{g}{l}(\cos x - 1) = E$ with different levels of the energy E. The constant $E = 0$ means the stable equilibria of center points $(\pm 2n\pi, 0)$, $n = 0, 1, 2, \ldots,$; the case $0 < E < \frac{g}{l}$ presents the family of oscillating solutions or rotations with small amplitudes; $E = \frac{g}{l}$ shows the unstable equilibria of saddle points $(\pm(2n-1)\pi, 0)$, $n = 1, 2, \ldots,$ with heteroclinic orbits (separatrix) connecting them; and $E > \frac{g}{l}$ gives the rotating solutions or oscillations with large amplitudes, as shown in Fig. 1.2b.

Very often, the pendulum (1.8) is linearized with $\sin x \approx x$ model oscillations of small amplitudes. For oscillations of larger amplitudes, it is accepted that a truncation resulting a Duffing oscillator [8] by letting $\sin x \approx x - \frac{1}{3!}x^3$ is appropriate. This is often done to connect oscillations with the heteroclinic orbit, to obtain the separatrix between oscillating and rotating solutions. This technique is quite common to transform an original nonlinear system into a Duffing oscillator, which is often seen as a good approximation. For example, Tufillaro [23] who investigated the oscillations of an elastic string by assuming that the string behaves as a Duffing type stiffness

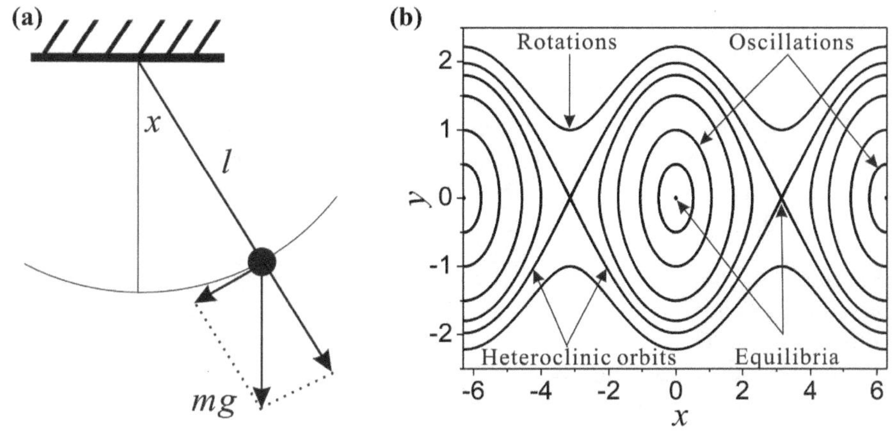

Fig. 1.2 **a** Mathematical pendulum and **b** phase portrait structure for pendulum: with oscillations, rotations and heteroclinic orbits for different choices of E., the details seen in the contents

and undergoes pseudo harmonic oscillations. In the paper by Holmes [24] a Duffing oscillator is introduced to approximate a double-well system called the Moon beam [25]. This system consists of a cantilever steel beam hung vertically, with a free end attracted by two magnets. Assuming that a single-mode approximation can be made for the motion, the benefits of this approach are apparent for oscillators since polynomial resistances can be easily treated by the Jacobian elliptic functions with simple formulations.

Then the fundamental question that can be addressed now is the following: what are the differences between the dynamics of the simple mathematical pendulum and that of an oscillator obtained by truncating at order resulting in a harmonic oscillator, or at order three yielding a Duffing oscillator truncated. Do the differences still remain for truncation at any order? Or in other words, in which way the nonlinear dynamics of the truncated system can be exactly the same as the one of the original system?

These questions have been studied extensively by a large number of scientists (see e.g. the books by Golubitsky et al. [26] and by Holmes [24]). The latest development in the theory of nonlinear dynamics, based upon an assumption of local smoothness, provides a theoretical framework to justify approximations by a polynomial expansion into so-called normal form, see [27, 28] for details. The kernel of this theory includes recognition, classification and finite determinacy of the singularities, which lead to the universal unfolding theory. These abundant analyzes lead to the answer that the truncated dynamics is equivalent to that of the original system if the truncation satisfies the following strict conditions:

1. smoothness assumption: the untruncated (original) system is differentiable to the required order;

2. locality assumption: the equivalence is valid locally even if global information is included in some cases;
3. finite determinacy: a finite truncation guarantees the equivalence qualitatively.

It is suggested that special care must be taken when it is not assured that all these conditions are satisfied. Now, these conditions stand for a strong motivation of our work presented in the following chapters. We have chosen two examples to provide a deeper understanding of these conditions.

Example 1. Consider the pendulum (1.8) again, with large angular motions, specifically, rotating solutions and separatrix. These nonlinear features are obviously much more interesting than the small oscillations, but they cannot be described by any finite order power expansion. As a matter of fact, for any finite $N = 2k - 1; k > 0$, the power expansion of $\sin x$ at the order N, gives

$$\ddot{x} - \frac{g}{\ell}\left(x - \frac{x^3}{3!} + \cdots + (-1)^k \frac{x^{2k-1}}{(2k-1)!}\right) = 0, \qquad (1.9)$$

but the dynamics of the truncated system may exhibit strong divergences with the dynamics of the original pendulum. Figure 1.3a–i shows the truncated dynamics staring from linear, cubic, to seventeenth order, which indicates that any truncated dynamics cannot model the rotating solutions and separatrix. This example illustrates that the truncated system does not include the global dynamics, even sometimes the global dynamics can be represented by the local dynamics.

Example 2. Now consider the following system which is differentiable infinitely many times.

$$\ddot{x} + f(x) = 0, \quad f(x) = \begin{cases} e^{-1/x^2}, x \neq 0 \\ 0, \qquad x = 0. \end{cases} \qquad (1.10)$$

Its Hamiltonian function is of the form given by Eq. 1.11 and the graph of the nonlinear function $f(x)$ is depicted in Fig. 1.4a.

$$H(x, y) = \frac{1}{2}y^2 + e^{-1/x^2}x + \sqrt{\pi}\,\mathrm{Erf}\left(\frac{1}{x}\right) = E, \;\; \mathrm{Erf}(x) = \frac{2}{\sqrt{\pi}}\int_0^x e^{-t^2}dt. \qquad (1.11)$$

Equation 1.11 determines the structure of the phase portraits shown in Fig. 1.4b and for which truncations always keep the same form as $\ddot{x} = 0$ no matter at which order the system is truncated.

It is worth emphasizing here that many dynamical systems in engineering are not smooth including Filippov systems [29, 30] which have discontinuous right hand sides, piecewise linear systems [31] or systems subjected to friction and impacts such as [32, 33]. For such systems the assumption of local smoothness fails and a power expansion does not converge to the original right hand side.

The SD oscillator, which is the centre of this book, can serve as an archetypal and inherently nonlinear dynamical system to challenge the classical approximation

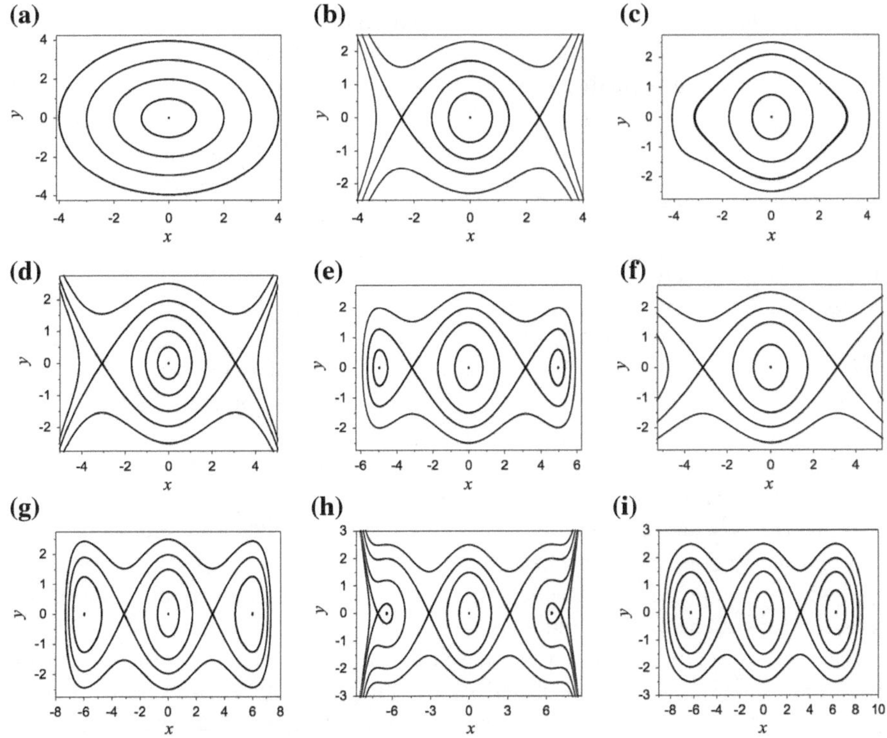

Fig. 1.3 Phase portraits for the mathematical pendulum which nonlinearity is modeled by Taylor expansions at successive orders: **a** first, **b** cubic, **c** quintic, **d** seventh, **e** ninth, **f** eleventh, **g** thirteen, **h** fourteen and **i** fifteenth

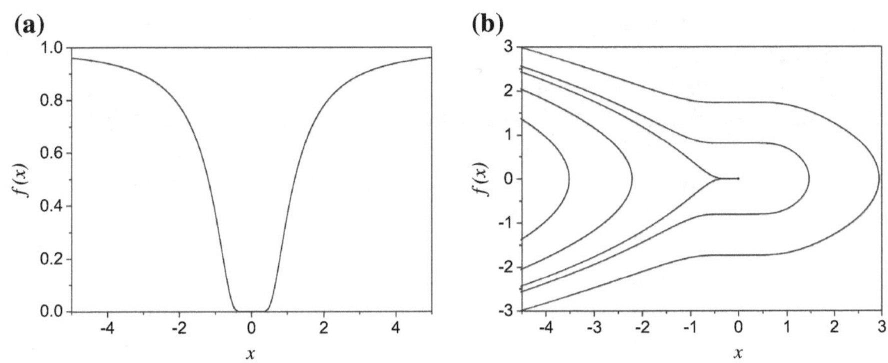

Fig. 1.4 **a** Graph for function $f(x)$ given in (1.10) and **b** phase portraits for (1.10), which indicates the saddle node equilibrium

approaches and to shed some light on the transition from smooth to non-smooth (discontinuous) dynamics. The latter is the newest development in nonlinear dynamics and feature some nonstandard bifurcation scenarios.

1.3 Scope of the Book

The smooth and discontinuous (SD) oscillator is introduced in this book to exhibit the new qualitative properties of the oscillator which can be either smooth or discontinuous [34–36]. When it is smooth it has an irrational nonlinearity for which new analytical treatment has been developed [34–36]. This oscillator has a simple geometrical nonlinearity allowing for a transition from smooth to discontinuous behaviour by controlling a geometrical parameter. It is made of a lump mass linked by a pair of linear springs pinned to rigid supports. Although each of the springs provides a linear resistance, the system is strongly nonlinear due to the geometrical configuration. This nonlinear behaviour could be interpreted in a similar way as large deformations in continuum mechanics.

The book comprises sixteen chapters arranged to gradually introduce and investigate this new archetypal oscillator.

Chapter 2 presents the SD oscillator and its first qualitative properties. After the description of its physical model and its equation of motion, it is shown that the system may have qualitatively different behaviours depending on the value of the geometrical control parameter.

Chapter 3 focuses on the discontinuous case. The dynamic responses are described for the damped or the undamped cases, where both free oscillations and the oscillations under an external forcing are investigated.

Chapter 4 deals with the resonances for the smooth and the discontinuous cases. The analysis requires an introduction of generalized winding number, which can be applied to other dynamical systems.

Chapters 5 and 6 are dedicated to the smooth case, where the complex behavior is explored. It is first proved via universal unfolding that the system exhibits a complex codimension two bifurcation. Then a very intricate structure of attractors is described within the framework of the so-called Wada structure.

The four following Chapters 7, 8, 9, 10 deal with questions of methodology. First a piecewise linear approach model of the restoring force is used, then the averaging method is presented and Hopf bifurcations are shown to occur near the equilibria. Then analytical calculations are performed using elliptic and hyperbolic functions. Finally, a cell mapping method is introduced to obtain new qualitative results.

Chapter 11 can be seen as an extension to the investigations of the previous sections. The composition of the system remains exactly the same, but the constraints are changed and involve unilateral contact and friction. This leads to the first qualitative analysis of interactions between the geometical nonlinearity with those arising from the contact and friction.

The four remaining main chapters are concerned with experiments and applications. Experiments in mechanics and electronics are presented in Chap. 12. Concerning applications a seismograph used in ancient China is presented in Chap. 13. Application in Chap. 14 a device for isolating large structures from ground vibrations or impacts is modeled and analyzed, whilst in Chap. 15, a secure communication is discussed.

Chapter 16 closes the book with comments about open problems and challenges.

References

1. Hooke, R. (1678). *Lectures de potentia restitutiva of a spring explaining the power of spring bodies*. London: John Martin.
2. Bernoulli, J. (1732). Meditations de chordis vibrantibus, cum pondusculis aequali a intervallo a se invicem dissitis ubi nimirum ex principio virium vivarum quaqeritur numerus vibrationum chordia pro una oscillatione penduli datae longitudinis D. *Comm. Acad. Sci. Petrop*, *3*(1728), 13–28.
3. Newton, I. (1687). *Philosphiae naturalis principia mathematica*. London.
4. Truesdell, C. (1960). *The rational mechanics of flexible or elastic bodies 1638-1788: Introduction to Leonhardi Euleri opera omnia, Vol. X et XI. seriei seeundae*. Orell Füssli Turiei, Switzerland.
5. Euler, L. (1750). E126, De novo genere oscillationum. *Comm. Acad. Sci. Petrop*, *11*(1139), 128–149.
6. Helmholtz, H. L. F. (1895). *On the Sensations of Tone as a Physiological Basis for the Theory of Music, App. XII, Theory of combinatorial tones* (3rd ed.). (A. J. Ellis, Trans.). New York: Longmans Green and Co. Reprinted by Dover publications.
7. Strutt, J. W., & Rayleigh, B. (1894). *The Theory of Sound* (2nd ed. Vols. I, 77). New York: Constable and Co. Ltd. Reprinted by Dover publications.
8. Duffing, G. (1918). *Erzwungene Schwingungen bei veränderlicher Eigenfrequenz und ihre technische Bedeutung*. Braunschweig: Vieweg and Sohn, Braunschweig Heft 41/42.
9. Chen, Y. S., & Leung, A. Y. T. (1998). *Bifurcation and chaos in engineering*. London: Springer.
10. Kovacic, I., & Brennan, M. J. (2011). *The duffing equation: nonlinear oscillators and their behaviour*. London: Wiley.
11. Ueda, Y. (2001). *The road to chaos-II*. Santa Cruz: Aerial Press.
12. Ueda, Y. (2006). Explosion of strange attractors exhibited by duffing's equation. *Annals of the New York Academy of Sciences*, *357*(1), 422–434.
13. Nayfeh, A. H., & Sanchez, N. E. (1989). Bifurcations in a forced softening duffing oscillator. *International Journal of Non-linear Mechanics*, *24*(6), 483–497.
14. Luo, A. C. J., & Huang, J. Z. (2014). On analytical routes to chaos in nonlinear systems. *International Journal of Bifurcation and Chaos*, *24*(4), 1430013.
15. Luo, A. C. J. (2006). Chaos and quasi-periodic motions on the homoclinic surface of nonlinear hamiltonian systems with two degrees of freedom. *Journal of Computational and Nonlinear Dynamics*, *1*(2), 135–142.
16. Guckenheimer, J., & Holmes, P. (1983). *Nonlinear oscillation, dynamical system and bifurcation of vector fields*. New York: Springer.
17. Van Der Pol, B. (1927). Vii. forced oscillations in a circuit with non-linear resistance (reception with reactive triode). *London Edinburgh and Dublin Philosophical Magazine and Journal of Science*, *3*(13), 65–80.
18. Lorenz, E. N. (1963). Deterministic non-periodic flow. *Journal of Atmospheric Sciences*, *20*(2), 130–141.

19. Chen, G. R., & Ueta, T. (1999). Yet another chaotic attractor. *International Journal of Bifurcation and Chaos*, *9*(7), 1465–1466.
20. Jinhu, Lu, & Chen, Guanrong. (2002). A new chaotic attractor coined. *International Journal of Bifurcation and Chaos*, *12*(3), 659–661.
21. Galileo, G. (1592) De motu (on motion). *Collection of Galileo Galilei's Manuscripts and Related Translations*, Retrieved.
22. Matthews, M. R. (2001). How pendulum studies can promote knowledge of the nature of science. *Journal of Science Education and Technology*, *10*(4), 359–368.
23. Nicholas, B. (1989). Tufillaro. Nonlinear and chaotic string vibrations. *American Journal of Physics*, *57*(5), 408–414.
24. Holmes, P. (1977). Bifurcation to divergence and flutter in flow induced oscillations: a finite dimensional analysis. *Journal of Sound and Vibration*, *53*(4), 471–503.
25. Moon, F. C., & Holmes, P. I. (1979). A magnetoelastic strange attractor. *Journal of Sound and Vibration*, *65*(5), 275–296.
26. Golubitsky, M., Schaeffer, D. G., & Stewart, I. (1985). *Singularities and groups in bifurcation theory* (Vol. I). New York: Springer.
27. Gamal, M. (1997). Periodic solutions of strongly nonlinear mathieu oscillators. *Journal of Non-linear Mechanics*, *32*(6), 1177–1185.
28. Chen, S. H., Yang, X. M., & Cheung, Y. K. (1998). Periodic solutions of strongly quadratic non-linear oscillators by the elliptic perturbation method. *Journal of Sound and Vibration*, *212*(5), 771–780.
29. Filippov, A. F. (1991). Differential equations with discontinuous right-hand sides. *Journal of Mathematical Analysis & Applications*, *154*(2), 377–390.
30. Filippov, A. F. (1988). *Differential equations with discontinuous right-hand sides: Control systems* (Vol. 18)., Mathematics and its applications The Netherlands: Springer. Originally published in Russian.
31. Santhosh, B., Padmanabhan, C., & Narayanan, S. (2014). Numeric-analytic solutions of the smooth and discontinuous oscillator. *International Journal of Mechanical Sciences*, *84*, 102–119.
32. Leine, R. I., Van Campen, D. H., De Kraker, A., & Van den Steen, L. (1998). Stick-slip vibrations induced by alternate friction models. *Nonlinear Dynamics*, *16*(1), 41–54.
33. Léger, A., & Pratt, E. (2016). *Qualitative analysis of nonsmooth dynamics*. London: ISTE Editions.
34. Cao, Q. J., Wiercigroch, M., Pavlovskaia, E. E., Grebogi, C., & Thompson, J. M. T. (2006). Archetypal oscillator for smooth and discontinuous dynamics. *Physics Review E*, *74*(046218), 1–5.
35. Cao, Q. J., Wiercigroch, M., Pavlovskaia, E. E., Grebogi, C., & Thompson, J. M. T. (2008). Piecewise linear approach to an archetypal oscillator for smooth and discontinuous dynamics. *Philosophical Transactions of the Royal Society, A-Mathematical, Physical and Engineering Sciences*, *366*(1865), 635–652.
36. Cao, Q. J., Wiercigroch, M., Pavlovskaia, E. E., Grebogi, C., & Thompson, J. M. T. (2008). The limit case response of the archetypal oscillator for smooth and discontinuous dynamics. *International Journal of Non-linear Mechanics*, *43*(6), 462–473.

Chapter 2
The SD Oscillator

This chapter introduces briefly the smooth and discontinuous (SD) oscillator which is a simple mechanical model or a geometrical oscillator with both smooth and discontinuous dynamics depending on the value of a geometrical parameter. Like the traditional harmonic oscillator, this model is also a simple mass-spring system comprising a lumped mass linked by a pair of linear springs pinned to its rigid supports vibrating along the perpendicular bisector of the supports. The overview of the system with the definitions and the fundamental properties are given here without detailed explanation, and will be expanded and discussed in the following chapters from Chaps. 3 to 6.[1]

2.1 Introduction

As mentioned in the introduction, Chap. 1, in addition to the smooth systems like Duffing oscillator, Lorenz system and van de Pol oscillator as well, many examples arise from mechanical and civil engineering [1, 2], electronics [3, 4], control [5, 6], computer graphics [7, 8], biology [9, 10], and others, which are non-smooth or discontinuous. This oscillator is motivated by the growing interest in the system which is not valid for the conventional nonlinear dynamics, especially. Although some theoretical foundations have been laid in the work by Filippov [11], Feigin [12], Kunze [13], Peterka [14], Shaw and Holmes [15] and Nordmark [16], there is a large disparity between development and understanding of smooth and discontinuous (non-smooth) systems.

[1] This chapter involves a common work with Professors Marian Wiercigroch, Ekaterina Pavlovskaia, Celso Grebogi, J. Michael T. Thompson, Center for Applied Dynamics Research, Aberdeen University, King's College, Aberdeen AB24 3UE, Scotland, UK.

Q. Cao and A. Léger, *A Smooth and Discontinuous Oscillator*, Springer Tracts in Mechanical Engineering, DOI 10.1007/978-3-662-53094-8_2

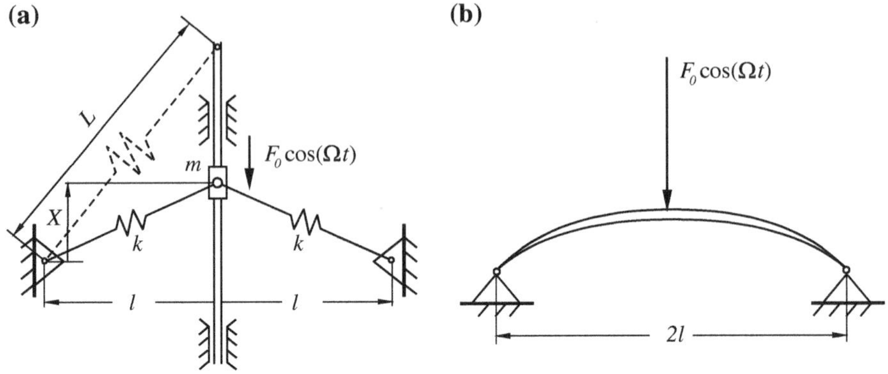

Fig. 2.1 **a** A mechanical oscillator consisting of a lumped mass m linked by a pair of linear springs with stiffness k and the equilibrium length L pinned to rigid supports abstracted from the first mode of a simple elastic arch demonstrated in **b**

An archetypal oscillator whose nonlinearity can be either smooth or discontinuous depending on the value of the smoothness parameter α is proposed and studied. As the considered oscillator has properties of both a smooth and a discontinuous system (at the limit), potentially a wealth of knowledge can be drawn from the well developed theory of continuous dynamics. Physically (as shown in Fig. 2.1a) this oscillator is similar to a snap-through truss system. It comprises a mass, m, linked by a pair of inclined elastic springs which are capable of resisting both in tension and compression; each spring of stiffness k is pinned to a rigid support. This model is inspired by the elastic arch described by Thompson and Hunt in [17] (see Fig. 2.1b). Although the springs themselves provide a linear restoring resistance, the resulting vertical force on the mass is strongly nonlinear because of changes to the geometric configuration. In the smooth regime, the system bears significant similarities to the Duffing oscillator, exhibiting the standard dynamics of hyperbolic structure patterns associated with equilibrium bifurcation of a supercritical pitchfork. The dynamics might be transformed from a single-well into a double-well behaviour at the degenerate value of the geometrical parameter. In the discontinuous case however, there is a substantial departure in the dynamics from the standard one. The system behaves a nonstandard dynamics caused by the loss of local hyperbolicity due to the collapse of the stable and unstable manifolds of the stationary state. In the presence of damping and external excitation, the system bears co-existence of attractors and a chaotic saddle, which becomes a chaotic attractor when the smoothness parameter drops to zero. This attractor may bifurcate to a large period periodic attractor or to a chaotic sea with islands of quasi-periodic attractors depending on the amplitude of the damping. All along this book the oscillator will be referred to as the SD oscillator and the attractors as the SD attractors.

2.2 Unperturbed Oscillator

The mechanical model or a mass-spring system dealt with all along the present book is very simple, which is inspired by the first mode of an elastic arch described in [17, 18], seen in Fig. 2.1b, and is similar to a snap-through truss system. As shown in Fig. 2.1a, it is constructed with a lumped mass m linked by a pair of linearly elastic springs with stiffness k and equilibrium length L being capable of resisting both in tension and in compression pinned to a rigid frame. Although the springs themselves provide a linear restoring resistance, the resulting vertical force supplied to the mass is nonlinear because of the geometrical configuration, which is a model for large deformations of the arch in Fig. 2.1b.

The equation of the motion of mass m is obtained according to the geometrical configuration of the springs, due to the Pythagorean theorem [19], and the Hooke's Law [20, 21].

$$m\ddot{X} + 2kX\left(1 - \frac{L}{\sqrt{X^2 + l^2}}\right) = 0, \tag{2.1}$$

where L is the natural length of each spring, X is the displacement of the mass and l is the half distance between the rigid supports. System (2.1) can be made dimensionless by letting $\omega_0^2 = \frac{2k}{m}$, $x = \frac{X}{L}$ and $\alpha = \frac{l}{L} \geq 0$,

$$\ddot{x} + \omega_0^2 x\left(1 - \frac{1}{\sqrt{x^2 + \alpha^2}}\right) = 0, \tag{2.2}$$

where α stands for a geometrical parameter, which defines the geometry of the oscillator as shown in Fig. 2.1a with its full physical practice. From the mechanical point of view, for $\alpha > 0$ the system represents a discrete elastic string presenting an archetypal smooth oscillator. If $\alpha = 0$, the model corresponds to an oscillating mass supported by two parallel vertical springs, as shown in Fig. 3.1a. It is worth reiterating here that the discontinuous dynamics can be obtained by decreasing the smoothness parameter $\alpha \to 0$. Equation (2.2) is now written as the following

$$\ddot{x} + \omega_0^2(x - \mathrm{Sign}(x)) = 0, \tag{2.3}$$

where Sign(x) is the sign of x. System (2.3) is the limit case of system (2.2) when $\alpha = 0$ building up an archetypal discontinuous oscillator consisting of a pair of opposite directed harmonic oscillators. Usually system (2.2) and (2.3) is called the SD oscillator throughout the book: S means smooth and D refers to discontinuous.

It is addressed that even the equation of motion (2.2) is derived by the simple Hooke's Law of a single linear spring [21], the resistance of our system is nonlinear due to the geometrical configuration, which can be regarded as the generalized Hooke's Law where the stiffness might be positive, negative or zero (quasi-zero) depending on the geometrical configuration of the system. The restoring force reads

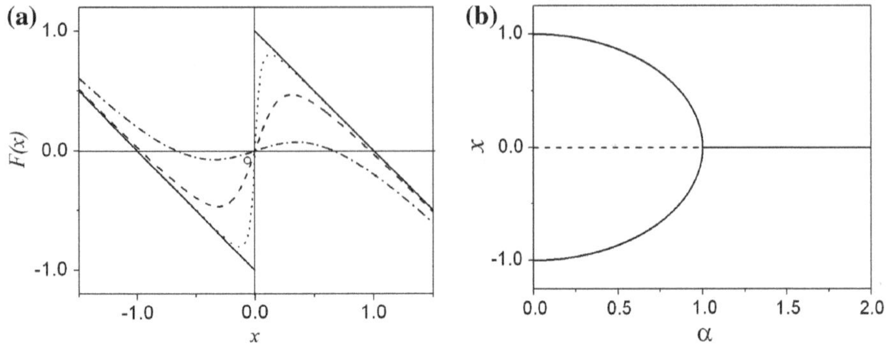

Fig. 2.2 **a** Nonlinear restoring force, $F(x)$; *solid line* marks the discontinuous case for $\alpha = 0$, *dotted*, *dash-dotted* and *dashed curves* are the smooth cases for $\alpha = 0.1, 0.5$ and $\alpha = 0.75$, respectively, and **b** The equilibrium bifurcation: the *solid* marks the stable and the *dashed* indicates the unstable branches with bifurcation point at $\alpha = 1$

$F(x) = -\omega_0^2 x \left(1 - \dfrac{1}{\sqrt{x^2 + \alpha^2}}\right)$, which is plotted for $\omega_0 = 1$ in Fig. 2.2a for different values of parameter α. The solid line represents the discontinuous case $\alpha = 0$, the dotted, the dashed dotted and the dashed lines mark the smooth cases, for $\alpha = 0.01$, 0.5, and 0.75 respectively.

To examine the influence of parameter α on the dynamics of (2.2) we construct the bifurcation diagram, depicted in Fig. 2.2b. The system undergoes a supercritical pitchfork bifurcation at $\alpha = 1$ where the stable branch $x = 0, \alpha \geq 1$ bifurcates into two stable branches at $x = \pm\sqrt{1 - \alpha^2}, 0 \leq \alpha < 1$, while the state branch $x = 0$ now becomes unstable. The Hamiltonian for system (2.2) can be written as

$$H(x, y) = \frac{1}{2}y^2 + \frac{1}{2}\omega_0^2 x^2 - \omega_0^2\sqrt{x^2 + \alpha^2} + \omega_0^2 \alpha, \tag{2.4}$$

where $\dot{x} = y$. With the help of the Hamiltonian function (2.4), the trajectories can be classified and analyzed. For both continuous and discontinuous cases, the phase portraits of systems (2.2) or (2.4) are plotted for different values of the Hamiltonian $H(x, y) = E$. For instance, for the smooth nonlinearity, $\alpha = 0.5$, the dynamical behaviour of double-well is similar to that of the Duffing oscillator [22], shown in Fig. 2.3a. For $\alpha = 0$, the behaviour is singular, as shown in Fig. 2.3b: the orbits for $E > 0$ are composed of two large segments of circles with their centers located at $(-1, 0)$ and $(1, 0)$ connected at $x = 0$. The case of $E < 0$ is represented by two families of circles. It is very interesting that the level $E = 0$ for the discontinuous system (2.3) is made up of two circles centered at $(\pm 1, 0)$ connecting at the singular point $(0, 0)$, which form a special singular homoclinic-like orbits. Even system (2.3) is discontinuous at $(0, 0)$ without eigenvalues and eigenvectors., the structure around it indicates a saddle-like behaviour, which is named the saddle-like equilibrium. The loss of hyperbolicity at $(0, 0)$ is caused by the tangency of the stable and unstable eigen directions. The pair of circles excluding the point $(0, 0)$ are not the manifolds of the singularity, but the flow along these circles approaches the point as $x \to 0$,

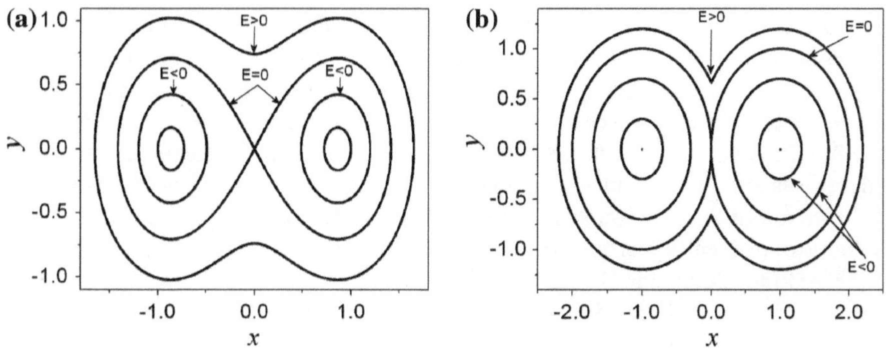

Fig. 2.3 Phase portraits; **a** smooth case for $\alpha = 0.5$ and **b** discontinuous case for $\alpha = 0$

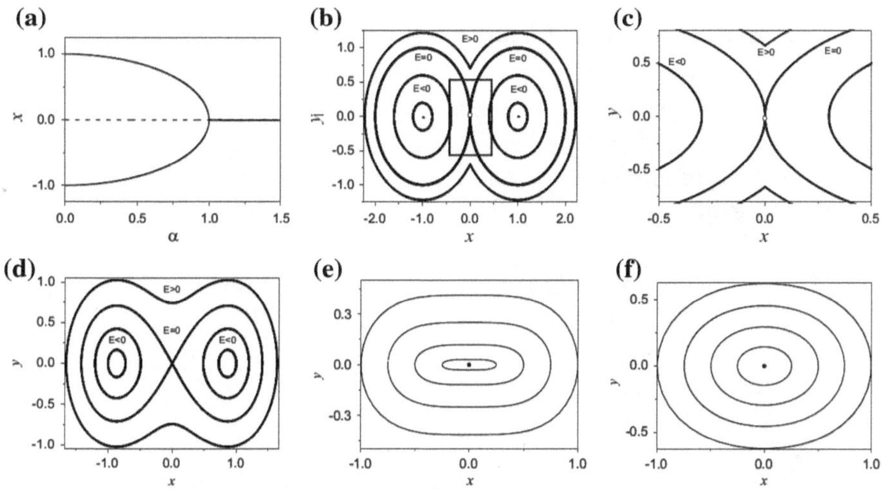

Fig. 2.4 **a** Equilibrium bifurcation and the Phase portraits of the SD oscillator: **b** phase portrait for $\alpha = 0$ with a homoclinic-like orbit composting a pair of cycles connected at a saddle-like singularity, the box area, **c** locally amplified structure, and **d** the portraits for double-well behaviour when $0 < \alpha < 1$ and **f** the single-well characteristic when $\alpha > 1$ transformed at the degenerate case **e** when $\alpha = 1$

and it will be trapped by the singularity. The solution of the special homoclinic-like orbits can be formulated as

$$\Gamma = \{(x_\pm(t), y_\pm(t)), t \in (-\frac{\pi}{\omega_0}, \frac{\pi}{\omega_0})\} \bigcup \{(0,0)\}, \tag{2.5}$$

where $(x_\pm(t), y_\pm(t)) = (\pm 1 \pm \cos \omega_0 t, \mp \sin \omega_0 t)$.

Figure 2.4 summaries all the patterns of the unperturbed dynamics of the SD oscillator for the equilibrium bifurcations with the supercritical pitchfork in Fig. 2.4a and the corresponding phase portraits as the geometrical parameter α varies. For

$\alpha > 0$ the system admits the standard smooth dynamics transforming the single-well dynamics from $\alpha > 1$, shown in Fig. 2.4f, into double-well behaviour to $0 < \alpha < 1$, shown in Fig. 2.4d, which results in the so-called single stability and bistablity shifting at $\alpha = 1$ with degenerate singularity, shown in Fig. 2.4e, which will be investigated in the following chapters. As the geometrical parameter α decreases to $\alpha = 0$, a sudden change happens leading to a nonstandard dynamics, as shown in Fig. 2.4b and the local amplified boxed area in Fig. 2.4c, which is caused by the loss of local hyperbolicity due to the collapse of the eigen direction of the stable and unstable manifolds at the stationary state. This discontinuous behaviour results in the homoclinic-like orbit consisting of a pair of cycles connected at the saddle-like equilibrium $(0, 0)$ centered at $(\pm 1, 0)$ respectively.

2.3 The Transition from Smooth to Discontinuous Dynamics

Now suppose system (2.1) is perturbed by a viscous damping and an external harmonic excitation of amplitude F_0 and frequency Ω. This leads to the following perturbed system,

$$m\ddot{X} + \delta\dot{X} + 2kX(1 - \frac{L}{\sqrt{X^2 + l^2}}) = F_0 \cos \Omega t. \tag{2.6}$$

Again system (2.6) can be written in a dimensionless form by letting $\tau = \omega_0 t$, $f_0 = \frac{F_0}{2kL}$, $\xi = \frac{\delta}{2m\omega_0}$ and $\omega = \frac{\Omega}{\omega_0}$,

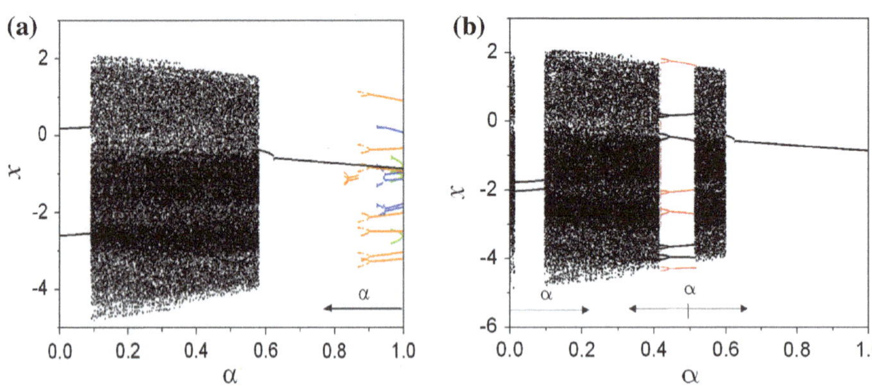

Fig. 2.5 Bifurcation diagrams for x versus α constructed for: **a** Decreasing from $\alpha = 1$ and following the attractors starting with period 1 (*black*), period 3 (*green*), 5 (*blue*) and 7 (*red*) respectively, **b** first, increasing from $\alpha = 0$ to 0.3 following the attractor starting with the initial condition (1,0). Second, decreasing α from 0.5 to 0.3 and increasing from α from 0.5 to 1, following the attractors starting with two period 4 solutions respectively

$$x'' + 2\xi x' + x(1 - \frac{1}{\sqrt{x^2 + \alpha^2}}) = f_0 \cos \omega\tau, \tag{2.7}$$

where "′" denotes the differentiation with respect to τ.

Numerical simulations are carried out by using Runge–Kutta method for system (2.7) for only the smooth case for $\alpha > 0$, while in the discontinuous case for $\alpha = 0$ the analytical technique is developed, the details will be given in the following chapters. It is always assumed $f_0 = 0.8$, $\xi = 0.01\sqrt{2}$ and $\omega = 0.75\sqrt{2}$ in the following calculations. Figure 2.5 shows bifurcation diagrams constructed for x sampled stroboscopically at phase zero versus control parameter α as α decreases from 1 to 0 (Fig. 2.5a) and α increases from 0 to 0.3 and from 0.5 to 1 and decreases from 0.5 to 0.3 (Fig. 2.5b). The system has co-existing periodic attractors and a strange chaotic attractor for $\alpha > 0$. However, for $\alpha = 0$ the system exhibits chaotic and periodic solutions with unusual high periods. These behaviours can be controlled by

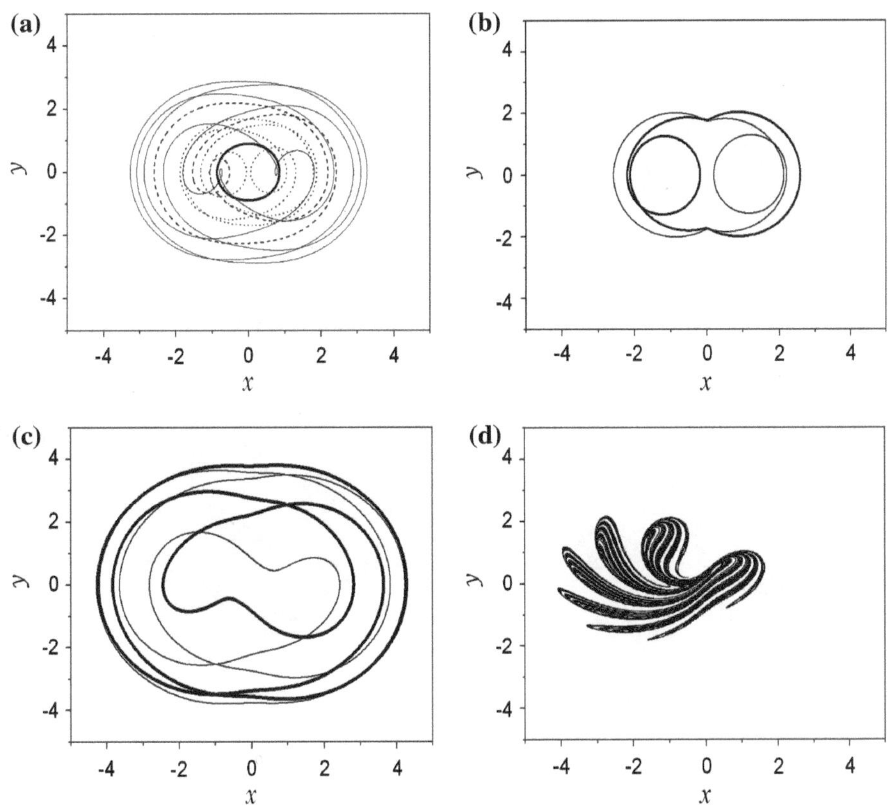

Fig. 2.6 Co-existence of periodic motions: **a** Period 1 (*thick line*), period 3 (*dash line*), 5 (*dotted line*) and 7 (*thin line*) respectively for $\alpha = 1.0$, **b** a pair of period two solutions for $\alpha = 0.05$, **c** and **d** represent the pair of period four and the chaotic attractor for $\alpha = 0.5$ with the largest Lyapunov exponent 0.1065

the strength of the damping. The main feature of these attractors is their topological similarity since they are associated with the hyperbolic stationary state. This similarity also holds for $\alpha > 0$. The chaotic attractor becomes a chaotic sea with islands representing a quasi periodic behaviour for $\alpha = 0$ and $\xi = 0$.

Figure 2.5a also shows the co-existing periodic solution of period one, three, five and seven at $\alpha = 1$ (orbits are shown in Fig. 2.6a) and their bifurcations under decreasing α. Other co-existing periodic solutions are found for $\alpha \in (0, 0.1)$, as shown in Fig. 2.5a, b. These two co-existing period 2 solutions are symmetrical and the orbits for $\alpha = 0.05$ are presented in Fig. 2.6b. For $\alpha \in (0.41, 0.52)$ a chaotic attractor co-exists with two period 4 solutions and the corresponding trajectories and the chaotic attractor are shown in Fig. 2.6c, d, respectively.

In addition to the co-existence of different attractors, the system exhibits a chaotic transient behaviour characterized by chaotic saddles, see [23–25]. The transient and the final periodic attractor are shown in Fig. 2.7b, c for $\alpha = 0.01$ and $\alpha = 0.001$

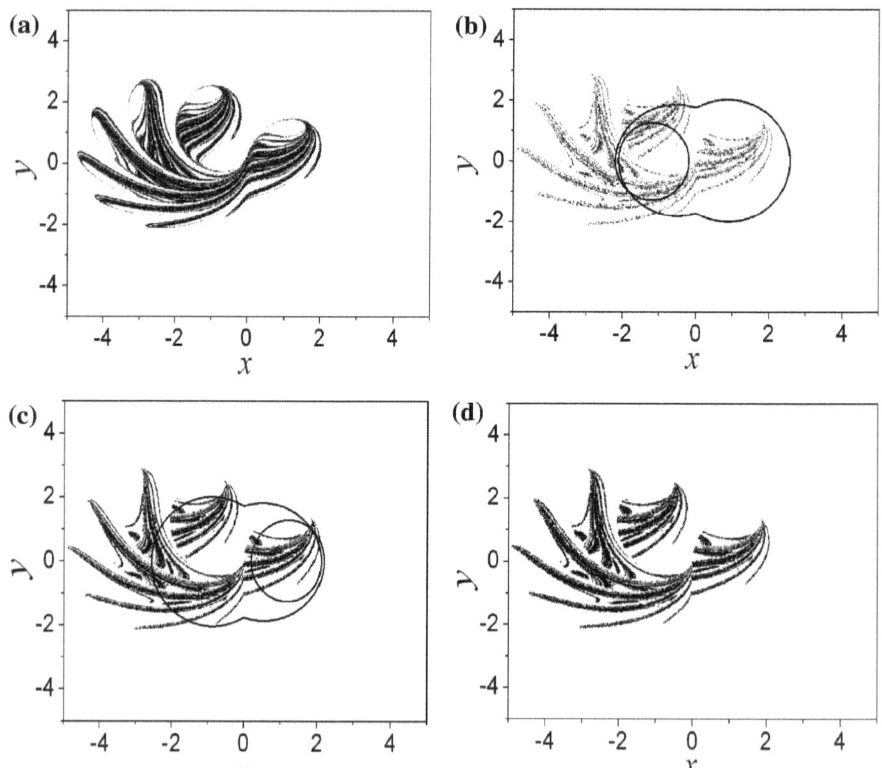

Fig. 2.7 **a** Chaotic attractor for $\alpha = 0.1$ with the largest Lyapunov exponent 0.0812, **b** chaotic saddle leading to period 2 solution for $\alpha = 0.01$ and **c** chaotic saddle leading to period 2 solution for $\alpha = 0.001$ **d** chaos for $\alpha = 0$ with the largest Lyapunov exponent 0.0480

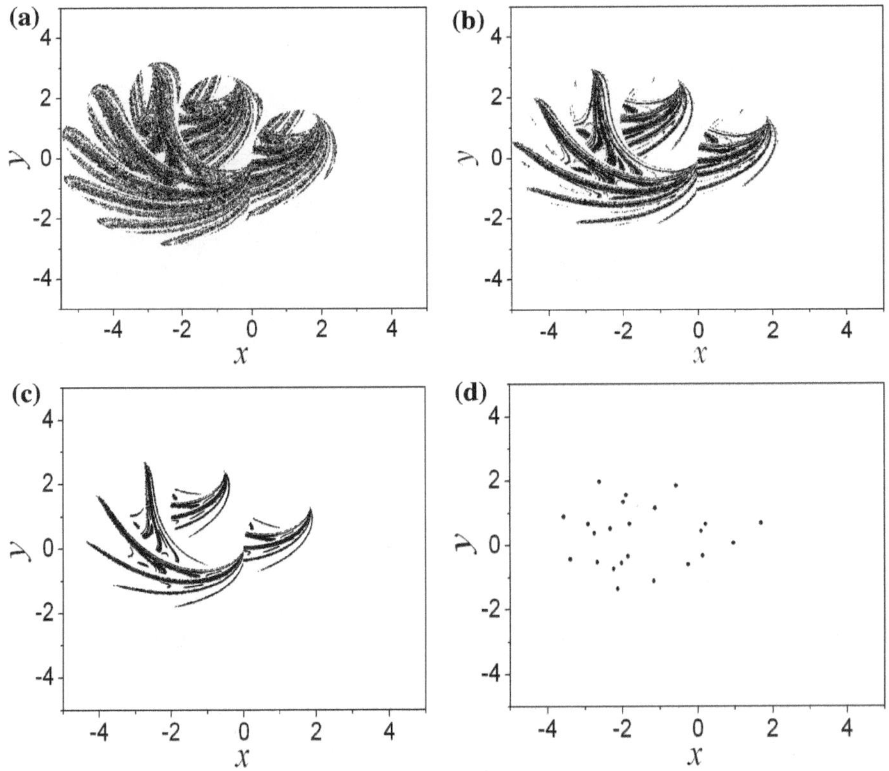

Fig. 2.8 Poincaré sections for $f_0 = 0.8$, $\omega = 0.75\sqrt{2}$, $\alpha = 0$; **a** chaotic attractor for $\xi = 0.005$ with the largest Lyapunov exponent 0.0525, **b** chaotic motion for $\xi = 0.0125$ with the largest Lyapunov exponent 0.0505, **c** chaotic motion for $\xi = 0.025$ with the largest Lyapunov exponent 0.0470 and **d** high periodic motion of period 23 for $\xi = 0.028$

respectively. The set of Poincaré maps shown in Fig. 2.7 for $\alpha = 0.1, 0.01, 0.001$ and 0 shows the topological similarity now associated with the discontinuity at the origin. The co-existing period 2 solutions persist for $\alpha = 0$ and the chaotic saddle becomes a chaotic attractor for $\alpha = 0$ as shown in Fig. 2.7d. The behaviour of this chaotic attractor can be controlled by the damping ratio ξ. Other attractors are presented in Fig. 2.8a for $\xi = 0.005$, Fig. 2.8b for $\xi = 0.0125$ and Fig. 2.8c for $\xi = 0.025$. A high period periodic attractor is shown in Fig. 2.8d having period 23 for $\xi = 0.028$. For both $\alpha = 0$ and $\xi = 0$, a chaotic sea and quasi-periodic behaviour can also be inferred via a semi-analytical method, see [26]. Figure 2.9 shows the chaotic sea together with a pair of quasi-period 2 solutions, a pair of quasi-period 5 solutions, the quasi-period 9 and the quasi-period 11 solutions respectively. The largest Lyapunov exponent for all the chaotic attractors presented here have been calculated using the chaos synchronization method, see [27] for instance, as shown in the captions for the corresponding figures.

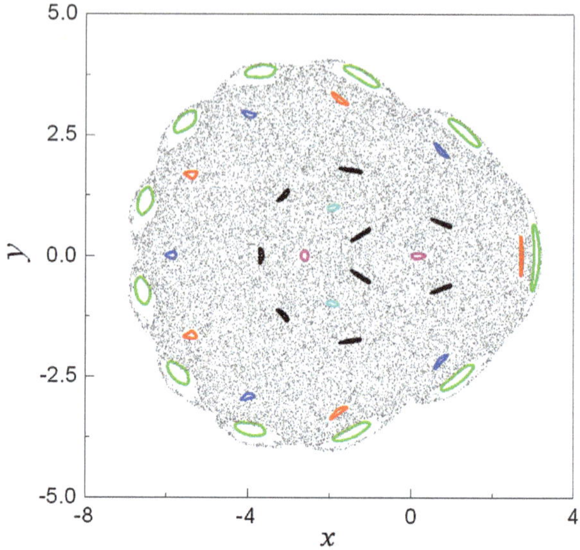

Fig. 2.9 Poincaré section for $f_0 = 0.8$, $\omega = 0.75\sqrt{2}$, $\alpha = 0$: chaotic sea or stochastic web and quasi-periodic trajectories (islands) for $\xi = 0$

Even only the primary characteristics of the perturbed oscillator are presented in this section, the main features of the system show that decreasing α further creates chaotic transients around a chaotic saddle, which leads to periodic orbits as the time increases. This chaotic transient becomes a chaotic attractor for $\alpha = 0$. The attractor can deform in shape or bifurcate to a high period periodic attractor depending on the strength of damping, or even can become a chaotic sea or stochastic web with islands of quasi-periodic trajectories for $\xi = 0$. It is also interesting to compare the finger like topology of the chaotic attractors found here with that observed by Thompson et al. in [28, 29]. The attractors shown in this chapter (2.2) and (2.3) are called the SD attractors: attractors of system (2.7) for both smooth ($\alpha > 0$) and discontinuous stage ($\alpha = 0$).

2.4 Summary

Let us close this chapter by a brief survey to list the main features that arise from this archetypal SD oscillator. At the first point, one must keep in mind that the unperturbed smooth oscillator has a single well property for $\alpha > 1$ shifting at $\alpha = 1$ to a double-well characteristics for $0 < \alpha < 1$, which is similar to Duffing oscillator and the system can be consequently integrated using classical methods for smooth systems such as Runge–Kutta method. However, it is different from the Duffing oscillator since there is no analytical solutions depending upon the elliptic functions. Furthermore, in the limit case of the SD oscillator, that is the discontinuous oscillator, classical methods no longer apply. The global qualitative analysis will be presented in the following chapters in detail.

References

1. Woo, K. C., Pavlovskaia, E., Wiercigroch, M., & Rodger, A. A. (2003). Modelling of ground moling dynamics by an impact oscillator with a frictional slider. *Meccanica, 38*(1), 85–97.
2. Karpenko, E. V., Pavlovskaia, E., & Wiercigroch, M. (2004). Nonlinear dynamic interactions of a jeffcott rotor with a preloaded snubber ring. *Journal of Sound and Vibration, 276*(1–2), 361–379.
3. Banerjee, S., & Chakrabarty, K. (1998). Nonlinear modeling and bifurcations in the boost converter. *IEEE Transactions on Power Electronics, 13*(2), 252–260.
4. Banerjee, S., Ott, E., Yorke, J. A., & Yuan, G. H. (1997). Anomalous bifurcations in dc-dc converters: boderline collisions in piecewise smooth maps. In *28th annual IEEE power electronics specialists' conference, PESC '97 record*, (Vol. 2), 22–27 Jun 1997, St. Louis, MO, pp. 1337–1344. IEEE.
5. Slotine, J. J., & Sastry, S. S. (1983). Tracking control of non-linear systems using sliding surfaces with application to robot manipulators. *International Journal of Control, 38*(2), 465–492.
6. Richard, P. Y., Cormerais, H., & Buisson, J. (2006). A generic design methodology for sliding mode control of switched systems. *Nonlinear Analysis, Theory, Methods and Applications, 65*(9), 1751–1772.
7. Luo, A. C. J., & Burkhardt, H. (1995). Intensity-based cooperative bidirectional stereo matching with simultaneous detection of discontinuities and occlusions. *International Journal of Computer Vision, 15*(3), 171–188.
8. Nishita, T., Sederberg, T. W., & Kakimoto, M. (1990). Ray tracing trimmed rational surface patches. *Computer Graphics, 24*(4), 337–345.
9. Kribs-Zaleta, C. M. (2004). To switch or taper off: The dynamics of saturation. *Mathematical Biosciences, 192*(2), 137–152.
10. Drulhe, S., Ferrari-Trecate, G., De Jong, H., & Viari, A. (2006). Reconstruction of switching thresholds in piecewise-affine models of genetic regulatory networks. In J. P. Hespanha & A. Tiwari (Eds.), *Hybrid systems: computation and control* (Vol. 3927, pp. 184–199)., Lecture notes in computer science Heidelberg: Springer.
11. Filippov, A. F. (1988). *Differential equations with discontinuous right-hand sides: Control systems* (Vol. 18)., Mathematics and its applications The Netherlands: Springer. Originally published in Russian.
12. Feigin, M. I. (1966). Resonance behaviour of a dynamical system with collisions. *Journal of Applied Mathematics and Mechanics, 30*(5), 1118–1123.
13. Kunze, M. (2000). *Non-smooth dynamical systems*. Berlin: Springer.
14. Peterka, F. (1981). *Introduction to vibration of mechanical systems with internal impacts*. Prague: Academia.
15. Shaw, S. W., & Holmes, P. J. (1983). A periodically forced piecewise linear oscillator. *Journal of Sound and Vibration, 90*(1), 129–155.
16. Nordmark, A. B. (1991). Non-periodic motion caused by grazing incidence in an impact oscillator. *Journal of Sound and Vibration, 145*(2), 279–297.
17. Thompson, J. M. T., & Hunt, G. W. (1973). *A general theory of elastic stability*. London: Wiley.
18. Savi, M. A., & Pacheco, P. M. C. L. (2003). Transient chaos in an elasto-plastic beam with hardening. *Journal of Brazilian Society of Mechanical Sciences and Engineering, 25*(2), 189–193.
19. Maor, E. (2007). *The Pythagorean theorem: A 4,000-year history*. Princeton: Princeton University Press.

20. Drake., E. T. (1996). *Restless Genius: Robert Hooke and his earthly thoughts*. Oxford: Oxford University Press.
21. Lewin, W. (1999). *Hooke's Law, simple harmonic oscillator.*, MIT course 8.01: Classical mechanics, lecture 10. (ogg) (videotape) Cambridge: MIT OCW.
22. Guckenheimer, J., & Holmes, P. (1983). *Nonlinear oscillation, dynamical system and bifurcation of vector fields*. New York: Springer.
23. Lai, Y. C., & Lerner, D. (1998). Effective scaling regime for computing the correlation dimension from chaotic time series. *Physica D-Nonlinear Phenomena, 115*, 1–18.
24. Tel, T. (1986). Characteristic exponents of chaotic repellers as eigenvalues. *Physics Letters A, 119*(2), 65–68.
25. Grebogi, C., Ott, E., & Yorke, J. A. (1983). Crises, sudden changes in chaotic attractors, and transient chaos. *Physica D-Nonlinear Phenomena, 7*(1–3), 181–200.
26. Pavlovskaia, E., & Wiercigroch, M. (2004). Analytical drift reconstruction for visco-elastic impact oscillators operating in periodic and chaotic regimes. *Chaos, Solitons and Fractals, 19*(1), 151–161.
27. Stefanski, A., & Kapitaniak, T. (2003). Estimation of the dominant lyapunov exponent of non-smooth systems on the basis of maps synchronization. *Chaos Solitons and Fractals, 15*(2), 233–244.
28. Thompson, J. M. T., & Ghaffari, R. (1983). Chaotic dynamics of an impact oscillator. *Physical Review A, 27*(3), 1741–1743.
29. Thompson, J. M. T., & Stewart, H. B. (2002). *Nonlinear dynamics and chaos* (2nd ed.). Chichester: Wiley and Sons.

Chapter 3
Discontinuous Dynamics

In this chapter we focus our attention mainly on the limit case as the geometrical parameter $\alpha \to 0$. The content of this chapter presents a new type of discontinuous oscillator with bistability which is a piecewise linear system with the discontinuity at $x = 0$. Due to the limitation of the numerical approximation, an analytical method to formulate the solutions of the discontinuous oscillator is proposed avoiding the barriers encountered in the conventional calculations due to the discontinuity. This scheme made it possible to obtain a chaotic solution theoretically, which gives the chaotic attractor analytically for this discontinuous oscillator. All the results presented here in this chapter and thereafter in this book for discontinuous dynamics are given by using this analytical procedure.[1]

3.1 Introduction

The oscillator presented in the above chapter may be either as a smooth oscillator or a discontinuous one. Of course, as long as the problem involves a single parameter which is a real number α, the smooth case could be seen as the generic case since it is associated with any strictly positive α, while the discontinuous case corresponds exactly to $\alpha = 0$. Due to the irrational form of the right hand side, it will be necessary to investigate the case $\alpha \neq 0$ in detail, this is an important aim of the present book. From general aspects of the theoretical foundations, the smooth case looks very much like Duffing equation, but the discontinuous case $\alpha = 0$ involves very unusual features which are the deepest specificities of the SD oscillator, and we shall consequently present the discontinuous dynamics of this limit case first. It is

[1] This chapter involves a common work with Professors Marian Wiercigroch, Ekaterina Pavlovskaia, Celso Grebogi, J. Michael T. Thompson, Center for Applied Dynamics Research, Aberdeen University, King's College, Aberdeen AB24 3UE, Scotland, UK.

Q. Cao and A. Léger, *A Smooth and Discontinuous Oscillator*, Springer Tracts in Mechanical Engineering, DOI 10.1007/978-3-662-53094-8_3

important that the classical tools for the calculation of smooth differential systems are no more applicable at the origin (0, 0) due to the discontinuity. A semi-analytical method, see [1], needs to be developed to overcome this barrier for discontinuous system.

Discontinuous dynamical systems have largely attracted the interests of nonlinear dynamical studies, see [2–4] for instance. Various types of discontinuities have been studied, see [5–8] for example. Discontinuous dynamical systems are of wide applicability both in engineering and in mathematics. Di Bernardo [9–11] studied the case of sliding systems with a border collision bifurcation. Banerjee [12, 13] investigated bifurcations of discontinuous maps for power circuits with discontinuous switches. Chin et al. [14], examined the bifurcations of a piecewise smooth map that captures the universal properties of impact oscillators near grazing point. Whiston [15] investigated the singularities of a vibro-impact system. These studies range from theoretical formulation and methodology to applications in engineering.

After recalling the model in the discontinuous case and its basic equation, the standard dynamics of the equilibrium state hyperbolicity of double-well and the loss of local hyperbolicity at the saddle-like equilibrium is presented for the unperturbed system.

Meanwhile, in the non-dissipative driven case, a resonance analysis is depicted with the KAM (Kolmogorov–Arnold–Moser) structure of the Poincaré section, see [16–18], exhibiting a series of periodic fixed points, related islands and chaotic connections for the islands. A special chaotic orbit, filling a finite area with periodic fixed points and corresponding islands, forms a chaotic sea (see [19–22]). Then numerical results for the system with weak dissipation and external excitation show the relevant dynamics, which involves co-existence of attractors, period doubling bifurcation, subharmonic and super harmonic responses and chaos. This is followed by an analysis of resonances.

3.2 Static Response of the SD Oscillator

We consider the limit case for the smooth and discontinuous oscillator presented in the previous chapters, see also [23, 24] for details. The mechanical model has been presented in Chap. 2, see Fig. 2.1a. Recall that it is built up with a lumped mass m and two linearly elastic springs of stiffness k by pinning them together at one end (the crown) and to two rigid foundations at the other end. Each spring has the equilibrium (unstressed) length, L, and is assumed to remain straight either under tension or compression. The distance between these two rigid foundations, often called abutments in engineering, is denoted by $2l$. Usually $L > l$ is assumed, which gives an unloaded symmetric arch of height

$$X_0 = \pm\sqrt{L^2 - l^2}. \tag{3.1}$$

Here the minus sign corresponds to the inverted arch which is of course a valid equilibrium state. Notice that if $L < l$, the unloaded springs will lie horizontally under tensile stress, and the behaviour will be that of a stretched spring. Of course in such a case there still remains the vertical geometrical nonlinearity even there is only a unique equilibrium state without loss of hyperbolicity. In this case the nonlinear behaviour may result in the harding spring behaviour.

The equilibrium response, $F(X_c)$ under a prescribed and time independent vertical loading P, in terms of the current height of the crown, X_c, is given by

$$F(X_c) = -2kX_c\left(1 - \frac{L}{\sqrt{X_c^2 + l^2}}\right), \tag{3.2}$$

the picture of which is omitted here, one can draw the response curve easily as the limit case later. Notice that m is an inertial mass only, which means that its weight is not taken into account without loss of generality.

We are interested in the limit case of the model shown in Fig. 2.1a as l decreases to zero, i.e. where the distance $2l$ between the abutments vanishes, which can be conceptually considered as the mechanism shown as in Fig. 3.1a. If the arch was carrying a load P, the springs will undergo a shortening s governed by the linearly elastic constitutive law $P = 2ks$ and the stored strain-energy will be

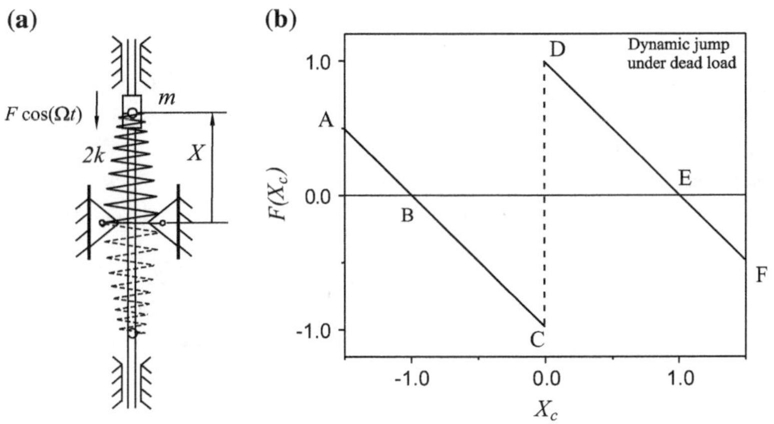

Fig. 3.1 **a** The limit case, discontinuous oscillator with lump mass and a conical spring of stiffness $2k$ capable of moving both above and below its pivot point, **b** limit case ($l = 0$) response $F(X_c)$ for $m = 1$, $k = \frac{1}{2}$ and $L = 1$

$$U = \frac{1}{2}(2k)s^2. \tag{3.3}$$

Observe that the choice of a linearly elastic constitutive law is a good approximation to reality when $s \ll L$ where it corresponds to Hooke's Law for an elastic body [25, 26]. The approximation is reasonably physical under tensile loads when P and s are both negative. However, under compressive loads, this idealization means that one is allowed to shrink the spring down to zero length ($s = L$) by applying the finite compressive load $P = 2kL$.

The load-deflection characteristics, $F(X_c)$ for the limit case of $l = 0$ is shown in Fig. 3.1b. Note that using $F(X_c)$ as defined in Eq. (3.2) gives us the convention plot for an arch with stable paths of positive slope. This diagram can be used to plot the response of the limiting arch with $l = 0$. On the stable equilibrium path ABC the arch points move upwards with the s vertical prings, and its height varies linearly with the applied load. Meanwhile on the stable equilibrium path DEF the arch points move downwards with the springs again vertical, and with its depth varying linearly with the applied load. At normal physical loading a sequence under a slow increase of the dead load, P, the response F would start at the unloaded state B and proceed quasi-statically along the stable equilibrium path to C. Here the system experiences a degenerate saddle-node fold, from which it would jump dynamically to an upside-down configuration as shown by the horizontal arrow. The degenerate equilibrium path CD is unstable everywhere, which might induce a buckling or a snap-through behaviour.

3.3 Equation of Motion

This section will now focuss on the discontinuous case of the SD oscillator using Hamiltonian for the unperturbed dynamics and the semi-analytical method, which will be developed to investigate the damped and forced oscillations in the following sections. The developed semi-analytical method would be used in the following of the book concerning the discontinuous dynamics.

From the point of view of the mechanical device, the springs depicted in Fig. 2.1a act as a conical compression spring with doubled stiffness $2k$ hinged at $X = 0$ as suggested by Fig. 3.1a. The mass is hinged to the springs in such a way that it could move inside. So that if the mass is forced, it can pass the fully compressed state of the spring at $X = 0$, causing the spring to flip or buckle. Thus the resistance of the spring could force the mass to oscillate either around upper or lower equilibrium positions.

The equation of motion for the limit case of the oscillator is rewritten here again for the convenience of the reader:

$$m\ddot{X} + 2k(X - \mathrm{Sign}(X)L) = 0, \mathrm{Sign}(x) = \begin{cases} 1, & \text{if } x > 0 \\ 0, & \text{if } x = 0 \\ 1, & \text{if } x < 0 \end{cases} \tag{3.4}$$

Using the same change of variables as in the smooth case, system (3.4) is changed into the following dimensionless one

$$\ddot{x} + \omega_0^2(x - \mathrm{sign}(x)) = 0, \tag{3.5}$$

by letting $\omega_0^2 = \frac{2k}{m}$, $x = \frac{X}{L}$. In the presence of viscous damping and external harmonic excitation of amplitude f_0 and frequency Ω, system (3.5), it is modified as

$$x'' + 2\xi x' + (x - \mathrm{sign}(x)) = f_0 \cos \omega\tau, \tag{3.6}$$

where the variables are now defined as

$$t = \frac{\tau}{\omega_0}, \quad f_0 = \frac{F_0}{\omega_0^2}, \quad \omega = \frac{\Omega}{\omega_0},$$

and where $x' = \dfrac{dx}{d\tau}$. Figure 3.2a shows the discontinuous oscillator of Duffing type restoring force and the potential energy with double well for $\omega_0^2 = 1$. The potential energy curve is non-smooth at $x = 0$ for the discontinuity. The damping free system (3.5) can be written as two first order differential equations by letting $x' = y$:

$$\begin{cases} x' = y, \\ y' = -(x - \mathrm{sign}(x)). \end{cases} \tag{3.7}$$

The equilibria of the system (3.7) are obtained as $(0, 0)$ and $(\pm 1, 0)$. Although the Jacobian matrix at point $(0, 0)$ does not exist due to the discontinuity, the structure of the phase portrait near $(0, 0)$ is similar to the hyperbolic one, the smooth case for $\alpha > 0$. So that the equilibrium point $(0, 0)$ is referred to as the saddle-like singularity. This is because both the right and left hand-side derivative exist having the same value 1. From the mathematical point of view, this singularity is a removable discontinuity. However, the Jacobian matrix at $(\pm 1, 0)$ does exist and it is calculated as

$$J_{(\pm 1,0)} = \begin{pmatrix} 0 & 1 \\ -1 & 0 \end{pmatrix},$$

with eigenvalues $\lambda_{1,2} = \pm i$. Thus the points $(\pm 1, 0)$ are center equilibria. The Hamiltonian and the potential can be obtained by multiplying both sides of the second equation of Eq. (3.7) by the first and integrating over $[0, x]$, written as

$$P(x) = \frac{1}{2}x^2 - |x|, H(x, y) = \frac{1}{2}x^2 + \frac{1}{2}y^2 - |x|. \tag{3.8}$$

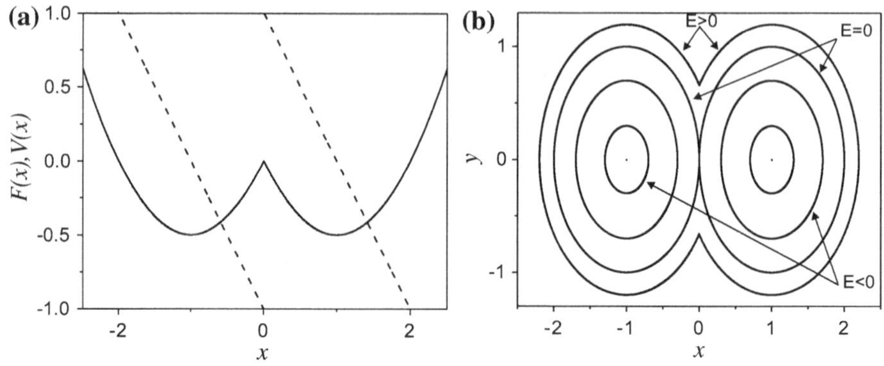

Fig. 3.2 **a** Potential energy marked by *solid curves* and restoring force marked by *dashed lines*. **b** Phase portraits of the discontinuous oscillator for $H(x, y) = E : E = 0.5, 0.0, -0.175, -0.35,$ and $E = -\frac{1}{2}$, respectively

Although the trajectories for Hamiltonian (3.8) has been plotted in the previous chapters, we would also like to repeat this procedure to address the discontinuous oscillator with double well. Figure 3.2a plots the discontinuous restoring force and also the discontinuous double potential well. In the same way, the trajectories can be classified by taking the different values of $H(x, y) = E$ and the corresponding phase portrait of the system is shown in Fig. 3.2b. This nonstandard discontinuous double well dynamical behaviour is similar to that of the standard Duffing oscillator for the double well, see [22] for details, with homoclinic-like structure. The Hamiltonian (3.8) represents two center points $(\pm 1, 0)$ for $E = -\frac{1}{2}$. The trajectories for $-\frac{1}{2} < E < 0$ are two families of circles with the centers at $(-1, 0)$ and $(1, 0)$ respectively, while the trajectories for $E > 0$ are composed of two large segments of circles with their centers located at $(-1, 0)$ and $(1, 0)$ connected at the vertical straight line $x = 0$. In the most interesting case, $E = 0$, the phase portrait of the system is depicted by two circles centered at $(\pm 1, 0)$ and connected to each other at the singularity point $(0, 0)$, forming a special singular homoclinic-like orbit. The structure near this point looks like a saddle. As discussed above, the Jacobian at $(0, 0)$ does not exit, and this isolated singularity has neither eigenvalue nor eigenvector. The two circles with exception of point $(0, 0)$, are not the manifolds of the singularity, but the flow along these circles approaches the point $(0, 0)$ as $x \to 0$ with the rate *dy*/*dx* tending to infinity. The flow will be trapped by the singularity when entering it in a finite time. This homoclinic-like orbit for $E = 0$, is theoretically expressed in Eq. (2.5).

3.4 Perturbed Discontinuous Dynamics

In this section we will now investigate the perturbed discontinuous dynamics using the developed semi-analytical procedure to the damped oscillator, undamped forcing oscillator and also to the forced system with both damping and external harmonic excitations.

3.4.1 Semi-analytical Analysis

A semi-analytical analysis, can now be developed for the calculation of the solutions of Eq. (3.6). For convenience, the perturbed system (3.6) can be written as two first order differential equations by letting $x' = y$:

$$\begin{cases} x' = y, \\ y' = -2\xi y - (x - \text{sign}(x)) + f_0 \cos \omega\tau. \end{cases} \tag{3.9}$$

The analytical solution for the perturbed system (3.9) is expressed as follows.

$$\begin{cases} x(\tau) = x_0(\tau) + x_g(\tau), \\ y(\tau) = y_0(\tau) + y_g(\tau), \end{cases} \tag{3.10}$$

where

$$x_0(\tau) = \frac{(1-\omega^2) f_0 \cos(\omega\tau) + 2 f_0 \xi \omega \sin(\omega\tau)}{4\xi^2\omega^2 + (1-\omega^2)^2},$$

$$y_0(\tau) = \frac{-\omega(1-\omega^2) f_0 \sin(\omega\tau) + 2 f_0 \xi \omega^2 \cos(\omega\tau)}{4\xi^2\omega^2 + (1-\omega^2)^2},$$

with

$$\begin{aligned} x_g(\tau) &= e^{-\xi(\tau-\tau_0)} \left(C_1 \cos \omega_d (\tau - \tau_0) + C_2 \sin \omega_d (\tau - \tau_0)\right) + \text{sign}(x(\tau_0)), \\ y_g(\tau) &= e^{-\xi(\tau-\tau_0))} \left((\omega_d C_2 - \xi C_1) \cos \omega_d (\tau - \tau_0) - (\omega_d C_1 + \xi C_2) \sin \omega_d (\tau - \tau_0)\right), \end{aligned}$$

where $x(\tau_0)$ and $y(\tau_0)$ are the initial values at the initial time τ_0, $\omega_d = \sqrt{1-\xi^2}$ is the damped frequency with a slight decreasing comparing with the natural frequency of system (3.6) vibrating around the centers $(\pm 1, 0)$.

The constants C_1 and C_2 are obtained due to initial conditions.

$$\begin{aligned} C_1 &= x(\tau_0) - x_0(\tau_0) - \text{sign}(x(\tau_0)) \\ C_2 &= \tfrac{1}{\omega_d} \left[y(\tau_0) - y_0(\tau_0)\right] + \tfrac{\xi}{\omega_d} \left[x(\tau_0) - x_0(\tau_0) - \text{sign}(x(\tau_0))\right] \end{aligned}$$

so that the solution finally reads:

$$\begin{aligned} x(\tau) = x_0(\tau) + \frac{e^{-\xi(\tau-\tau_0)}}{\omega_d} &\left[(y(\tau_0) - y_0(\tau_0) + \xi \left(x(\tau_0) - x_0(\tau_0) - \text{sign}(x(\tau_0))\omega_2\right)\right) \sin \omega_d (\tau - \tau_0) \\ &+\omega_d \left(x(\tau_0) - x_0(\tau_0) - \text{sign}(x(\tau_0))\omega_2\right) \cos \omega_d (\tau - \tau_0)\big] + \text{sign}(x(\tau_0)), \end{aligned}$$

$$\begin{aligned} y(\tau) = y_0(\tau) + \frac{e^{-\xi(\tau-\tau_0)}}{\omega_d} &\{\omega_d \left(y(t_0) - y_0(t_0)\right) \cos \omega_d (\tau - \tau_0) \\ &- \left[\left(\xi^2 + \omega_d^2\right) \left(x(\tau_0) - x_0(\tau_0) - \text{sign}(x(t_0))\right) + \xi \left(y(\tau_0) - y_0(\tau_0)\right)\right] \sin \omega_d (\tau - \tau_0)\}. \end{aligned}$$

In this discontinuous case, the phase space is divided into two regimes $x < 0$ and $x > 0$ connected at $x = 0$. The constants above depend on the initial conditions and also on the transient point when the trajectory crosses the boundaries $x = 0$.

3.4.2 Dissipative Oscillations

In the dissipative case with $\xi \neq 0$, still in the unforced case i.e. $f_0 = 0$, Eq. (3.6) becomes

$$x'' + 2\xi x' + (x - \mathrm{Sign}(x)) = 0. \tag{3.11}$$

where the damping ratio ξ is always assumed to be small. The equilibria of this damped system are the same as that of the system (3.7). The equilibria $(\pm 1, 0)$ of system (3.11) are stable centers with eigenvalues having negative real parts, $\lambda_{1,2} = -\xi \pm i\sqrt{1-\xi^2}$. The property of the saddle-like singularity $(0, 0)$ remains unchanged as that of system (3.7). The basin of attractions of the stable singularities $(\pm 1, 0)$ are shown in Fig. 3.3a and the basin boundary is plotted in Fig. 3.3b. The point $(0, 0)$ is located in the basin boundary which is the separatrix curve to separate the basin of attractions of equilibria $(\pm 1, 0)$. Along this separatrix, the flow approaches the point $(0, 0)$ as $x \to 0$ with the rate dy/dx tending to infinity. The flow will be trapped by the singularity when entering it at a finite time, which is similar with the unperturbed situation discussed in Sect. 2.3.

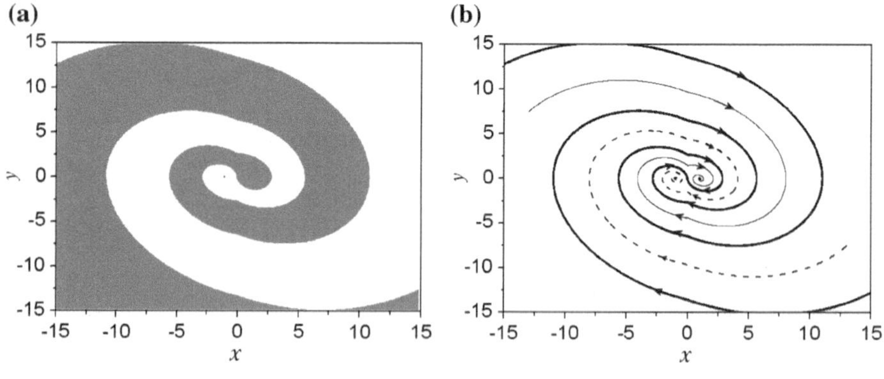

Fig. 3.3 Damped behaviour for $\xi = 0.2$: **a** The basins of attractions are shown in *grey* for the center $(1, 0)$ and in *white* for $(-1, 0)$, **b** the separatrix on the border of the basin of attractions, is marked as the *bold solid curve*, and the trajectories converging to the stable centers $(1, 0)$ and $(-1, 0)$ are shown by *thin solid* and *dashed curves* respectively

3.4.3 Undamped Forcing Oscillations

Introducing a forcing opens the analysis of resonances. This will be studied later. Studying undamped forced oscillations means that Eq. (3.6) is taken with $\xi = 0$ and $f_0 \neq 0$, so that the dynamics is governed by

$$x'' + (x - \text{sign}(x)) = f_0 \cos \omega\tau. \tag{3.12}$$

This system is non-integrable due to the explicit presence of time τ in the forcing term. The time-dependent Hamiltonian is obtained and written as

$$H_\tau(x, y, \tau) = \frac{1}{2}y^2 + \frac{1}{2}x^2 - |x| - f_0 x \cos \omega\tau, \tag{3.13}$$

where y stands for x'.

In the following, a semi-analytical method is used to solve this time dependent Hamiltonian system. We will firstly present the quasi-periodic behaviours according to Poincaré section regardless the rotation of the natural frequency avoiding the use of terminology of resonance, which will be discussed later.

Figure 3.4 plots the Poincaré section for $\alpha = 0, \xi = 0, f_0 = 0.8$ and $\omega = 1.05$. KAM structure, see [16, 19] for instance, shows a series of fixed periodic points encircled by the quasi-periodic islands surrounded by the chaotic orbit. A special chaotic orbit is found to be distributed to fill the finite area surrounding the isolated

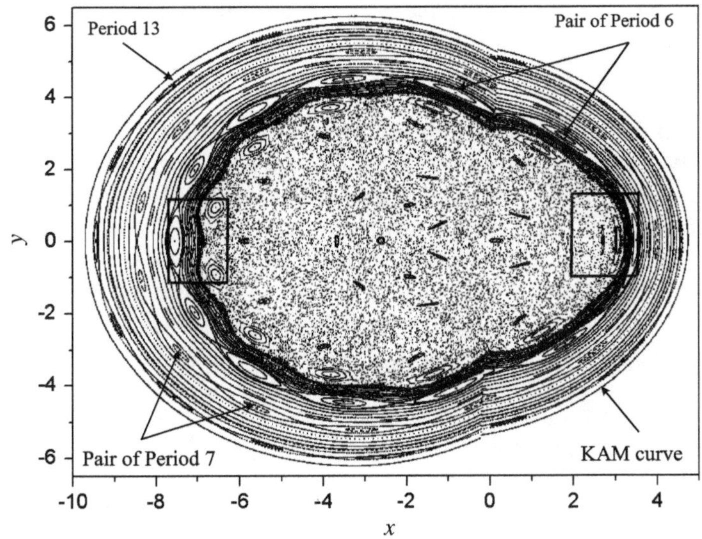

Fig. 3.4 Global KAM structure on Poincaré section with KAM orbits, series of fixed points with the related quasi period islands chained by the chaotic trajectories

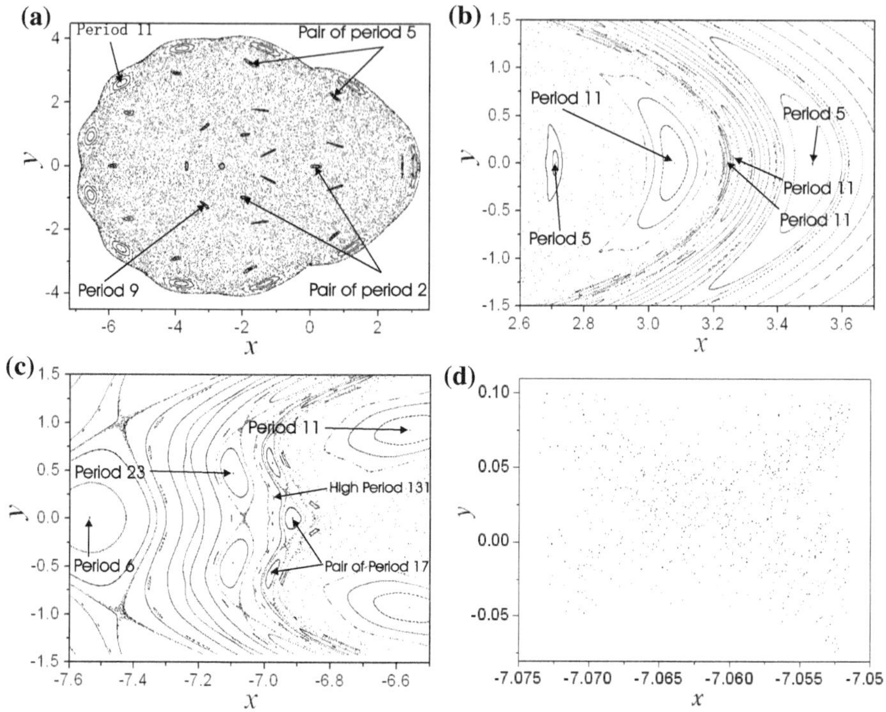

Fig. 3.5 Local amplifications of Fig. 3.4: **a** The chaotic trajectory filling the finite region and connecting islands encircling series of series of fixed points. **b** Local amplification for the right rectangle area in Fig. 3.4, **c** local amplification for the left rectangle area in Fig. 3.4, **d** local amplification for chaotic orbit connecting the quasi-periodic island chain, in the rectangle area in c

islands which encircles the related series resonance fixed points. As shown in the outside of the finite region, series of quasi-periodic islands consist of a pair of quasi-period 6, a pair of quasi-period 7 and a quasi-period 13 islands connected by the related chaotic orbits, respectively, which are separated by KAM curves.

Figure 3.5a shows the dynamics within the finite area filled by the special chaotic trajectory which is known as the chaotic sea or stochastic web, see details in [16, 19]. This chaotic trajectory connects quasi-periodic islands, [21], encircling the series of pair of period 2, a pair of period 5, one period 9 and one period 11 fixed points. The structure of these dynamics forms a chaotic sea, [19–21], with quasi-periodic solutions. Figure 3.5b, c are the locally amplified structure in the rectangle area near the right- and left-hand side of the dark looped area in Fig. 3.4. Various series of resonant solutions are displayed. Examples for these resonance solutions are marked: series of resonance periodic fixed points of period 17, period 23 and a high order fixed points of period 131 being visible with the related islands and the chaotic orbits. Figure 3.5d shows the locally amplified chaotic orbit connecting the island chain, in the rectangle area in Fig. 3.5c.

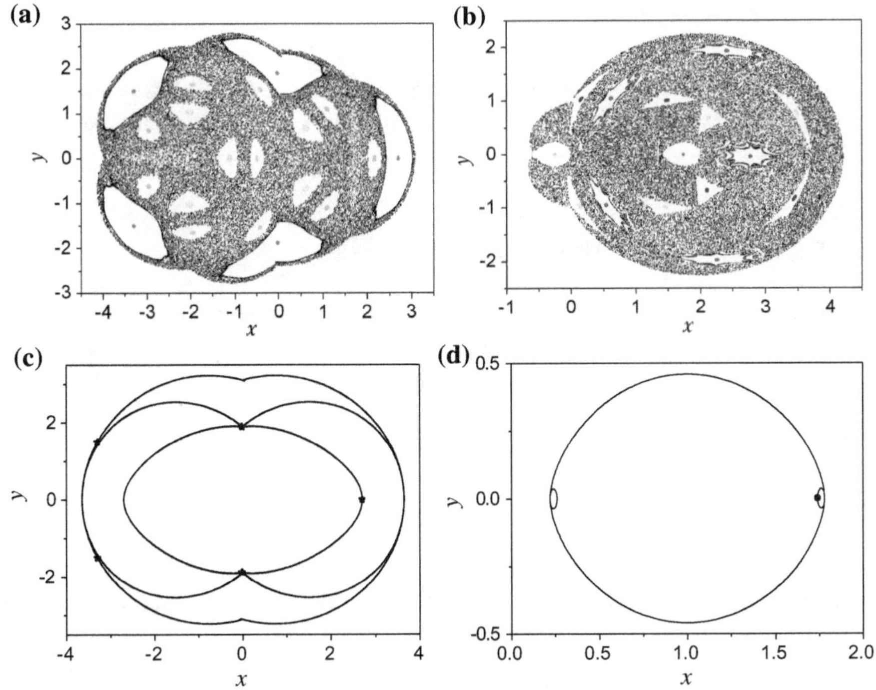

Fig. 3.6 Chaotic seas and periodic solutions for $f_0 = 0.8$: **a** For $\omega = 1.3$, the series of fixed points of period 5 are marked *gray*, *light gray*, *green*, *yellow*, respectively, **b** for $\omega = \frac{1}{3}$, the series of fixed points are marked *gray* and *light gray* for the pair of period one, *dark gray* and *yellow* for the pair of period 4, and the red for the period 5, **c** and **d** are the trajectories the Poincaré sections for one of the series of fixed points of period 5 in **a** and one of the period on in **b** respectively

More chaotic seas and periodic behaviours can be found by varying the external frequency ω. Figure 3.6 presents two other examples of chaotic sea for $f_0 = 0.8$ and $\omega = 1.3$ in Fig. 3.6a and $\omega = \frac{1}{3}$ in Fig. 3.6b respectively. In Fig. 3.6a, four series of fixed points of period 5 are found and marked with color gray, light gray, green, and blue respectively. In Fig. 3.6b, the pair of fixed points of period one are marked with color gray and light gray, the pair fixed points of period 4 are marked with color dark gray and yellow and the series fixed points of period 5 is marked with red and the fixed points of period 5 are marked with blue. Trajectories and Poincaré sections for one of the series of period 5 in Fig. 3.6a and one of the pair of period one in Fig. 3.6b are displayed in Fig. 3.6c and d respectively.

3.4.4 Forced Dynamics

In this subsection, the forced dynamics with dissipation, the forced dynamics without damping and the forced system with both damping and external harmonic excitations are investigated.

3.4.4.1 Influence with Weak Dissipation

As discussed in previous sections, the discontinuous system without dissipation exhibits primary resonant solutions and corresponding quasi-periodic solutions. In this subsection, we first study the behaviour of these solutions in the presence of weak dissipation.

Bifurcation diagrams are given for velocity y versus the weak dissipation ξ in Fig. 3.7a and b, starting from quasi periodic solutions for $\xi = 0$. These diagrams show the bifurcation from quasi periodic to periodic solutions and chaotic solutions in the case of weak dissipation.

From Fig. 3.7a and b, the co-existence of periodic solutions can be seen with a weak dissipation. In Fig. 3.7a three bifurcation diagrams are plotted starting from the quasi-periodic solutions as ξ changes. The one marked by black, starts from a quasi-periodic solution of period 9 which bifurcates to the periodic solution of the same period 9 as ξ increases. The blue starts from a quasi-periodic orbit of period 2 which bifurcates to period 2 and the red starting from quasi-periodic of period 5 bifurcates to period 5. The similarity can be seen in Fig. 3.7b, the bifurcation diagram marked blue starts from periodic solution of period 11 bifurcating to peri-

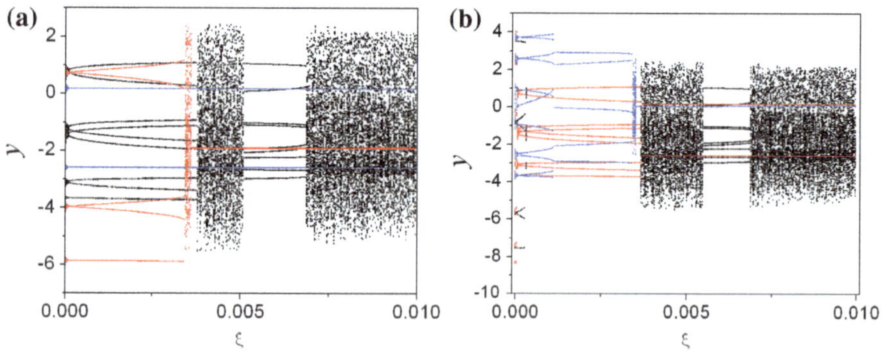

Fig. 3.7 Resonance or quasi-periodic solution bifurcation diagrams for y versus ξ: **a** The *black plots* mark the bifurcation diagram starting from quasi-periodic orbit of period 9, the *blue* starting from quasi-periodic orbit of period 2 and the *red* starting from quasi-periodic orbit of period 5, respectively, **b** the *black plots* mark the bifurcation diagram starting from quasi-periodic orbit of period 6, the *blue* starting from quasi-periodic orbit of period 11 and the *red* starting from quasi-periodic orbit of period 13, respectively

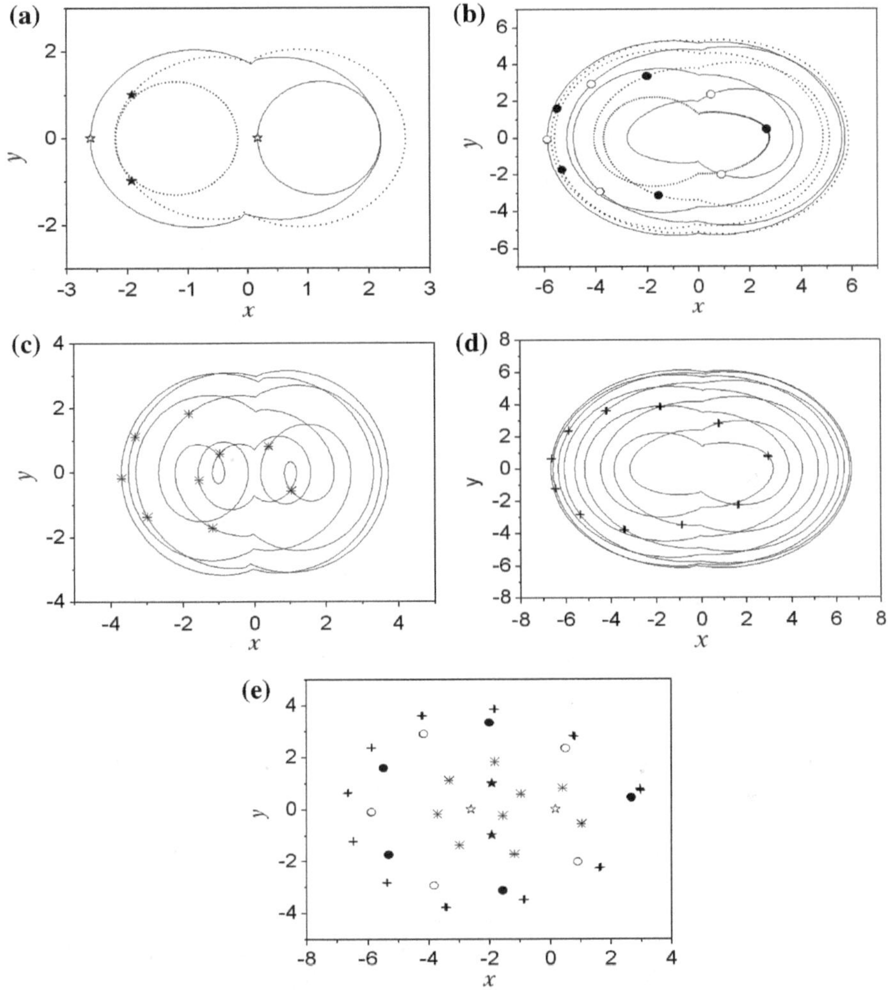

Fig. 3.8 The co-existence of periodic orbits for $\xi = 0.001$, $f_0 = 0.8$, $\omega = 1.05$: **a** A pair of period 2 solutions, **b** a pair of period 5 solutions, **c** period 11 solution, **d** period 11 solution, **e** Poincaré sections for all the co-existed period solutions

odic solution of period 11 as well as ξ varies the small value. This concludes that the co-existence of the primary resonance solutions within the finite region, as shown in Fig. 3.5, governed by the special chaotic trajectory remains unchanged while the corresponding quasi-periodic solutions bifurcate to the periodic solutions in the weak dissipation. Figure 3.8 shows this co-existence of periodic orbits for $\xi = 0.001$, $f_0 = 0.8$, $\omega = 1.05$; (a) A pair of period 2 solutions, (b) a pair of period 5 solutions, (c) a period 9 solution and (d) a period 11 solution, (e) the Poincaré section for co-existing periodic orbits.

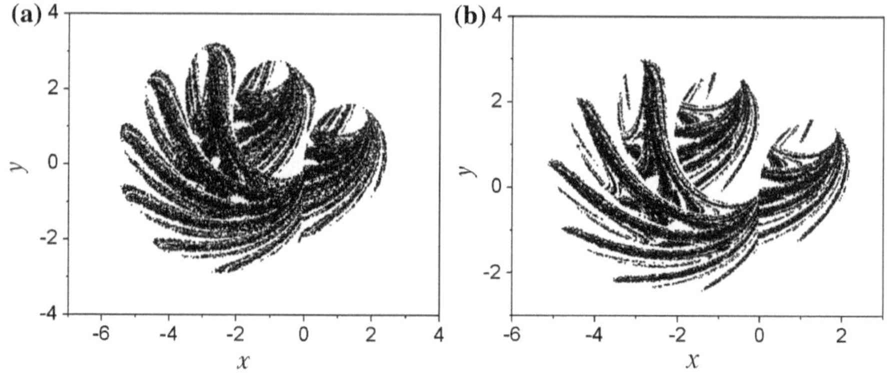

Fig. 3.9 Chaotic orbits under weak dissipative for $f_0 = 0.8$, $\omega = 1.05$: **a** $\xi = 0.005$, **b** $\xi = 0.01$

The primary resonance solutions in the outer area of the finite area filled by the special chaotic trajectory, as shown in Fig. 3.5, together with the related quasi-periodic solutions are nonpersistent. In Fig. 3.7b, the bifurcation diagram marked by black starts from quasi-periodic of period 6 which bifurcates to period 9 periodic solution as ξ increase and the bifurcation diagram marked by red points starts from a quasi-periodic solution of period 13 which bifurcates to period 9 as ξ increases as well.

In addition to the quasi- periodic bifurcations, other periodic windows and chaotic regions can be seen from Fig. 3.7a, b. There are two main intervals in which chaotic motions occur. The chaotic attractors for $f_0 = 0.8, \omega = 1.05$, are presented in Fig. 3.9a for $\xi = 0.005$ and Fig. 3.9b for $\xi = 0.01$, respectively.

Semi-analytical studies also show the long time transient from a chaotic saddle leading to periodic solution. This behaviour can be characterized by chaotic saddles, see [27–29]. The chaotic transient and the final period 2 attractor are shown in Fig. 3.10a, for $\xi = 0.0035$. Figure 3.10b shows the bifurcation diagram for y versus time τ, which shows the transient process from the chaotic saddle to the period 2 attractor.

The chaotic attractors and the chaotic saddle have the same topology. Similar behaviour was observed by Thompson and Ghaffari, see also Thompson and Stewart, [30, 31]. The latter also arose in a model of linear oscillator with impacts, [31], which has apparent similarities with the present model.

3.4.4.2 Amplitude Response

Figure 3.11a shows the bifurcation diagram of the system as f_0 changes for $\omega = 1.05$ and $\xi = 0.01$. From this bifurcation diagram, multiple periodic windows and chaotic regions can be seen. As an example, the period doubling leading to chaos is presented in Fig. 3.12 for $\omega = 1.05$ and $\xi = 0.01$. Figure 3.12a–c shows the periodic orbits

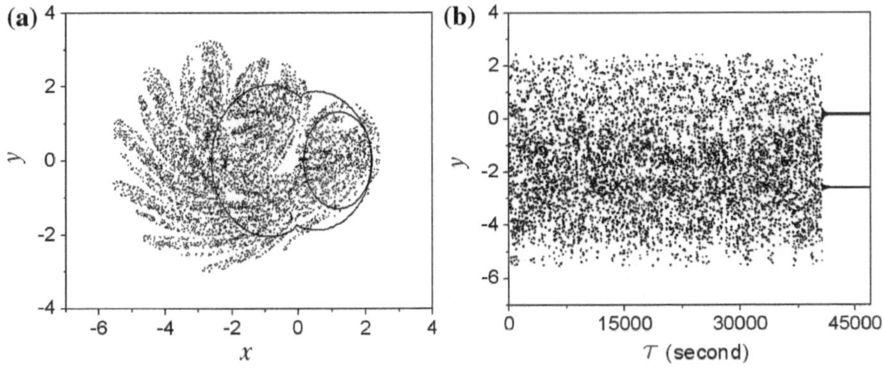

Fig. 3.10 State space plots for $f_0 = 0.8$, $\omega = 1.05$ and $\xi = 0.0035$: **a** Chaotic saddle leading to period 2 solution, **b** bifurcation diagram for y versus time τ

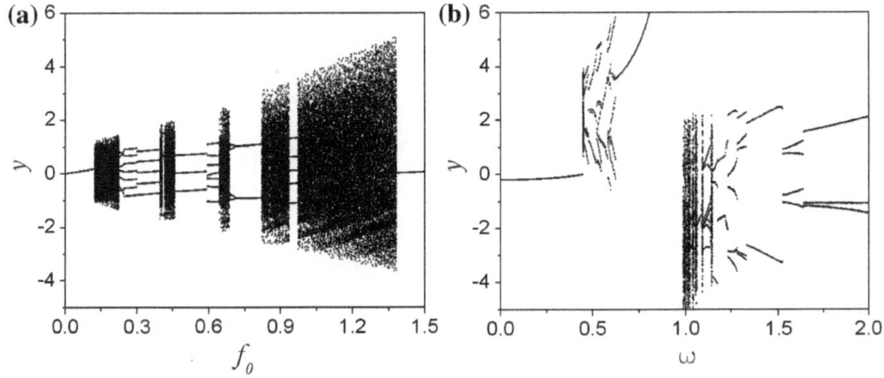

Fig. 3.11 Bifurcation diagrams: **a** for y versus f_0 for $\omega = 1.05$ and $\xi = 0.01$, **b** for y versus ω for $f_0 = 0.8$ and $\xi = 0.001$

and the Poincaré sections for period 2, $f_0 = 0.75$, period 4, $f_0 = 0.7$, and period 8, $f_0 = 0.6875$. Figure 3.12d shows the chaotic attractor, $f_0 = 0.675$, led by period doubling.

3.4.4.3 Frequency Response

By contrast, Fig. 3.11b shows the bifurcation diagram for velocity y versus the external frequency ω for $f_0 = 0.8$ and $\xi = 0.001$. From this bifurcation diagram, multiple periodic windows and chaos can be seen. Some special periodic orbits are shown in the following figures. Figure 3.13a, b show the trajectories for the subharmonic periodic solutions of period 3 and period 2, for $\omega = 0.5$ and $\omega = 0.56$, respectively. Figure 3.14a, b present the primary harmonic trajectories of period 7 and period 3, for $\omega = 1.125$ and $\omega = 1.64$, respectively and Fig. 3.15a, b display the trajectories

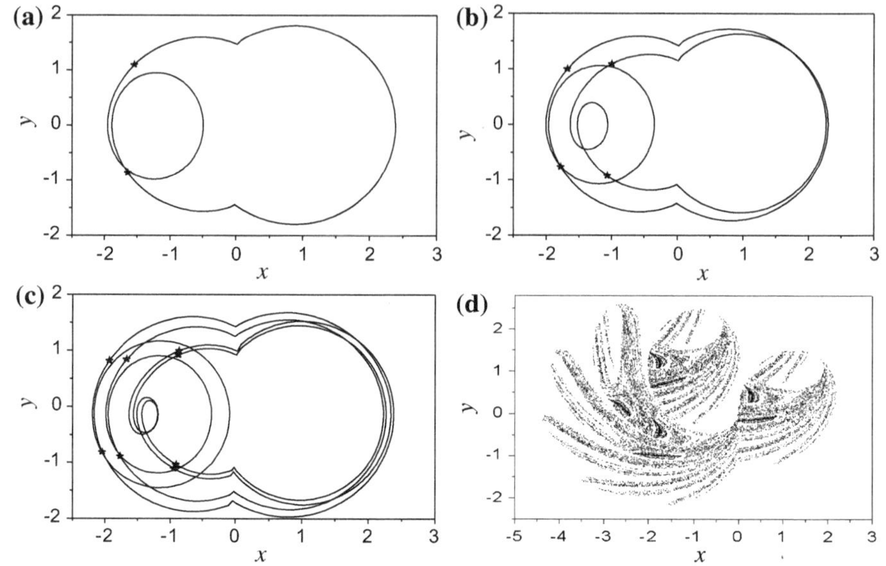

Fig. 3.12 Period doubling bifurcation to chaos for $\omega = 1.05$ and $\xi = 0.01$: **a** periodic solution of period 2 for $f_0 = 0.75$, **b** periodic solution of period 4 for $f_0 = 0.7$, **c** periodic solution of period 8 for $f_0 = 0.6875$, **d** chaotic attractor led from period doubling for $f_0 = 0.675$

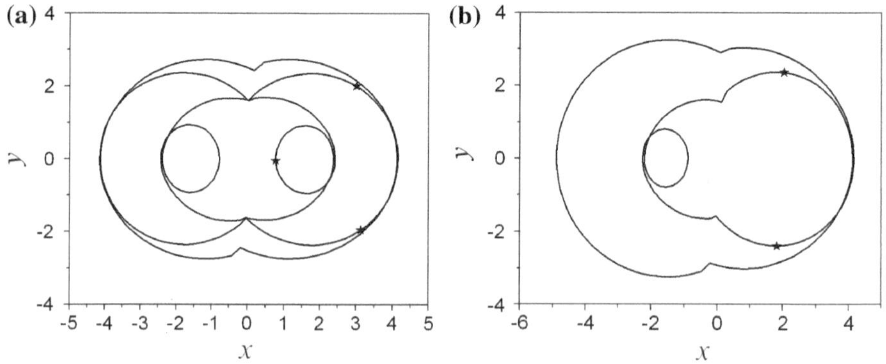

Fig. 3.13 Subharmonic periodic solutions for $f_0 = 0.8$ and $\xi = 0.001$: **a** Period 3 for $\omega = 0.5$ and **b** period 2 for $\omega = 0.56$

of one super harmonic periodic solution of period 2 and super harmonic periodic solution of period 3 for $\omega = 1.4$ and $\omega = 2$, respectively. The solid stars in the corresponding figures mark the corresponding Poincaré sections.

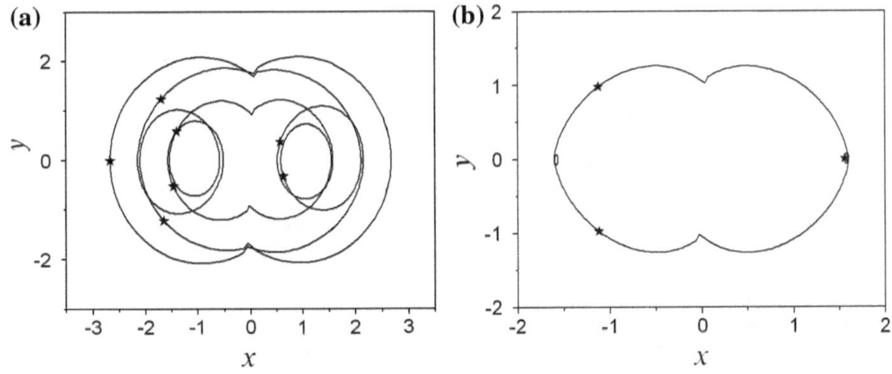

Fig. 3.14 Primary resonance for $f_0 = 0.8$ and $\xi = 0.001$: **a** Period 7 for $\omega = 1.125$ and **b** period 3 for $\omega = 1.64$

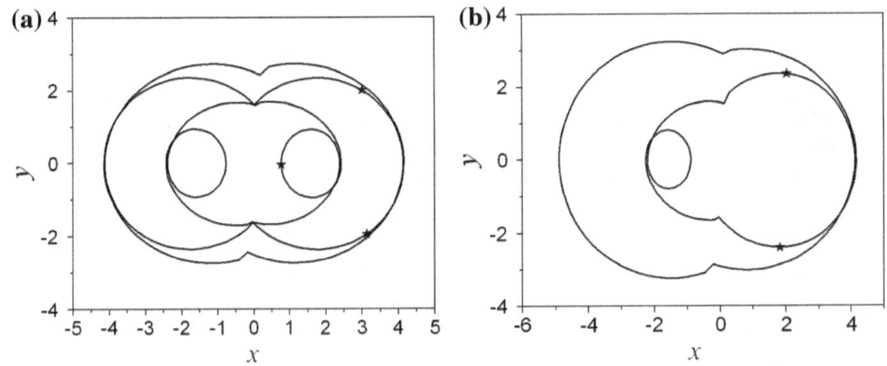

Fig. 3.15 Super harmonic periodic solutions for $f_0 = 0.8$ and $\xi = 0.001$: **a** Period 2 for $\omega = 1.4$ and **b** period 3 for $\omega = 2.0$

3.5 Summary

This chapter has presented a new type of discontinuous oscillator with bistability which is the limit case of the SD oscillator. The complicated dynamics has been presented through the introduction of an analytical method to formulate the solutions of the discontinuous oscillator, avoiding the barriers encountered in the conventional calculations due to the discontinuity. This scheme made it possible to obtain chaotic solutions or chaotic attractors analytically for this discontinuous oscillator.

References

1. Pavlovska, E., & Wiercigroch, M. (2004). Analytical drift reconstruction for visco-elastic impact oscillators operating in periodic and chaotic regimes. *Chaos, Solitons and Fractals, 19*(1), 151–161.
2. Leine, R. I., & Van, D. H. (2002). Campen. Discontinuous bifurcations of periodic solutions. *Mathematical and Computer Modelling, 36*(3), 259–273.
3. Leine, R. I., Van Campen, D. H., & Van de Vrande, B. L. (2000). Bifurcations in nonlinear discontinuous systems. *Nonlinear Dynamics, 23*(23), 105–164.
4. Leine, R. I., Van Campen, D. H., De Kraker, A., & Van den Steen, L. (1998). Stick-slip vibrations induced by alternate friction models. *Nonlinear Dynamics, 16*(1), 41–54.
5. Filippov, A. F. (1988). *Differential equations with discontinuous right-hand sides: Control systems* (Vol. 18). Mathematics and its applications The Netherlands: Springer. Originally published in Russian.
6. Filippov, A. F. (1991). Differential equations with discontinuous right-hand sides. *Journal of Mathematical Analysis & Applications, 154*(2), 377–390.
7. Xu, L., Lu, M. W., & Cao, Q. J. (2002). Nonlinear vibrations of dynamical systems with a general form of piecewise-linear viscous damping by incremental harmonic balance method. *Physics Letters A, 301*(1–2), 65–73.
8. Cao, Q. J., Xu, L., Djidjeli, K., Price, W. G., & Twizell, E. H. (2001). Analysis of period-doubling and chaos of a non-symmetric oscillator with piecewise-linearity. *Chaos Solitons and Fractals, 12*(10), 1917–1927.
9. Di Bernardo, M., & Kowalczyk, P. (2003). Nordmark a sliding bifurcations: A novel mechanism for the sudden onset of chaos in dry friction oscillators. *International Journal of Bifurcation and Chaos, 13*(10), 2935–2948.
10. Di Bernardo, M., Garofalo, F., Iannelli, L., & Vasca, F. (2002). Bifurcations in piecewise-smooth feedback systems. *International Journal of Control, 75*(16–17), 1243–1259.
11. Di Bernardo, M., Kowalczyk, P., & Nordmark, A. (2002). Bifurcations of dynamical systems with sliding: derivation of normal-form mappings. *Physica D-Nonlinear Phenomena, 170*(3–4), 175–205.
12. Banerjee, S., & Grebogi, C. (1999). Border collision bifurcations in two-dimensional piecewise smooth maps. *Physical Review E, 59*(4), 4052–4061.
13. Banerjee, S., Karthik, M. S., Yuan, G. H., & Yorke, J. A. (2000). Bifurcations in one-dimensional piecewise smooth maps - theory and applications in switching circuits. *IEEE Transactions on Circuits and Systems-I: Fundamental Theory & Applications, 47*(3), 389–394.
14. Chin, W., & Grebogi, C. (1996). Universal grazing bifurcations in impact oscillators. *Physical Review E, 53*(1), 134–139.
15. Whiston, G. S. (1992). Singularities in vibro-impact dynamics. *Journal of Sound and Vibration, 152*(3), 427–460.
16. Lichtenberg, A. J., & Lieberman, M. A. (1992). *Regular and chaotic dynamics*. New York: Springer.
17. Arnold, V. I. (1983). *Mathematical methods of classical mechanics*. New York: Springer.
18. Jackson, E. A. (1990). *Persentives on nonlinear dynamics 1,2*. Cambridge: Cambridge University Press.
19. Chernikov, A. A., Sagdeev, R. Z., Usikov, D. A., Yu Zakharov, M., & Zaslavsky, G. M. (1987). Minimal chaos and stochastic webs. *Nature, 326*, 559–563.
20. Pekarsky, S., & Rom-Kedar, V. (1997). Uniform stochastic web in lowdimensional hamiltonian systems. *Physical Letters A, 225*(4–6), 274–286.
21. Luo, A. C. J. (2006). Chaos and quasi-periodic motions on the homoclinic surface of nonlinear hamiltonian systems with two degrees of freedom. *Journal of Computational and Nonlinear Dynamics, 1*(2), 135–142.
22. Guckenheimer, J., & Holmes, P. (1983). *Nonlinear oscillation, dynamical system and bifurcation of vector fields*. New York: Springer.

23. Cao, Q. J., Wiercigroch, M., Pavlovskaia, E. E., Grebogi, C., & Thompson, J. M. T. (2006). Archetypal oscillator for smooth and discontinuous dynamics. *Physics Review E*, *74*(046218), 1–5.
24. Cao, Q. J., Wiercigroch, M., Pavlovskaia, E. E., Grebogi, C., & Thompson, J. M. T. (2008). Piecewise linear approach to an archetypal oscillator for smooth and discontinuous dynamics. *Philosophical Transactions of the Royal Society, A-Mathematical, Physical and Engineering Sciences*, *366*(1865), 635–652.
25. Lewin, W. (1999). *Hooke's Law, simple harmonic oscillator.* MIT course 8.01: Classical mechanics, lecture 10. (ogg) (videotape) Cambridge: MIT OCW.
26. Struganova, I. (2005). A spring, hooke's law, and archimedes' principle. *Physics Teacher*, *43*(8), 516–518.
27. Lai, Y. C., & Lerner, D. (1998). Effective scaling regime for computing the correlation dimension from chaotic time series. *Physica D-Nonlinear Phenomena*, *115*, 1–18.
28. Tel, T. (1986). Characteristic exponents of chaotic repellers as eigenvalues. *Physics Letters A*, *119*(2), 65–68.
29. Grebogi, C., Ott, E., & Yorke, J. A. (1983). Crises, sudden changes in chaotic attractors, and transient chaos. *Physica D-Nonlinear Phenomena*, *7*(1–3), 181–200.
30. Thompson, J. M. T., & Ghaffari, R. (1983). Chaotic dynamics of an impact oscillator. *Physical Review A*, *27*(3), 1741–1743.
31. Thompson, J. M. T., & Stewart, H. B. (2002). *Nonlinear dynamics and chaos* (2nd ed.). Chichester: Wiley.

Chapter 4
Resonances

Attention is now paid to the resonance behaviour of the discontinuous case of the oscillator. Although the method and the results presented here are for the discontinuous case, they are also valid for the smooth dynamics. To clarify the resonant phenomena, we introduce the concepts of generalized winding number and the generalized canonical generating functions to derive the complicated resonant dynamics for the discontinuous system. The time dependent Hamiltonian for the discontinuous system enables us to understand such a resonant motion by a composed winding number consisting of multiple phases, each of which is confined within one of the separate resonant area, satisfying the regular resonant condition. The complex resonant structures of stochastic webs or chaotic sea with stochastic trajectories entering different resonant layers of resonant island chains observed are considered as the combination of the separated linear system according to the composed winding behaviour. This might suggest further investigations into resonant synchronization and mechanisms of vibrational power flow transfer from nonlinear smooth to discontinuous dynamical systems.[1]

4.1 Introduction

Resonance phenomena of a harmonically excited system with multiple potential well plays an important role in nonlinear dynamics research. In this chapter, we present the resonant behaviours of a discontinuous dynamical system derived from the SD oscillator to gain a better understanding of the mechanism of the resonance transitions. Time dependent Hamiltonian is introduced for the oscillator with bistability. This system comprises two subsystems connected at $x = 0$ where the system is

[1]This chapter relies on a common work with Professor Yeping Xiong, School of Engineering, University of Southampton, Southampton SO17 1BJ, UK.

Q. Cao and A. Léger, *A Smooth and Discontinuous Oscillator*, Springer Tracts in Mechanical Engineering, DOI 10.1007/978-3-662-53094-8_4

discontinuous. We construct a series of generating functions and canonical transformations to obtain the canonical form with action variables [1] of the system of which the resonant behaviours will be investigated. A composed winding number is introduced to explore resonant phenomena. The formulation given in this section for resonant phenomena generalizes the formula $n\omega_0 = m\omega$ used in the regular perturbation theory, where n and m are relative prime numbers, ω_0 and ω respectively the natural and external frequencies. Understanding the resonant behaviour of the SD oscillator at the discontinuous phase enables us to reveal the vibrational energy transfer mechanism [2] between smooth and discontinuous nonlinear dynamical systems and the use of chaos controlling and synchronization [3, 4] in security communication using nonlinear systems.

Much attention has been paid to the resonance phenomena of harmonically excited forcing system with multiple well dynamics, [5–7], which plays an important role in the nonlinear dynamics research, e.g. [8–10]. The formulation of resonances by regular perturbation theory, Hamiltonian and KAM theory [11, 12], is inconvenient to study the resonant phenomena exhibited in a harmonically excited forcing system with single well potentials, as it is impossible to find a pair of co-prime integers m and n which satisfies the condition $m\omega_0 = n\omega$, where ω_0 and ω are the natural frequency and the external excitation frequency [13, 14], respectively. This system comprises two subsystems connected at $x = 0$, where each of them exhibits regular resonance, while the overall system generates complex resonances due to its discontinuous nature. Although the formulation is presented in the particular case of the discontinuous system, it is valid for both smooth and discontinuous cases.

4.2 Time Dependent Hamiltonian

Consider again the forced discontinuous oscillator of frequency ω and amplitude f_0 without damping discussed in Chaps. 2 and 3. Here for convenience, it is rewritten as follows

$$x'' + (x - \text{sign}(x)) = f_0 \cos \omega\tau, \tag{4.1}$$

for which the time-dependent Hamiltonian can be obtained by letting $y = x'$,

$$H_\tau(x, y, \tau) = \frac{1}{2}y^2 + \frac{1}{2}x^2 - |x| - f_0 x \cos \omega\tau. \tag{4.2}$$

The first order term $|x|$ can be incorporated into the above Hamiltonian by shifting the origin from $(0, 0)$ to the equilibria $(\pm 1, 0)$, by means of the time dependent canonical transformation, [1, 15]. New variables $(x, y, \tau) \to (\overline{p}, \overline{q}, \tau)$ are defined as $\overline{p} = x - \text{sign}(x)$ and $\overline{q} = y$, and the canonical generating function is constructed as

$$F_2(\bar{p}, \bar{q}, t) = y\bar{p} = y(x - \text{sign}(x)). \tag{4.3}$$

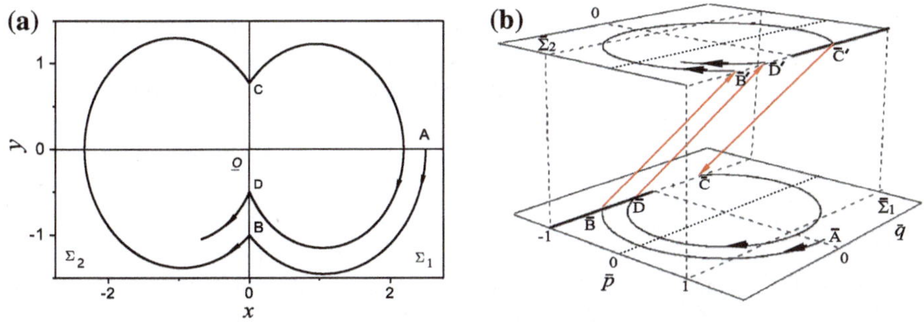

Fig. 4.1 **a** Trajectory in xy plane, **b** trajectory and the transient at $\bar{p} = \pm 1$: the *solid curve* marks the part of the trajectory in $\bar{\Sigma}_1$ and $\bar{\Sigma}_2$. The trajectory starts from $\bar{A} \in \bar{\Sigma}_1$ and the transients or the jumps at $\bar{B} \to \bar{B}'$; $\bar{C}' \to \bar{C}$ and $\bar{D} \to \bar{D}'$ such that the arc $\widehat{\bar{A}\bar{B}} \in \bar{\Sigma}_1$, $\widehat{\bar{B}'\bar{C}'} \in \bar{\Sigma}_2$ and $\widehat{\bar{C}\bar{D}} \in \bar{\Sigma}_1$

This transformation maps two half-planes Σ_1 and Σ_2 in the xy plane onto two half-planes $\bar{\Sigma}_1$ and $\bar{\Sigma}_2$ in the $\bar{q}\bar{p}$ plane, as shown in Fig. 4.1, and defined below

$$\begin{cases} \Sigma_1 = \{(x, y)|x > 0\} \bigcup \{(x, y)|x = 0, y > 0\}, \\ \Sigma_2 = \{(x, y)|x < 0\} \bigcup \{(x, y)|x = 0, y < 0\}, \\ \bar{\Sigma}_1 = \{(\bar{p}, \bar{p})|\bar{p} > -1\} \bigcup \{(\bar{p}, \bar{q})|\bar{p} = -1, \bar{q} > 0\}, \\ \bar{\Sigma}_2 = \{(\bar{p}, \bar{q})|\bar{p} < 1\} \bigcup \{(\bar{p}, \bar{q})|\bar{p} = 1, \bar{q} < 0\}. \end{cases} \tag{4.4}$$

As it can be seen from the Fig. 4.1, the overlap for $\bar{p}\bar{q}$ plane in the region $\bar{p} \in (-1, 1)$ has to be taken into account for an appropriate reconstruction of the trajectory.

Hamiltonian (4.2) can be transformed into the following form in term of the new variables $\overline{p}, \overline{q}$.

$$H_\tau(\bar{p}, \bar{q}, \tau) = \begin{cases} \frac{1}{2}\bar{p}^2 + \frac{1}{2}\bar{q}^2 - (\bar{p} + 1) f_0 \cos\omega\tau, \ (\bar{p}, \bar{q}) \in \bar{\sum}_1, \\ \frac{1}{2}\bar{p}^2 + \frac{1}{2}\bar{q}^2 - (\bar{p} - 1) f_0 \cos\omega\tau, \ (\bar{p}, \bar{q}) \in \bar{\sum}_2. \end{cases} \tag{4.5}$$

We can treat time τ and $E = -\bar{H}_\tau$ as additional coordinates in an extended phase space [1], by introducing an auxiliary parameter ξ playing the role of time [1, 16]. The new Hamiltonian of the forced system with the canonical generating function $F_2 = \bar{p}\bar{q} + E\tau$ in the extended phase space is

$$\bar{H}(\bar{p}, \bar{q}, E, \tau) = \begin{cases} \frac{1}{2}\bar{p}^2 + \frac{1}{2}\bar{q}^2 + E - (\bar{p} + 1) f_0 \cos\omega\tau, \ (\bar{p}, \bar{q}) \in \bar{\sum}_1, \\ \frac{1}{2}\bar{p}^2 + \frac{1}{2}\bar{q}^2 + E - (\bar{p} - 1) f_0 \cos\omega\tau, \ (\bar{p}, \bar{q}) \in \bar{\sum}_2. \end{cases} \tag{4.6}$$

Each of the branches of Eq. (4.6) is similar to an autonomous two degrees-of-freedom and it is non-integrable due to a lack of the integral factor. The first two terms of the extended Hamiltonian represent a simple harmonic oscillator, whose corresponding action and angle variables (J, θ) are introduced and defined as $\bar{p} = \sqrt{2J}\sin\theta, \bar{q} = \sqrt{2J}\cos\theta$, see [1] for details. This maps $\bar{\Sigma}_1 \to \Xi_1$ and $\bar{\Sigma}_2 \to \Xi_2$, which leads to

$$H_J(J, E, \theta, \tau) = \begin{cases} J + E - (\sqrt{2J}\sin\theta + 1) f_0 \cos\omega\tau, & (J, \theta) \in \Xi_1, \\ J + E - (\sqrt{2J}\sin\theta - 1) f_0 \cos\omega\tau, & (J, \theta) \in \Xi_2. \end{cases} \tag{4.7}$$

To obtain the action and angle variables, a canonical transformation is made via generating function $\bar{F}_2(J_\theta, J_\varphi; \theta, \varphi) = \varphi J_\varphi + \theta J_\theta$, such that $(J, \theta, E, \tau) \to (J_\theta, \theta, J_\varphi, \varphi)$ defined by $J_\theta = J, \theta = \theta, J_\varphi = \frac{E}{\omega}$ and $\varphi = \omega\tau$. The Hamiltonian is then of the form

$$\begin{aligned} \hat{H}_J(J_\theta, J_\varphi; \theta, \varphi) &= \hat{H}_J^0(J_\theta, J_\varphi) + \hat{H}_J(J_\theta, J_\varphi; \theta, \varphi) \\ &= \omega_\theta J_\theta + \omega J_\varphi - \begin{cases} \left(\sqrt{2J_\theta}\sin\theta + 1\right) f_0 \cos\varphi, & (J_\theta, \theta) \in \Xi_1, \\ \left(\sqrt{2J_\theta}\sin\theta - 1\right) f_0 \cos\varphi, & (J_\theta, \theta) \in \Xi_2. \end{cases} \end{aligned} \tag{4.8}$$

This extended four dimensional system has only three independent variables. One of them is the angle $\varphi = (\omega\tau) \bmod (2\pi)$. The Poincaré section for $\varphi = 0$ is equivalent to $T = 2\pi/\omega$. Two frequencies for the two angle functions (θ, φ) are introduced to characterize the extended tori for the unperturbed part of the Hamiltonian and written as

$$\omega_\theta = \frac{\partial \hat{H}_J^0}{\partial J_\theta} = \omega_0 = 1, \quad \omega_\varphi = \frac{\partial \hat{H}_J^0}{\partial J_\varphi} = \omega, \tag{4.9}$$

where ω_φ and ω_θ are the forcing and natural frequencies, respectively. The dynamic behaviour of the forced system depends on the ratio of these two frequencies. If there are co-prime integers m and n, such that the frequencies are commensurate, satisfying $m\omega_\varphi - n\omega_\theta = 0$, the ratio $\alpha = \omega_\theta/\omega_\varphi = m/n$ is defined as the winding number, or an $m : n$ primary resonance occurs [16, 17].

A $m : n$ resonance is the closed orbit on the torus (φ, θ), m turns in the short way and n turns in the long way. On the Poincaré section (or stroboscopic map, $t = t_0 + k\frac{2\pi}{\omega}$), this resonance is observed as a collection of fixed points. The stable fixed points are surrounded by the corresponding closed orbits in the neighborhood of the fixed points which make island chains representing the quasi-periodic motions. The unstable series of fixed points are always associated with chaotic orbits connecting the corresponding islands. In addition to the resonances, the system also exhibits generic KAM curves densely covering the closed curve in a long period of time. There exists a special chaotic orbit filling the finite region between the separatrices of the lower order of resonances forming the stochastic web [1, 18].

4.3 Generalized Winding Number and Resonances

Even the resonance and the winding number has been defined in regular perturbation theory, one cannot find the relative prime integers m and n such that $m\omega_\varphi - n\omega_\theta = 0$ for each resonance in system (4.7). As shown in Fig. 4.2, the resonances cannot be described as $m\omega_\varphi - n\omega_\theta = 0$ for any co-prime integers m and n for $\omega_\theta = 1$ and $\omega_\varphi = \frac{1}{3}$. To understand these resonances, the resonant trajectories in this system can be divided into two types: the first type consists of trajectories which belong to either of the two half planes, while the other type is located in the opposite half plane, respectively. For the perturbation $\hat{H}_J(J_\theta, J_\varphi; \theta, \varphi) \neq 0$ in Hamiltonian (4.7), the periodic trajectory can be divided into pieces by the pair of successive transient points. Each of these pieces is composed of two half circles located in their half planes. Suppose that there are $2k$ transient points and i circles and j twists for angle variables ω and θ in each piece of the trajectory. We say that (i, j) is the piece winding number for the corresponding piece of the trajectory. The composed winding number for the combined resonant orbit can be formulated in the following way.

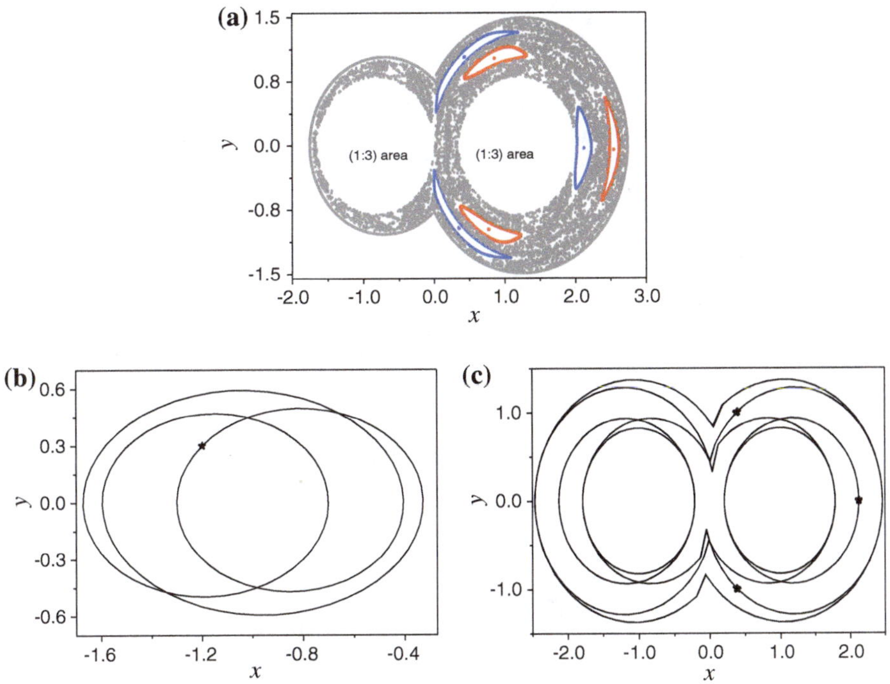

Fig. 4.2 Stochastic web and trajectories for $f_0 = 0.2$, $\omega = \frac{1}{3}$: **a** Stochastic web showing a pair of harmonic motion areas of (1 : 3) for each of the single system, and the pair of composite resonant solution with the winding number (3 : 7) marked with *blue* and *red*, respectively, **b** harmonic trajectory for the (1 : 3) vibration and **c** the resonant solution of (3 : 7) respectively

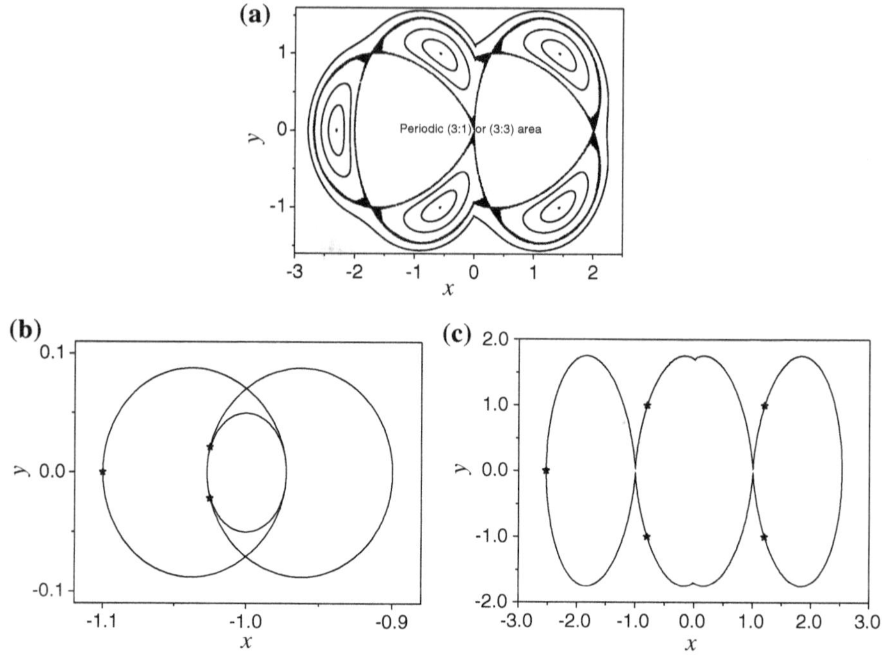

Fig. 4.3 **a** Resonant layer separatrix for $f_0 = 0.4$, $\omega = 3$ showing the pair of harmonic motion areas for both $(3:1)$ and $(3:3)$ and the resonant solution with winding number $(5:1)$, **b** harmonic trajectory of $(3:3)$ and **c** the resonant trajectories of winding number $(5:1)$ respectively

Assuming that $(m:n)$ involves co-prime integers satisfying $m\omega_\varphi - n\omega_\theta = 0$, we split both m and n into $m_1, m_2, \ldots, m_k$ and $n_1, n_2, \ldots, n_l$. All the possible piece winding numbers are $(m_{k_i} : n_{l_j})$, $m_{k_i} = 0, 1, \ldots, m$ and $n_{l_j} = 0, 1, \ldots, n$. The composed resonant co-relative number, or the winding number, is then defined as:

$$(M:N) = \sum_{i=1}^{k}(m_{k_i} : n_{l_i}) = \left(\sum_{i=1}^{k} m_{k_i} : \sum_{i=1}^{k} n_{l_i}\right) \tag{4.10}$$

where $(m_{k_i} : n_{l_i})$ are the segment resonant number of the piece of the trajectory.

Figure 4.2a for parameters taken as $f_0 = 0.2$, $\omega_\varphi = \frac{1}{3}$ shows the stochastic or resonant web [18, 19] with a chaotic orbit, marked grey, surrounding a pair of harmonic motion areas of $(1:3)$ for each of the subsystem, and the pair of composed resonant solutions of $(3:7)$, the corresponding islands for the quasi-periodic solutions are marked in blue and red respectively. Figure 4.2b displays one of the $(1:3)$ and Fig. 4.2c is one of the $(3:7)$ resonant trajectories. Figure 4.3a shows a resonant separatrix for $f_0 = 0.2$, $\omega_\varphi = 3$ for the harmonic motion areas for both $(3:1)$ and $(3:3)$ and the resonant solution of winding number $(5:1)$. Figure 4.3b shows one of the $(3:3)$ harmonic solutions, and Fig. 4.3c depicts the trajectory of resonance of

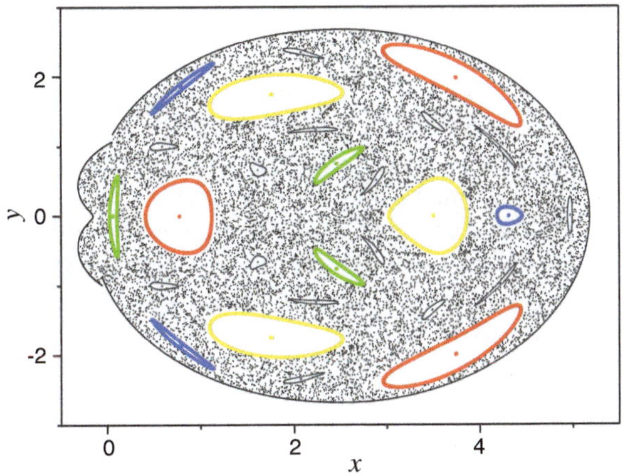

Fig. 4.4 Chaotic sea comprising a chaotic trajectory, marked *grey*, the island chains of four resonant solutions with winding number (3 : 8) for parameter $f_0 = 1.2$, $\omega_\varphi = \frac{1}{3}$ marked *blue* and *green*, *red* and *yellow*, respectively and the islands chain of the second resonance of (15 : 39) plotted *dark grey*

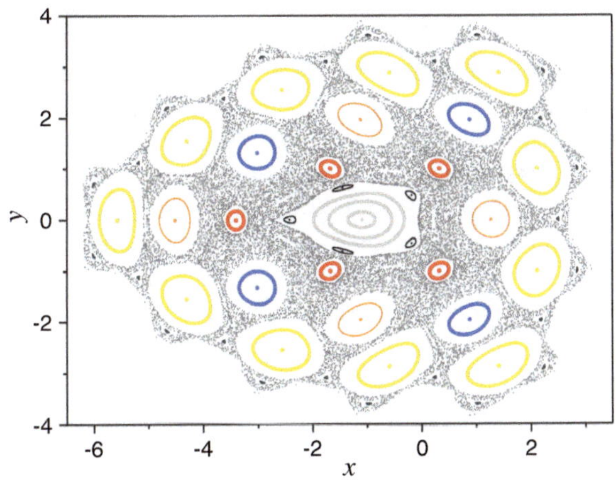

Fig. 4.5 Chaotic sea comprising a chaotic trajectory, marked *grey*, and the island chains for $f_0 = 10$, $\omega_\varphi = 3$ with a prime resonance of winding number (1 : 1), marked *light grey*, a (5 : 5) resonant island chain, marked *red*, a pair of (4 : 4) resonant island chains marked *thin blue* and *orange*, respectively and the islands of (11 : 11) marked *yellow*, the second resonant island chains of (5 : 5) near the center of (1 : 1) plotted black and the outer one of (17 : 25) plotted with *dark grey*

winding number (5 : 1). Figure 4.4 plots the structures of the resonant webs or a chaos sea [18–20] with a chaotic trajectory plotted grey entering different resonant layers of the corresponding resonant islands, which is computed for $f_0 = 1.2$, $\omega_\varphi = \frac{1}{3}$ and presents two pairs of prime resonances with winding number (3 : 8), the corresponding islands are marked in blue and green, red and yellow, respectively, and the islands of the second resonance of (15 : 39) are marked dark grey. In Fig. 4.5 the structures of the resonance web or a chaos sea for $f_0 = 10$, $\omega_\varphi = 3$ is demonstrated with a prime resonance of winding number (1 : 1), the corresponding isolated islands marked in light grey, the (5 : 5) resonant islands chain, marked in red, a pair of (4 : 4) resonant island chains, marked as blue and orange, respectively, the resonance of (11 : 11) and the corresponding islands plotted with yellow. In addition to the prime resonance, the second resonances are also formulated with the composed winding number, which is not multiple of the prime resonance. The second resonant islands of (5 : 5) near the (1 : 1) resonance are colored in black, and the outer second resonance of (17 : 25) is plotted in dark grey.

4.4 Summary

This chapter has presented a number of resonances which occur in the discontinuous oscillator derived from the SD oscillator by introducing the generalized winding number and the generalized canonical generating functions. The time dependent Hamiltonian for the discontinuous system enables us to understand such a resonant behaviour by a composed winding number consisting of multiple phases, each of which is confined within one of the separate resonant area (half plane, as described in Fig. 4.1), satisfying the regular resonant condition. Although the formulation derived in this chapter concerns the particular discontinuous system, it is valid for conventional nonlinear systems with multiple well potentials. Here the complex resonant structures of stochastic webs or chaotic sea with stochastic trajectories entering different resonant layers of resonant island chains observed may suggest further investigations into resonant synchronization and mechanisms of vibrational power flow transfer from nonlinear smooth to discontinuous dynamical systems.

References

1. Lichtenberg, A. J., & Lieberman, M. A. (1992). *Regular and chaotic dynamics*. New York: Springer.
2. Xiong, Y. P., Xing, J. T., & Price, W. G. (2005). Interactive power flow characteristics of an integrated equipment-nonlinear isolator-travelling flexible ship excited by sea waves. *Journal of Sound and Vibration*, *287*(1–2), 245–276.
3. Ueda, Y. (2001). *The road to chaos-II*. Santa Cruz: Aerial Press.
4. Ott, E., Grebogi, C., & Yorke, J. A. (1990). Controlling chaos. *Physical Review Letters*, *64*(11), 1196–1199.
5. Landau, L. D., & Lifschitz, E. M. (1977). *Quantum mechanics: Non-relativistic theory*. Oxford: Pergamon Press.
6. Frasca, M. (1998). Duality in perturbation theory and the quantum adiabatic approximation. *Physics Review A*, *58*(5), 3439–3442.
7. Frasca, M. (2006). A strongly perturbed quantum system is a semiclassical system. *Proceedings of the Royal Society A Mathematical Physical and Engineering Sciences*, *463*(2085), 2195–2200.
8. Arnold, V. I. (1983). *Mathematical methods of classical mechanics*. New York: Springer.

9. Chernikov, A. A., Sagdeev, R. Z., Usikov, D. A., Yu Zakharov, M., & Zaslavsky, G. M. (1987). Minimal chaos and stochastic webs. *Nature*, *326*, 559–563.
10. Jackson, E. A. (1990). *Persentives on nonlinear dynamics 1,2*. Cambridge: Cambridge University Press.
11. Zaslavsky, G. M. (2005). *Hamiltonian chaos and fractional dynamics*. Oxford: Oxford University Press.
12. Treschev, D., & Zubelevich, O. (2010). *Introduction to the perturbation theory of Hamiltonian system.*, Springer monographs in mathematics Heidelberg: Springer.
13. Luo, A. C. J., & Keqin, G. (2003). On resonant separatrix bands of a duffing oscillator with a twin-well potential. *Chaos, Solitons and Fractals*, *273*(4), 771–782.
14. Luo, A. C. J. (2004). On the symmetry of solutions in non-smooth dynamical systems with two constraints. *Journal of Sound and Vibration*, *273*(4–5), 1118–1126.
15. DE Freitas, M. S. T., Viana, R. L., & Grebogi, C. (2004). Multistability, basin boundary structure, and chaotic behavior in a suspension bridge model. *International Journal of Bifurcation and Chaos*, *14*(3), 927–950.
16. Nayfeh, A. H., & Sanchez, N. E. (1989). Bifurcations in a forced softening duffing oscillator. *International Journal of Non-linear Mechanics*, *24*(6), 483–497.
17. Wiggins, S. (1990). *Introduction to applied nonlinear dynamical systems and chaos*. New York: Springer.
18. Kapitaniak, T., & Bishop, S. R. (1999). *The illustrated dictionary of nonlinear dynamics and chaos*. Chichester: Wiley.
19. Evan-Iwanowski, R. M. (1976). *Resonance oscillations in mechanical systems*. Amsterdam, Oxford: Elsevier Scientific Publishing Company.
20. Cao, Q. J., Wiercigroch, M., Pavlovskaia, E. E., Grebogi, C., & Thompson, J. M. T. (2006). Archetypal oscillator for smooth and discontinuous dynamics. *Physics Review E*, *74*(046218), 1–5.

Chapter 5
Co-Dimension Two Bifurcation

This chapter investigates the fundamental properties of the perturbed behaviour of the SD oscillator in the degenerate case. As we have already seen, there is a degenerate singularity due to the change of hyperbolicity when parameter α varies crossing $\alpha = 1$, where the system is transformed from a single well dynamics to that of a double-well, in other words from a single stability to a bistability. This singularity can also be treated as the transition from non snap through buckling ($\alpha > 1$) to a snap through buckling ($\alpha < 1$) under a static load. We turn our attention to the interesting perturbed behaviour of the complex codimension two bifurcations of the oscillator at the degenerate equilibrium point $(0, 0)$ near $\alpha = 1$. It is found that the universal unfolding with two parameters is the perturbed SD oscillator with a nonlinear visco-damping. This universal unfolding reveals the complicated codimension two bifurcation phenomena in the parameter space with homoclinic bifurcation, closed orbit bifurcation, Hopf bifurcations and also the pitchfork bifurcations demonstrated at the same time when the geometrical parameter varies.[1]

5.1 Introduction

From the dynamical point of view, the interesting point is the degeneration of the singularity which results in the double zero eigenvalue problem admitting codimension two bifurcation behaviour. This phenomenon attracts the research for decades [1–3]. Codimension two bifurcations can also be found in many areas from engineering [4, 5] to physics [6] and have been widely investigated in the literature [7–9]. Universal unfolding is a powerful tool to unfold the degenerate nonlinear behaviour [3, 10] which is folded by the degeneration of the system. We are willing to derive

[1]This chapter relies on a common work with Professors Ruilan Tian and Shaopu Yang, Center for Nonlinear Dynamics Research, Shijiazhang Tiedao University, Shijiazhuang 050043, China.

Q. Cao and A. Léger, *A Smooth and Discontinuous Oscillator*, Springer Tracts in Mechanical Engineering, DOI 10.1007/978-3-662-53094-8_5

the complex dynamic behaviour of the SD oscillator near the degenerate point $\alpha = 1$ by using universal unfolding particularly for the codimension two bifurcation when the oscillator involves a nonlinear viscous damping with two parameters which will be treated as the unfolding parameters to show the global bifurcations including the bifurcation of closed orbit bursting out limit cycles. The universal property may offer an opportunity to the exact bifurcation depending upon the physical parameters of the oscillator.

5.2 Equilibrium Analysis

Recall the unperturbed dimensionless form of the SD oscillator:

$$x'' + x\left(1 - \frac{1}{\sqrt{x^2 + \alpha^2}}\right) = 0, \tag{5.1}$$

in which the parameter α is taken equal to zero in the previous chapter. This leads to the so-called non-smooth or discontinuous dynamics while it is now simply assumed to be strictly positive, which gives the smooth case. The case $\alpha > 1$ represents a pre-tensioned discrete elastic string, see for example in [11–13], the case of $0 < \alpha < 1$ represents a pre-compressed stage, which behaves a double well potential, while $\alpha = 1$ corresponds to the transition of codimension two bifurcations.

The first step of the equilibrium analysis for system (5.1) is to study the set of equilibria of the system with its stability.

Let $y = x'$, then system (5.1) can be rewritten in the following form

$$\begin{cases} x' = y, \\ y' = -x\left(1 - \dfrac{1}{\sqrt{\alpha^2 + x^2}}\right), \end{cases} \tag{5.2}$$

whose equilibria are obtained and written as

$$(x_0, y_0) = (0, 0), (x_{1,2}, y_{1,2}) = (\pm\sqrt{1 - \alpha^2}, 0),$$

which consist of three branches of $x = 0$ and $x = \pm\sqrt{1 - \alpha^2}$, which will be merged into a single branch $x = 0$ when α increases crossing $\alpha = 0$ to $\alpha > 0$. The Jacobian for the equilibrium (0, 0) is obtained for $\alpha > 0$, written as

$$J_{(0,0)} = \begin{bmatrix} 0 & 1 \\ \dfrac{1 - \alpha}{\alpha} & 0 \end{bmatrix}. \tag{5.3}$$

It is noticed that in the special case when $\alpha = 1$, the $J_{(0,0)}$ is the so called double zero eigenvalue problem, which are critical interested in nonlinear dynamics research. The Jacobian and the eigenvalues for (0, 0) are listed as follows.

1. When $0 < \alpha < 1$, the equilibrium $(0, 0)$ is a saddle point for the eigenvalues of the Jacobian $\lambda_{1,2} = \pm\sqrt{(1-\alpha)/\alpha}$.
2. When $\alpha = 0$, the singularity $(0, 0)$ of the discontinuous system is the limit of saddle point of the smooth system as $\alpha \to 0$. The hyperbolicity is broken as the eigenvalues $\lambda_{1,2} = \pm\sqrt{(1-\alpha)/\alpha}$ do not have finite limit as $\alpha \to 0$. Therefore, the singularity point $(0, 0)$ for the discontinuous system is a saddle-like singularity.
3. When $\alpha = 1$, the eigenvalues of the equilibrium point $(0, 0)$ are $\lambda_{1,2} = 0$, which implies that equilibrium $(0, 0)$ is a non-hyperbolic point, see [14, 15] for instance. This case introduces a codimension two bifurcation, which will be investigated in detail in the following section.
4. While for $\alpha > 1$, the equilibrium $(0, 0)$ of the system is a center with the eigenvalues of its Jacobian are $\lambda_{1,2} = \pm i\sqrt{(1-\alpha)/\alpha}$.

The Jacobian for the pair of equilibria $(\pm\sqrt{1-\alpha^2}, 0)$ for $0 \leq \alpha < 1$ is obtained as

$$J_{(0,0)} = \begin{bmatrix} 0 & 1 \\ -(1-\alpha^2) & 0 \end{bmatrix}, \tag{5.4}$$

with the characteristic equation and the eigenvalues written as

$$\lambda^2 + (1-\alpha^2) = 0, \lambda_{1,2} = \pm i\sqrt{1-\alpha^2}, \tag{5.5}$$

which implies that the pair of equilibria are center points. This will be studied analytically in subsequent chapters and leads to Hopf bifurcations [16, 17] near these non-trivial equilibria.

5.3 Universal Unfoldings

The section introduces the universal unfolding of the system to investigate the codimension bifurcation phenomenon at the equilibrium $(0, 0)$ near $\alpha = 1$ in the presence of a nonlinear viscous damping. The restoring force of system (5.1) reads as $F(x) = -x(1 - \frac{1}{\sqrt{x^2+\alpha^2}})$. Using a Taylor series up to the cubic term in a neighborhood of $x = 0$ it is approximated by

$$F(x) = \frac{(1-\alpha)}{\alpha}x - \frac{1}{2\alpha^3}x^3.$$

Thus, system (5.1) can be written close to $x = 0$ as

$$x'' - \frac{(1-\alpha)}{\alpha}x + \frac{1}{2\alpha^3}x^3 = 0. \tag{5.6}$$

- For $\alpha \geq 1$, let $\omega^2 = \dfrac{\alpha - 1}{\alpha}$, $\alpha_0 = \dfrac{1}{2\alpha^3}$, system (5.6) becomes

$$x'' + \omega^2 x + \alpha_0 x^3 = 0. \tag{5.7}$$

- For $\alpha < 1$, let $\omega^2 = \dfrac{1 - \alpha}{\alpha}$, $\alpha_0 = \dfrac{1}{2\alpha^3}$, system (5.6) becomes

$$x'' - \omega^2 x + \alpha_0 x^3 = 0. \tag{5.8}$$

Now suppose system (5.6) has nonlinear viscous damping with two parameters ξ and γ. The unforced dissipative oscillator is then the following:

$$x'' + (\xi + \gamma x^2)x' - \frac{1 - \alpha}{\alpha}x + \frac{1}{2\alpha^3}x^3 = 0, \tag{5.9}$$

where ξ and γ can take arbitrary real values. Let $x' = y$, system (5.9) becomes

$$\begin{cases} x' = y \\ y' = \dfrac{1 - \alpha}{\alpha}x - \xi y - \gamma x^2 y - \dfrac{1}{2\alpha^3}x^3. \end{cases} \tag{5.10}$$

System (5.10) has a trivial zero solution $(x_1, x_2) = (0, 0)$ at which the Jacobian is

$$\begin{bmatrix} 0 & 1 \\ \dfrac{1 - \alpha}{\alpha} & -\xi \end{bmatrix}. \tag{5.11}$$

The characteristic equation corresponding to the trivial zero solution is

$$\lambda^2 + \xi\lambda + \frac{\alpha - 1}{\alpha} = 0. \tag{5.12}$$

When $\xi = 0$ and $\alpha = 1$, system (5.10) has one non-semisimple double zero eigenvalues

$$\lambda_{1,2} = 0. \tag{5.13}$$

Let $\varepsilon_1 = \dfrac{1 - \alpha}{\alpha}$, $\varepsilon_2 = \xi$, system (5.10) becomes

$$\begin{cases} x' = y, \\ y' = \varepsilon_1 x - \varepsilon_2 y - ax^3 - bx^2 y, \end{cases} \tag{5.14}$$

where $a = \dfrac{1}{2}(1 + \varepsilon_1)^3$, $b = \gamma$, and $a > 0$. Assume $b \neq 0$ and apply a scaling transformation as follows

$$x \to \frac{\sqrt{a}}{b}x, \quad y \to \frac{a\sqrt{a}}{b^2}y, \quad t \to \frac{b}{a}t. \tag{5.15}$$

System (5.14) can then be written as

$$\begin{cases} x' = y, \\ y' = \mu_1 x + \mu_2 y - x^3 - x^2 y, \end{cases} \tag{5.16}$$

which is equivalent to the damped SD oscillator, system (5.10) according to transformation (5.15), with $\mu_1 = \frac{b^2}{a^2}\varepsilon_1,\ \mu_2 = -\frac{b}{a}\varepsilon_2$.

System (5.16) is the universal unfolding for the normal form of the SD oscillator with nonlinear damping and without forcing.

The folded dynamics near the origin of the parameter space is given by

$$\begin{cases} x' = y, \\ y' = -x^3 - x^2 y. \end{cases} \tag{5.17}$$

In order to get the unfolded dynamics with the bifurcation diagram for the codimension two bifurcations, the following analysis is carried out.

At first, when $\mu_1 \geq 0$, the equilibria for system (5.16) are

$$(\pm\sqrt{\mu_1}, 0), \ (0, 0),$$

where a pitchfork bifurcation occur for $\mu_1 = 0,\ \mu_2 \neq 0$.

Secondly, when $\mu_1 < 0$, $(0, 0)$ is the unique equilibrium. The Jacobian at this equilibrium is given by

$$Df(0,0) = \begin{bmatrix} 0 & 1 \\ \mu_1 & \mu_2 \end{bmatrix}. \tag{5.18}$$

Let $x = \bar{x},\ y = \sqrt{-\mu_1}\,\bar{y}$, it turns out that Hopf bifurcation occurs at $\mu_1 < 0,\ \mu_2 = 0$.

Furthermore, when $\mu_1 > 0$, re-scaling (5.16) by

$$\mu_1 = \delta^2, \ \mu_2 = \delta^2\zeta, \ x \to \delta x, \ y \to \delta^2 y, \ t \to \frac{t}{\delta}, \tag{5.19}$$

gives

$$\begin{cases} x' = y, \\ y' = x - x^3 + \delta(\zeta y - x^2 y), \end{cases} \tag{5.20}$$

For $\delta = 0$, system (5.20) is a Hamiltonian system

$$\begin{cases} x' = y, \\ y' = x - x^3, \end{cases} \tag{5.21}$$

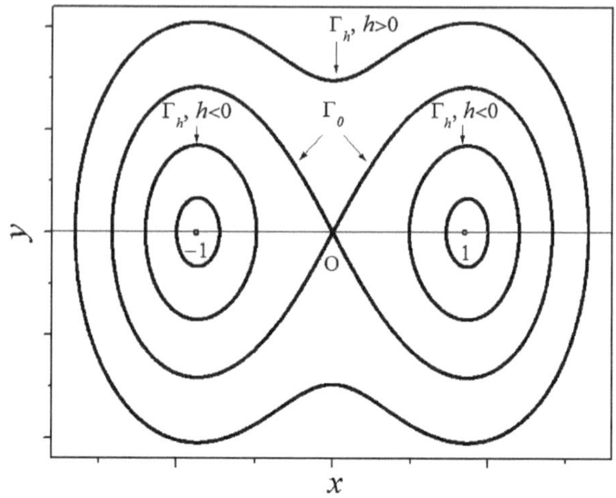

Fig. 5.1 Phase portrait for system (5.21): $\Gamma_h = \{(x, y)|H(x, y) = h\}$

the Hamiltonian of which is

$$H(x, y) = \frac{y^2}{2} - \frac{x^2}{2} + \frac{x^4}{4}, \tag{5.22}$$

for which the phase portrait is plotted in Fig. 5.1. Using this Hamiltonian (5.22), the trajectories of system (5.21) can be classified by $H(x, y) = h$.

Let

$$\Gamma_h = \{(x, y) \mid H(x, y) = h\}.$$

- When $-\frac{1}{4} < h < 0$, Γ_h represents a pair of closed orbits;
- when $h \to -\frac{1}{4}$, Γ_h shrinks to the pair of equilibria $(\pm 1, 0)$;
- when $h \to 0^-$, Γ_0 extends into a double homoclinic orbit connecting at the equilibria $(0, 0)$;
- while $h > 0$, Γ_h gives the large closed orbit encircling the equilibria $(\pm 1, 0)$, $(0, 0)$ and the double connected homoclinic orbit, seen in Fig. 5.1. This large closed orbit does shrink itself to the double homoclinic orbit as $h \to 0^+$.

This double connected homoclinic orbit, can be written as

$$\begin{cases} x_0 = \pm\sqrt{2}\,\text{sech}\, t, \\ y_0 = \mp\sqrt{2}\,\text{sech}\, t \tanh t. \end{cases} \tag{5.23}$$

5.4 Codimension Two Bifurcations

In the following subsections, the complicated codimension two bifurcations shall be discussed for near the parameter $\alpha = 1$. The first part is for the universal bifurcations and the second part is the SD oscillator bifurcations.

5.4.1 *Universal Codimension Bifurcations*

For $\delta \neq 0$, system (5.20) is dissipative, the double connected homoclinic orbit is preserved, *if and only if*

$$M(\zeta) = \int_{+\infty}^{-\infty} y_0(\zeta y_0 - x_0^2 y_0)dt = 0. \tag{5.24}$$

It is known that (the details of the proof can be seen in [18]) the double homoclinic orbit occurs for $\zeta = \frac{4}{5}$, i.e. $\mu_2 = \frac{4}{5}\mu_1 > 0$. Finally, to analyze the bifurcations of multiple limit cycles based on Eq. (5.20), we need to consider the number of zeros of subharmonic Melnikov function. Let $(\overline{x}(t),\ \overline{y}(t))$ represents the closed periodic orbits inside or outside a pair of homoclinic loops Γ_0. The Hamiltonian function, defined by Eq. (5.22), is now $H(\overline{x}, \overline{y}) = h$ with period T. Thus, the subharmonic Melnikov function can be obtained as

$$\begin{aligned} M(h) &= \textstyle\int_0^T y(T)[\zeta y(T) - x^2(T)y(T)]dt \\ &= \textstyle\int_{H=h} \zeta y - x^2 y dx \\ &= \zeta I_0(h) - I_1(h), \end{aligned} \tag{5.25}$$

where

$$I_0(h) = \int_{H=h} ydx,\ I_1(h) = \int_{H=h} x^2 ydx. \tag{5.26}$$

Let us introduce $P(h) = \dfrac{I_1(h)}{I_0(h)}$, then $P(h)$ satisfies a Ricatti equation [19]

$$4h(1 + 4h)P' = 5P^2 + 8hP - 4P - 4h. \tag{5.27}$$

We can define

$$\overline{M}(h) = \frac{M(h)}{I_0} = \zeta - P(h). \tag{5.28}$$

Since $I_0(h) \neq 0$, the number of the zeros for $\overline{M}(h)$ is equal to the number of the zeros of $M(h)$, so that we analyze the zeros of function $\overline{M}(h)$ instead of function $M(h)$.

Equation $\overline{M}(h) = 0$ yields

$$P(h) = \zeta. \tag{5.29}$$

From Eq. (5.27), we know $P(h)$ can also be determined from the solutions of following equations

$$\begin{cases} \dot{h} = 4h(1+4h), \\ \dot{P} = 5P^2 + 8hP - 4P - 4h. \end{cases} \tag{5.30}$$

Therefore, the zeros of $\overline{M}(h)$ can be determined by the intersection of $P(h)$ and the line $P(h) = \xi$. The results are the following

- When $\zeta \in (1, +\infty)$, $P(h)$ and the line $P(h) = \xi$ has only one intersection point for some $h_1 > 0$. This suggests that only one closed orbit exists, enclosing all the three singular points for $\zeta \in (1, +\infty)$.
- When $\zeta \in (\frac{4}{5}, 1)$, $P(h)$ and the line $P(h) = \xi$ has two intersection points in the interval $h_2 \in (-\frac{1}{4}, 0)$ and $h_3 \in (0, +\infty)$ respectively. This implies, for $\zeta \in (\frac{4}{5}, 1)$, that three orbits, two for $h_2 \in (-\frac{1}{4}, 0)$ and one for $h_3 \in (0, +\infty)$, are maintained.
- At $P(h) = \zeta = \frac{4}{5}$, $h_4 = 0$ and $h_5 > 0$. We have the homoclinic saddle connection already described, coexisting with a closed orbit lying outside the connection.
- When $\zeta \in (c, \frac{4}{5})$, $P(h)$ and the line $P(h) = \xi$ has two intersection points in the interval $h_5, h_6 \in (0, +\infty)$. Therefore, there exist two closed orbits, both enclosing all three fixed points.
- It can be proved that $P(h)$ has a unique minimum $\zeta \approx 0.752$, for details see [20]. Therefore, these orbits coalesce and vanish as ζ passes through c, the detail can be seen in Fig. 5.2.

The bifurcation diagram for system (5.16) in (μ_1, μ_2) plane can be obtained by using corresponding transformation given in formula (5.19) above, and is described below:

1. $\mu_1 = 0, \mu_2 \neq 0$, the pitchfork bifurcation;
2. $\mu_2 = 0, \mu_1 < 0$, the Hopf bifurcation;
3. $\mu_1 = \mu_2 > 0$, the (double) Hopf bifurcation;
4. $\mu_2 = \frac{4}{5}\mu_1 > 0$, a (double) homoclinic bifurcation;
5. $\mu_2 = c\mu_1 > 0, c \approx 0.752$, the closed orbit bifurcation.

The bifurcation diagram described above divides the two-dimensional parameter space (the (μ_1, μ_2) plane) into six persistent regions, in each of which the system is structurally stable, while the structural stability dissolves and the bifurcation occurs in the bifurcation sets, see [21, 22]. Bifurcation diagram, codimension two bifurcation and phase portraits are presented in Fig. 5.3.

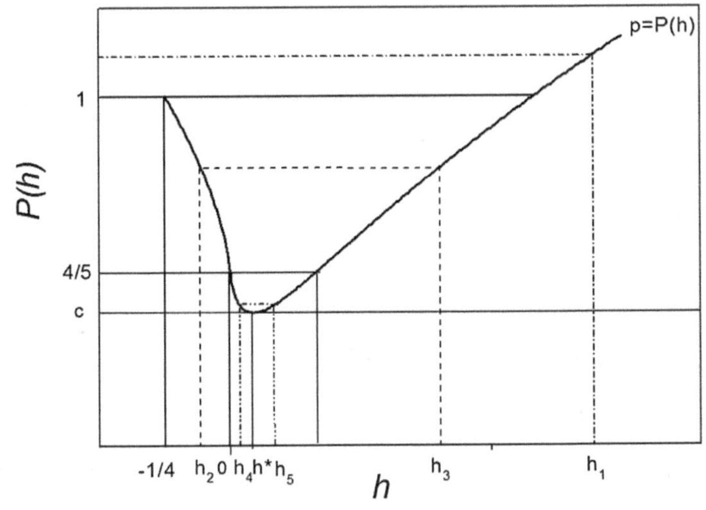

Fig. 5.2 P(h) curve: the topological classification of the orbits of the universal unfolding (5.20)

5.4.2 SD Oscillator Codimension Bifurcations

Coming back to the physical parameters of the damped SD oscillator, codimension two bifurcations appear as follows, with homoclinic bifurcation, closed orbit bifurcation, Hopf bifurcations and pitchfork bifurcations displayed at the same time.

Recalling that the parameters are

$$\varepsilon_1 = \frac{1-\alpha}{\alpha},\ \varepsilon_2 = \xi, a = \frac{1}{2}(1+\varepsilon_1)^3,\ b = \gamma,\ a > 0,\ \mu_1 = \frac{b^2}{a^2}\varepsilon_1, \mu_2 = -\frac{b}{a}\varepsilon_2,$$

the bifurcation diagram for system (5.10) in the original parameter space $(\alpha,\ \xi)$ reads, using the notations of Fig. 5.3,

1. $OR = OR^+ \bigcup\{(1,0)\} \bigcup OR^- = \{\alpha = 1, \xi \neq 0, \gamma \neq 0\}$, the pitchfork bifurcation;
2. $OH_1 = \{\xi = 0, \alpha > 1, \gamma \neq 0\}$, the Hopf bifurcation;
3. $OH_2 = \left\{\gamma\left(\frac{1}{\alpha} - 1\right) + \frac{1}{2\alpha^3}\xi = 0, 0 < \alpha < 1, \gamma\xi < 0\right\}$, the (double) Hopf bifurcation;
4. $OHL = \left\{4\gamma\left(\frac{1}{\alpha} - 1\right) + \frac{5}{2\alpha^3}\xi = 0, 0 < \alpha < 1, \gamma\xi < 0\right\}$, a (double) homoclinic bifurcation;
5. $OB = \left\{c\gamma\left(\frac{1}{\alpha} - 1\right) + \frac{1}{2\alpha^3}\xi = 0, 0 < \alpha < 1, \gamma\xi < 0, c \approx 0.752\right\}$, the closed orbit bifurcation.

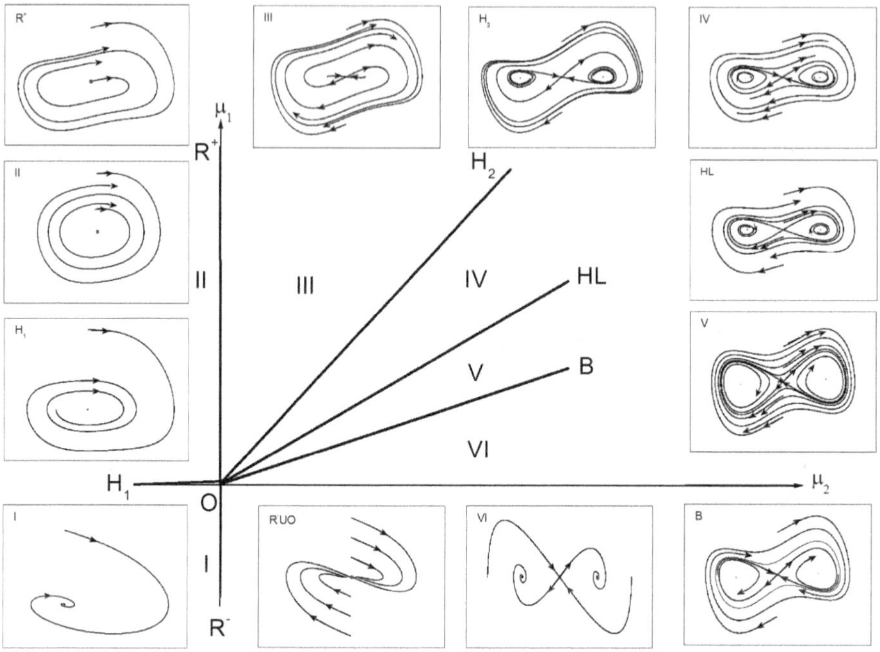

Fig. 5.3 Codimension two bifurcations: the bifurcation diagram described in (μ_1, μ_2) plane marked by I, II, III, IV, V, and VI, respectively, and the corresponding phase portraits plotted and marked with the same symbol including that of the bifurcation ones, the details see the contents

Let us define

$$\sum = OR \bigcup OH_1 \bigcup OH_2 \bigcup OHL \bigcup OB, \tag{5.31}$$

to exhibit the bifurcation diagram for system (5.10) in the original parameter space (the $(\alpha, \ \xi)$ plane).

For any fixed $\gamma \neq 0$, $\sum$ divides the two dimensional parameter space $\alpha - \xi$ into six persistent regions written as: OI, OII, OIII, OIV, OV and OVI. Inside each of these regions, the system is structurally stable, while it is structurally unstable when the parameters are on $\sum$.

Details for the bifurcation diagram $\sum$, are shown in Fig. 5.4. To understand the codimension two bifurcations, one can think to a point (α, ξ) moving clockwise along the circumference of half unit radius centered at $(1, 0)$ passing through the regions and crossing the branches of the bifurcation diagram successively, starting from OR^+, which shows the dynamics of system (5.10) changing from one stage to another.

- The system has a pitchfork bifurcation when (α, ξ) is on the branch OR^+ of $\sum$ with the unique stable proper node which becomes a sink, or a stable focus when (α, ξ) enters region OI;

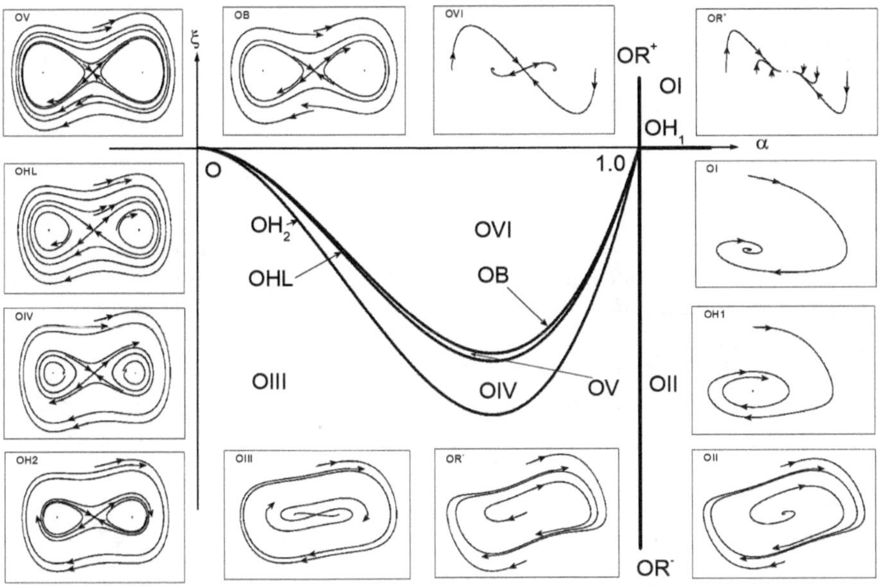

Fig. 5.4 Bifurcation diagram and the corresponding perturbed dynamics: $\sum$ the bifurcation diagram in the (α, ξ) plane marked by OI, OII, OIII, OIV, OV, and OVI, respectively, and the corresponding phase portraits plotted and marked with the same symbol including that of the bifurcation ones, the details see the contents

- When (α, ξ) touches the branch OH_1, there appears the Hopf bifurcation which bifurcates into a stable limit cycle encircling the unique source when (α, ξ) enters into region OII;
- The pitchfork bifurcation occurs again when (α, ξ) locates on OR^- with the stable limit cycle encircling an unstable proper node which breaks into one saddle and a pair of unstable proper nodes when (α, ξ) enters region OIII, while the stable limit cycle remains unchanged encircling the equilibria;
- This pair of nodes becomes a pair of unstable fine foci admitting Hopf bifurcations as soon as the point (α, ξ) meets OH_2, which bifurcates into a pair of unstable limit cycles and the encircled pair of stable foci, and these foci and cycles are encircled by the larger stable one when the point (α, ξ) enters region OIV;
- The pair of unstable limit cycles may extend into a double connected homoclinic orbit encircling the pair of stable foci in each branch when (α, ξ) tends to OHL from region OIV, this double homoclinic orbit bifurcates into an unstable large limit cycle encircling the pair of the stable foci and the saddle encircled by the larger stable one when (α, ξ) comes into region OV;
- Suppose that the point (α, ξ) moves again tending to the branch OB of $\sum$, then the pair of large limit cycles, the stable and the unstable one, meet together merging into a semi-stable limit cycle. This closed orbit disappears when (α, ξ) gets into region OVI, while the pair of stable foci change their property into a pair of stable

proper nodes; all of these three equilibria will merge together into a stable proper node admitting the pitchfork bifurcation again when (α, ξ) comes back to OR^+.

5.5 Summary

This chapter has presented the complex codimension two bifurcations of the SD oscillator at the degenerate equilibrium point (0, 0) near $\alpha = 1$. It is interesting that the universal unfolding with two parameters is found to be the SD oscillator with a nonlinear visco-damping. This universal unfolding reveals the complicated codimension two bifurcation phenomena in the physical parameter space with homoclinic bifurcation, closed orbit bifurcation, Hopf bifurcations and also pitchfork bifurcations demonstrated at the same time when the geometrical parameter varies. This provides a picturesque example with rich nonlinear phenomena which may happen in a practical mechanical model.

References

1. Ashwin, P., & Aston, P. J. (1998). Blowout bifurcations of codimension two. *Physics Letters A*, *244*(4), 261–270.
2. Han, M. A. (1999). Global behavior of limit cycles in rotated vector fields. *Journal of Differential Equations*, *151*(1), 20–35.
3. Holmes, P. (1977). Bifurcation to divergence and flutter in flow induced oscillations: A finite dimensional analysis. *Journal of Sound and Vibration*, *53*(4), 471–503.
4. Yagasaki, K. (1999). Codimension-two bifurcations in a pendulum with feedback control. *International Journal of Non-Linear Mechanics*, *34*(6), 983–1002.
5. Cline, J. M., Descheneau, J., Giovannini, M., & Vinet, J. (2003). Cosmology of codimension-two braneworlds. *Journal of High Energy Physics*, *2003*(6), 048.
6. Fatimah, S., & Ruijgrok, M. (2002). Bifurcation in an autoparametric system in 1:1 internal resonance with parametric excitation. *International Journal of Non-linear Mechanics*, *37*(2), 297–308.
7. Pellegrini, L., & Possio, C. T. (1996). A non-ideal CSTR: A high codimension bifurcation analysis. *Chemical Engineering Science*, *51*(11), 3151–3156.
8. Potzsche, C. (2013). Bifurcations in a periodic discrete-time environment. *Nonlinear Analysis Real World Applications*, *14*(1), 53–82.
9. Duan, L. X., & Lu, Q. S. (2006). Codimension-two bifurcation analysis on firing activities in Chay neuron model. *Chaos, Solitons and Fractals*, *30*(5), 1172–1179.
10. Golubitsky, M., Schaeffer, D. G., & Stewart, I. (1985). *Singularities and groups in bifurcation theory* (Vol. I). New York: Springer.
11. Tufillaro, N. B., Reilly, J., & Abbott, T. (1992). *An experimental approach to nonlinear dynamics and chaos*. Boston: Addison-Wesley.
12. Tufillaro, N. B. (1989). Nonlinear and chaotic string vibrations. *American Journal of Physics*, *57*(5), 408–414.
13. Tufillaro, N. B. (1990). Torsional parametric oscillations in wires. *European Journal of Physics*, *11*(2), 122–124.
14. Holmes, P., & Marsden, J. E. (2014). Bifurcations of dynamical systems and nonlinear oscillations in engineering systems. *Dynamics and Design Conference: D & D*, 561–582.

15. Wiggins, S. (1990). *Introduction to applied nonlinear dynamical systems and chaos*. New York: Springer.
16. Hopf, E. (1942). Abzweigung einer periodischen Lösung von einer stationären Lösung. *Berich. Sächs. Akad. Wiss., Leipzig, Math. Phys. Kl.*, *94*(19), 15–25.
17. Hopf, E. (1943). Abzweigung einer periodischer Lösung von einer stationären Lösung eines Differentialsystems. *Ber. Verh. Sächs. Akad. Wiss. Leipzig. Math-Nat. K1*, *95*, 3–22.
18. Horozov, E. I. (1979). Versal deformation of equivariant vector fields in the case of symmetry of order 2 and 3. *Trudy Sem Petrovsk*, *5*, 163–192.
19. Zhang, W., & Yu, P. (2000). A study of the limit cycles associated with a generalized codimension-3 lienard oscillator. *Journal of Sound and Vibration*, *231*(1), 145–173.
20. Carr, J. (1981). *Applications of centre manifold theory* (Vol. 35). Applied mathematical sciences. New York: Springer.
21. Alexander, J. C., & Yorke, J. (1978). Global bifurcations of periodic orbits. *American Journal of Mathematics*, *100*(2), 263–292.
22. Hermann, G. (1967). Stability of equilibrium of elastic systems subjected to nonconservative forces. *Applied Mechanics Review*, *20*(2), 103–108.

Chapter 6
Wada Basin Dynamics

This chapter investigates a specific point of the very intricate asymptotic behaviour of the SD oscillator, which is known as the Wada basin dynamics. The oscillator is subjected to a linear viscous damping and to a sinusoidal forcing. As described and already observed through direct numerical integration, this system may possess more than twenty coexisted low-period periodic attractors for a given set of parameters. The large number of stable orbits yields a complex structure of closely interwoven basins of attraction. We obtain the so-called Wada basins of which the boundaries are rigorously described. The consequence of this complex structure is the existence of a large number of final states and the difficulty to predict the final state, which addresses the question of non predictability, so that the qualitative fate of a given initial data cannot be given.[1]

6.1 Introduction

Many processes in nature do not possess only one long-term asymptotic state or attractor [1], but are rather characterized by multiple coexisting attractors for a given set of parameters. It is of fundamental importance to be able to determine which attractor a specified initial condition leads to. The collection of initial conditions whose trajectories approach a specified attractor is called the basin of an attractor and the boundary separating the attraction is called basin boundary. To assess the significance of a specific attractor, it is necessary to know the basin of attraction and

[1]This chapter is due to the contribution by Professor Yongxiang Zhang, Shenyang Agricultural University, College of Science, Shenyang, Liaoning, China, Professors Guanwei Luo, Gansu Key Laboratory for System Dynamics and Reliability of Railway Transportation Equipment, Lanzhou Jiaotong University, Lanzhou, China and Doctor Mei Lin, The Engineering Institute, Air Engineering Univesity, Xi'an, Shanxi, China.

Q. Cao and A. Léger, *A Smooth and Discontinuous Oscillator*, Springer Tracts in Mechanical Engineering, DOI 10.1007/978-3-662-53094-8_6

the geometry of the basin boundary [2, 3]. One of the goals of nonlinear dynamics is to determine the global structure of dynamical systems. One of these 'global' structures in dynamical systems are the boundaries of basins. The basin boundaries can be either smooth or fractal [2]. We say that a basin boundary is fractal if it contains a transversal homoclinic point (or its basin boundary is a Cantor set) [3]. When three attractors coexist in phase space, a strange phenomenon is possible, namely the Wada property [4]. A basin boundary satisfies the Wada property if every open neighborhood of any point on the boundary has a nonempty intersection with at least three different basins. Although it is hard to imagine, it is possible to have three or more regions sharing the same boundary in nonlinear dynamics. When this type of basin occurs, the basin boundaries have a complicated structure. The Wada property was firstly introduced in the physics literature by Kennedy and Yorke and it attracts a particular attention to the topological properties underlying them [4]. An interesting consequence of the Wada property is the existence of several possible final states and difficulty of predicting the finial state. Therefore, the unpredictability of a system with the Wada property is even stronger than the fractal basin boundaries (non-Wada property) [5, 6]. On the other hand, the unpredictability of the parameter plane can also be explained by depending whether the associated basins are Wada or not [7]. The occurrence of Wada basin boundaries in the real world is still a very interesting problem. Several applications to plasma physics [8], mechanics [9], ecology [10] and chaotic scattering problems [11] have been proposed. Although fractal basin boundaries have been shown to be very common in nonlinear dynamics, the number of examples displaying the Wada property is still quite limited. In particular, the results on the Wada property are limited to less than five basins. However, it is expected that some systems in nature have a large number of coexisting attractors [12]. The behaviour of such systems will be more complex since there is the added feature of interactions among the various attractors and their basins. The purpose of this chapter is to study the Wada basin dynamics of a shallow arch model which exhibits a large number of coexisting attractors and complex structure of basins.

The uncertain dynamics and unpredictability can occur when the dynamical systems present the Wada basin boundaries [13, 14]. Besides the Wada basin boundaries of the periodic attractors [15], the Wada basin boundaries of quasi-periodic attractors [16] and the Wada basin boundaries of strange non-chaotic attractors [17] can also be observed in quasi-periodically forced systems. Thus, different types of attractors can exhibit the unpredictability for the specific initial conditions on the basin boundaries. Here, we focus on some Wada basin boundaries of the oscillator with more than 20 coexisting low-period periodic attractors. This situation exists when the final destination of the initial conditions can take place through more different paths. The degree of unpredictability of destinations can be more severe than the usual case [14]. The practical consequences of encountering such a system are obviously very serious although the underlying equation of motion is strictly speaking deterministic. Due to measurement errors, simple prediction of the qualitative outcome often of a perfectly modeled experiment would be impossible.

The global Wada basin dynamics described here is different from the typical basin dynamics (the basins of less than five coexisting attractors) because of the complexity of basin boundaries. From the point of view of dynamics, the Wada basin boundaries are still not entirely clear for a large number of coexisting attractors. Here, the SD oscillator may exhibit more than 20 low-period periodic attractors and complex basin boundaries. It is easy to visualize from an intuitive point of view whether or not a system satisfies the Wada property. However, the goal is to verify the Wada property with some rigor for some typical basins. Though the basin boundary points are common to two or three basins, it is difficult to find the boundary points that are shared by more than four basins. Here, some boundary points are shared by more than four basins (even more than 20 basins). In order to get the number of coexisting periodic orbits and study the basin structure with high accuracy, the phase space is divided into several critical regions. For a certain set of parameters, the system can possess more than 20 coexisting low-period periodic attractors. Some results (Theorems 6.1 and 6.2, see in [18]) will be used to prove the Wada basin boundaries and the partially Wada basin boundaries.

6.2 Preliminaries

Recall that the SD oscillator is smooth as long as the geometrical parameter α is strictly positive, while it is discontinuous when $\alpha = 0$. Here, we focus on the Wada basin dynamics for the smooth oscillator i.e. $\alpha > 0$. The dimensionless form, the generic perturbed oscillator is rewritten in the following form.

$$\ddot{x} + \beta\dot{x} + x(1 - \frac{1}{\sqrt{x^2 + \alpha^2}}) = f_0 \cos \Omega t. \tag{6.1}$$

where $\beta = 2\xi$, as described in system (2.7)

For clarity, we repeat some basic notions briefly [18], adding important details. Let M denote either the plane or any other smooth 2-dimensional manifold without boundary. Let F be a dissipative diffeomorphism from M to itself, that is, both F and its inverse are continuous and differentiable. For an open set B, write $\partial B = \overline{B} - B$ for the boundary of B, where $\overline{B}$ is the closure of the set B. The point q is called a periodic point of period k if $F^k(q) = q$ and $F^i(q) \neq q$ for $1 \leq i \leq q$. A periodic point q is hyperbolic if the Jacobian matrix $DF(q)$ has no eigenvalues with absolute values 1. A periodic point q of period k is an attractor (respectively, saddle) if the eigenvalues e_1 and e_2 of $DF(q)$ satisfy $|e_1| < 1$ and $|e_2| < 1$ (respectively $|e_1| < 1$ and $|e_2| > 1$). For a periodic point q of period k, the stable manifold $W^s(q)$ and unstable manifold $W^u(q)$ are defined to be the following sets:

$$W^s(q) = \{x \in M | F^{kn}(x) \longrightarrow F^{kn}(q), \text{ as } n \longrightarrow \infty\},$$
$$W^s(q) = \{x \in M | F^{-kn}(x) \longrightarrow F^{-kn}(q), \text{ as } n \longrightarrow \infty\}$$

The basin of attraction of an attracting set is the set of all initial conditions whose trajectories tend asymptotically to the attracting set. A generalized attractor is the

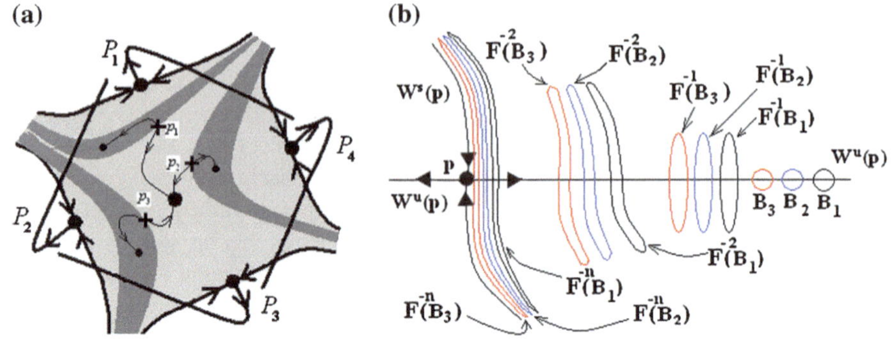

Fig. 6.1 **a** The *black point* (p_1, p_2 and p_3 marked by '+') is accessible from the *light grey* basin, because it is the first boundary point that a curve hits starting from the interior of the basin; **b** The lambda lemma when $W^u(p)$ intersects the basin $B_i (i = 1, 2, 3)$

union of any finite number of attractors, and a generalized basin is the basin of a generalized attractor. A point P on the basin boundary is a Wada point if every open neighborhood of P has a nonempty intersection with at least three different basins. A basin boundary is a Wada basin boundary if all of its points are Wada points, i.e. every open neighborhood of a point belonging to a Wada basin boundary has a nonempty intersection with a least three different basins. For systems that show coexistence of Wada points and non-Wada points in the same boundary we propose the name of partially Wada basin boundaries [5]. A point p in ∂B is accessible from basin B if a curve can be drawn, starting in B so that p is the first boundary point of B the curve hits (see Fig. 6.1a for a schematic explanation of this concept, p_1, p_2 and p_3 marked by '+' are three accessible points).

For two-dimensional diffeomorphisms, the mechanism for Wada basin boundaries is well understood, thanks to the rigorous mathematical work by Kennedy, Nusse, and Yorke [18]. Kennedy and Yorke proved [4] that, for a periodic point on the basin boundary, if the following two conditions are satisfied: (1) its unstable manifold intersects every basin, and (2) its stable manifold is dense in each of the basin boundaries, then the basins have the Wada property. Condition (1) is fundamental for verifying the Wada property by referring to Fig. 6.1b, where there are three coexisting basins. Suppose that B_1, B_2 and B_3 are small disks that belong to different basins. If we calculate the preimages of B_1, B_2 and B_3, that is their images when the time goes backward, we find that they successively approach the stable manifold of p. In other words, the preimages of B_1, B_2 and B_3 become exponentially stretched in the direction of the stable manifold and get arbitrarily close to it. In the limit, all points on the stable manifold of the periodic orbit are boundary points, that is, there are points that belong to all three basins arbitrarily close to then. Therefore all points on the stable manifold of p are Wada points. In fact, this is guaranteed mathematically by the lambda lemma due to Palis [19]. Since the stable manifold of p is dense in each of the basin boundaries, all boundaries must be common to all basins. This method is always an open possibility but in practice it is hard to do it rigorously. Here, we use the following theorems to verify our main results.

Theorem 6.1 *Let p be a B_1-accessible hyperbolic periodic point, and assume that the unstable manifold $W^u(p)$ intersects each B_k, where $1 \leq k \leq N$. (a) Every point z on the stable manifold $W^s(p)$ is a Wada point. w.r.t. $B_k(1 \leq k \leq N)$. (b) If the orbit of p is the only B_1-accessible periodic orbit, then ∂B_1 is a Wada basin boundary w.r.t. $B_k(1 \leq k \leq N)$.*

The following theorem is used to verify the partially Wada basin boundaries.

Theorem 6.2 *Let F be a dissipative diffeomorphism of the plane. Let p and q be two accessible hyperbolic periodic points, and assume that the unstable manifold $W^u(p)$ intersects at least three basins and the unstable manifold $W^u(q)$ intersects only two basins. Then ∂B_1 is a partially Wada basin boundary w.r.t. $B_k(1 \leq k \leq N)$.*

Proof Let p be as in the theorem. Then, by the Proposition 1 in [18], $W^s(p)$ is contained in each basin boundary $B_k(1 \leq k \leq N)$. Hence, every $z_1 \in W^s(p)$ is a Wada point w.r.t. $B_k(1 \leq k \leq N)$. Let the unstable manifold $W^u(q)$ intersects only two basins B_k and B_n $(1 \leq k \leq N)$, $(1 \leq n \leq N)$ and $k \neq n$. Then by the Proposition 2 [4] or the Proposition 1 [18], each point $z_2 \in W^s(q)$ is in the boundary $\partial B_k(1 \leq k \leq N)$. By the lambda lemma [19], this implies that every open neighborhood of z_2 contains points of two basins B_k and B_n (the point z_2 is not a Wada point). Since the stable manifolds $W^s(p)$ are contained in each basin boundary $B_k(1 \leq k \leq N)$. Hence, every $z_1 \in W^s(p)$ and $W^s(p)$ are contained in the basin boundary ∂B_k, the coexistence of Wada points and non-Wada points can exist in the same boundary ∂B_k. By the concept of partially Wada basin boundary [5], then ∂B_k is a partially Wada basin boundary.

The partially Wada basin boundaries can be explained by Fig. 6.1: A period-4 saddle-hyperbolic orbits (P_1, P_2, P_3, P_4) which are accessible from the light grey basin, their unstable manifolds cross more than three basins; while a period-3 saddle-hyperbolic orbits (p_1, p_2, p_3) which are accessible from the light grey basin, their unstable manifolds cross two basins.

6.3 The Attractors and Their Basins

The main focus of this section is to show that the system (6.1) has more than 20 low-period periodic attractors for a given set of parameters. Generally, the more important attractor may have larger basins of attraction, the higher probability that the final motion converges to this attractor. Identifying their basins can help us to get the larger stable (or safe) region. All the numerical computations were made using a fourth-order Runge–Kutta integrator with a fixed time step of 0.01. To generate numerically the basins of attraction, a grid of 900×900 points has been selected, which were taken as initial conditions for all figures in the $\{x, \dot{x}\}$ plane. Each initial condition was integrated until the corresponding trajectory approaches one of the attractors. The initial condition was then plotted in a color that corresponds to the attractor. Statistically speaking, the more important attractors may have larger basins of attraction. We use the relative size of the basin of attraction (i.e. the number of initial conditions converging to a particular attractor divided by the total number of initial conditions taken into account) to characterize a given basin. The following parameters have been fixed throughout the study for system (6.1): $\beta = 0.01, \alpha = 0.6$, $\omega = 2$. We vary the parameter f_0 and the study has been concentrated on the interval $0.7 \leq f_0 \leq 1.0$, where more than 20 low-period periodic attractors coexist. Here, we investigate all coexisting low-period periodic attractors for the Poincaré time-π map associated with system (6.1) with $= 1.0$ as a paradigm. All attractors have been detected as shown in Tables 6.1, 6.2, 6.3 and 6.4. Because of many basins and too few colors, we divided the region ($[-50, 50] \times [-50, 50]$) into six sub-regions (A–F) by different colors (see Fig. 6.2a). Figure 6.2b ($[-3.5, 3.5] \times [-3, 3]$) shows four critical regimes (A–D) and three color boxes (three enlargements shown in Fig. 6.2c–e). The region A (light gray) is the basin of a period-3 attractor (P3C marked by red '•' in Fig. 6.2b). Three plots of the periodic attractors (together with their basins) for the Poincaré time-π map are shown in Fig. 6.2c–e.

Note that different colors present different basins in every Fig. 6.2c–e, the same color may be used in the different figures because of the limited colors. Six different attractors are detected in the region B (P1L, P1R, P3L, P3R, P5 and P7). We call P1L (marked by dark red '∘') and P1R (marked by yellow '∘') the period-1 orbits located on the left and right, respectively. We call P3L (marked by dark red '+') and P3R (marked by dark blue '+') the period-3 orbits located on the left and right, respectively. We call P5 (marked by yellow '+') and P7 (marked by green '+') the period-5 orbits and period-7 orbits, respectively. Let us now look at the size of the basins of the periodic attractors in the region B. The basins of period-1 (P1L and P1R) are relatively large compared to the others. They make up about 76.6 % of the whole number of initial conditions (37.8 % for all P1L orbits and 38.8 % for all P1R orbits) in the space of initial conditions under consideration and the remaining 23.4 % of all initial points belong to the basins of other periodic orbits. The basins of P5 and P3 make up about 22.1 % of the whole number of initial conditions (11.0 % for all P5 orbits, 5.5 % for all P3L orbits and 5.6 % for all P3R orbits) in the space of initial conditions. The basin of attractor with period-7 covers only 1.3 % of the space of initial conditions. For the size of the basins within the P1(P1L and P1R) we obtain a decrease in the basin size with increasing value for the initial conditions being considered (the oscillations in the size of the basins of the primary P1 attractors is related to the existence of P7 attractors surrounding the primary ones. They eat up a part of the basins of the primary orbits they are connected with). Figure 6.2d shows a blowup ($[-0.6, 2] \times [2, 2.5]$) of the region C by a 900×900 grid. In this region,

Table 6.1 The periodic attractor in the region A

Attractors	p1	p2	p3
P3C	(−1.3687, 1.2634)	(−1.1963, −1.2837)	(1.8803, 0.0820)

Table 6.2 The periodic attractor in the region B

Attractors	p1	p2	p3	p4	p5
P1L	(−1.3687, 1.2634)				
P1R	(0.4753, 0.0035)				
P3L	(−0.6164, 0.1304)	(−1.4584, 0.0776)	(−0.7666, −0.1966)		
P3R	(0.4662, −0.3345)	(0.6290, 0.2991)	(0.0704, 0.0489)		
P5	(0.1275, −0.4079)	(−1.0431, −0.3756)	(0.9089, 0.1416)	(−0.0456, 0.3287)	(−1.3482, 0.3421)
P7	(0.0915, 0.2647)	(0.8355, 0.0677)	(0.1815, −0.3042)	(−0.2534, −0.0882)	(−0.3644, 0.1507)
Attractors	p6	p7			
P6					
p7	(−0.9134, −0.2932)	(−1.4204, 0.2277)			

Table 6.3 The periodic attractor in the region C

Attractors	p1	p2	p3	p4	p5	p6
P6A1	(0.7466, 2.2249)	(0.9478, 2.2811)	(0.8125, 2.3712)	(0.4583, 2.3865)	(0.4360, 2.2516)	(0.2757, 2.3030)
P6A2	(1.0769, 2.3490)	(0.5205, 2.2950)	(0.7166, 2.3806)	(0.9886, 2.2034)	(1.2018, 2.2461)	(0.6764, 2.2397)
P6A3	(1.4541, 2.3043)	(1.2963, 2.1192)	(0.3475, 2.2193)	(0.8198, 2.1952)	(0.3686, 2.3954)	(0.9237, 2.4384)
P6A4	(0.1934, 2.2338)	(0.0294, 2.3213)	(0.2054, 2.3938)	(0.5514, 2.3793)	(0.6994, 2.2998)	(0.5054, 2.2368)
P1A1	(0.6103, 2.3031)					
P1A2	(0.8607, 2.2895)					
P1A3	(0.3629, 2.3107)					
P1A4	(1.1141, 2.2617)					

eight period attractors (P6A1, P6A2, P6A3, P6A4, P1A1, P1A2, P1A3 and P1A4) are observed. Their basins are color coded light gray, yellow, black, white, dark red, dark green, light green, and light red, respectively. The basins of period-6 (P6A1, P6A2, P6A3 and P6A4) are relatively large compared to the others. They make up about 70.0 % [57.1 % for all P6A1 orbits] of the whole number of initial conditions in the space of initial conditions under consideration and 23.3 % [12.3 % for all P1A1 orbits] of all initial points belong to the basins of period-1 orbits (P1A1, P1A2, P1A3 and P1A4). The basins of the attractors in the region A and B make up about 6.8 % of the whole number of initial conditions. There are additional attractors, which together attract less than 0.9 % of the points in the picture (shown in blue). Hence the periodic attractors with higher period periodic orbits are more difficult to find because of their small basins. Similarly, Fig. 6.2e shows a blowup ($[1, 3] \times [-2.5, -1]$) of the region D by a 900×900 grid. In this region, at least eight period attractors (P6A1′, P6A2′, P6A3′, P6A4′, P1A1′, P1A2′, P1A3′ and P1A4′) are observed. Their basins are color coded by eight colors and the attractors are denoted by color '+' and '∘'. The basins of period-6 (P6A1′, P6A2′, P6A3′ and P6A4′) are relatively large compared to the others. They make up about 68.4 % [52.7 % for all P6A1′ orbits] of the whole number of initial conditions in the space of initial conditions under consideration and the 24.8 % [14.5 % for all P1A1′ orbits] of all initial points belong to the basins of period-1 orbits (P1A1′, P1A2′, P1A3′ and P1A4′). The basins of the attractors in the region A and B make up about 6.0 % of the whole number of initial conditions. There are additional attractors, which together attract less than 0.8 % of

Table 6.4 The periodic attractor in the region D

Attractors	p1	p2	p3	p4	p5	p6
P6A1′	(2.0785, −1.9718)	(2.1605, −1.8387)	(2.6072, −1.5125)	(2.4362, −1.6325)	(2.3043, −1.9139)	(2.5860, −1.6656)
P6A2′	(2.2364, −1.7980)	(2.4276, −1.7009)	(1.9374, −1.9683)	(2.0765, −2.0654)	(1.8380, −2.0926)	(2.3880, −1.8556)
P6A3′	(1.5860, −2.0736)	(2.6559, −1.5968)	(2.3096, −2.0002)	(1.7580, −2.2416)	(2.0600, −1.8736)	(2.4620, −1.5606)
P6A4′	(2.6280, −1.3696)	(2.8129, −1.3827)	(2.2737, −1.7308)	(2.5220, −1.8196)	(2.0120, −2.1256)	(1.8260, −1.9876)
P1A1′	(2.3700, −1.7659)					
P1A2′	(2.1575, −1.9238)					
P1A3′	(2.5592, −1.5834)					
P1A4′	(1.1255, −2.0544)					

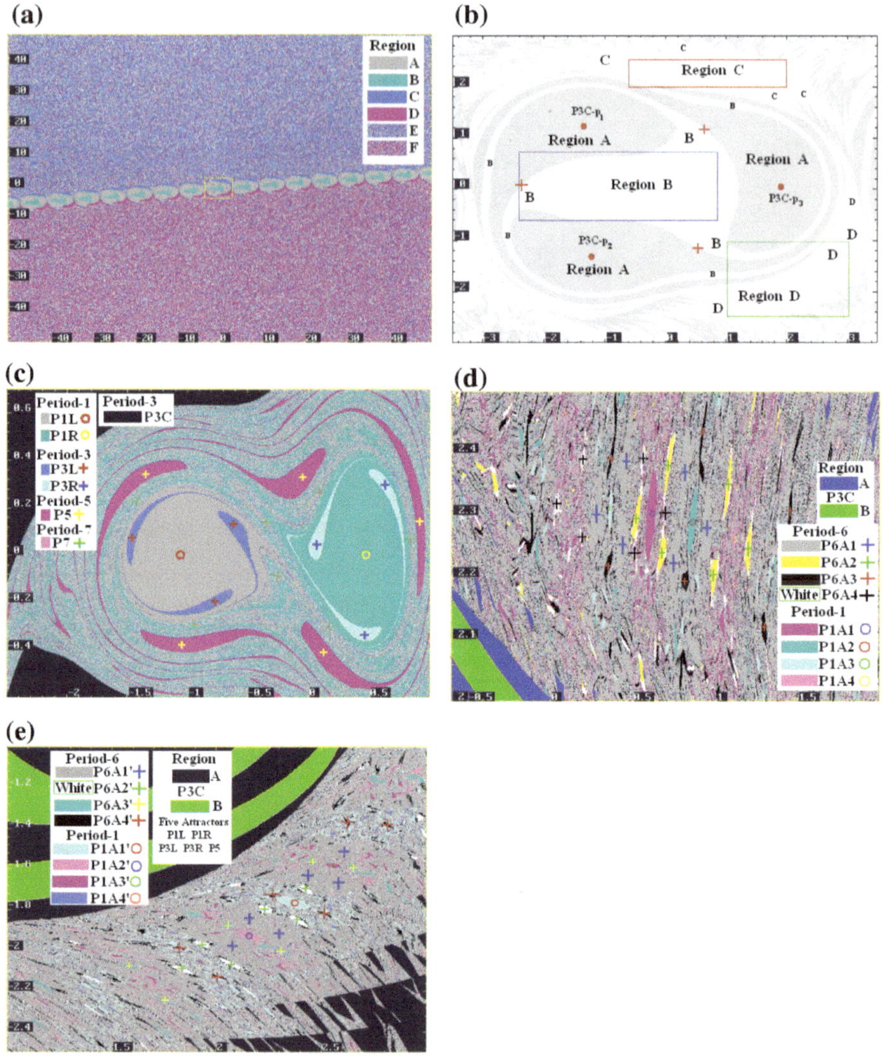

Fig. 6.2 Basin of attraction and coexisting attractors for the time-π map of the SD oscillator (6.1) for $\beta = 0.01$, $f_0 = 1$ and $\omega = 2$. **a** a generalized basin with four regions (A–D) in the (x, x') plane; **b** a schematic view for an enlargement shown in (**a**); **c** six attractors and their basins in the region B; **d** eight attractors and their basins in the region C; **e** eight attractors and their basins in the region D

the points in the picture (dark blue). See Table 6.5 for the basin area of the attractors. The most interesting observation is that the region E and the region F yield a complex interwoven structure of the basins. At least 20 basins of the coexisting attractors are in these regions. We will focus on the region E and F in the following section.

Table 6.5 The basin area of the attractor

Region A	Basin area	Region B	Basin area	Region C	Basin area	Region D	Basin area
P3C	100 % Sa	P1L	37.8 % Sb	P6A1	57.1 % Sc	P6A1′	52.7 % Sd
		P1R	38.8 % Sb	P6A2	5.3 % Sc	P6A2′	5.2 % Sd
		P3L	5.5 % Sb	P6A3	3.9 % Sc	P6A3′	7.8 % Sd
		P3R	5.6 % Sb	P6A4	3.7 % Sc	P6A4′	2.7 % Sd
		P5	11.0 % Sb	P1A1	12.3 % Sc	P1A1′	14.5 % Sd
		P7	1.3 % Sb	P1A2	3.8 % Sc	P1A2′	3.8 % Sd
				P1A3	3.4 % Sc	P1A3′	2.7 % Sd
				P1A4	3.8 % Sc	P1A4′	3.8 % Sd

6.4 Wada Property and the Uncertainty Exponents

The main focus of this section is to show some evidences that the system (Eq. 6.1) has the Wada property (totally Wada and partially Wada). We use three numerical methods. The first numerical method is for finding accessible saddle periodic orbits on basin boundaries. Usually, the accessible trajectories converge to a periodic trajectory which is an accessible saddle. For those periodic trajectories, one branch of its local unstable manifold is an accessible curve. This method is called ABST method [18]. In practice, the ABST method works fine for many dynamical systems and it is used to find an accessible saddle periodic point. It is easy to verify that it has found an accessible hyperbolic periodic point [20]. The second numerical method is for plotting one dimensional unstable (or stable) manifold. One may expect that plots of unstable manifolds would be inaccurate due to the sensitivity of initial condition near a saddle periodic orbit, but plots are usually very accurate. The method for plotting unstable (or stable) manifolds is called UM (or SM) method [18], see [21, 22] for a detailed justification of the algorithm and more details. The third method is for finding periodic orbits of some specified period and this method is called RP method [18]. The RP method attempts to plot some periodic points of a specified period. This routine has a basic step in which it chooses a point in the screen area at random and then applies a Quasi-Newton method up to 50 times. The routine plots when it has found a point which is within a distance of 10^{-11} from a periodic point of a specified period. Therefore, one can use the UM method to verify that the periodic orbits are indeed accessible from basin B. If a branch of the unstable manifold of the periodic point P on the basin boundary goes to the regional attractor of its basin, then this branch of the unstable manifold of P is a curve in the basin that hits the basin boundary at the point P, so that the point P is accessible. We use the UM method to plot the other branch of the unstable manifold of P, although we would only need a small piece of what is plotted to show that it intersects other basins. Firstly, we are interested in the basin boundaries for some typical basins of attraction. In the region A, we obtain a period-3 saddle-hyperbolic orbit S3C (Fig. 6.2a) on the basin boundary which is accessible from the basin of P3C (marked by red '+' in Fig. 6.2b).

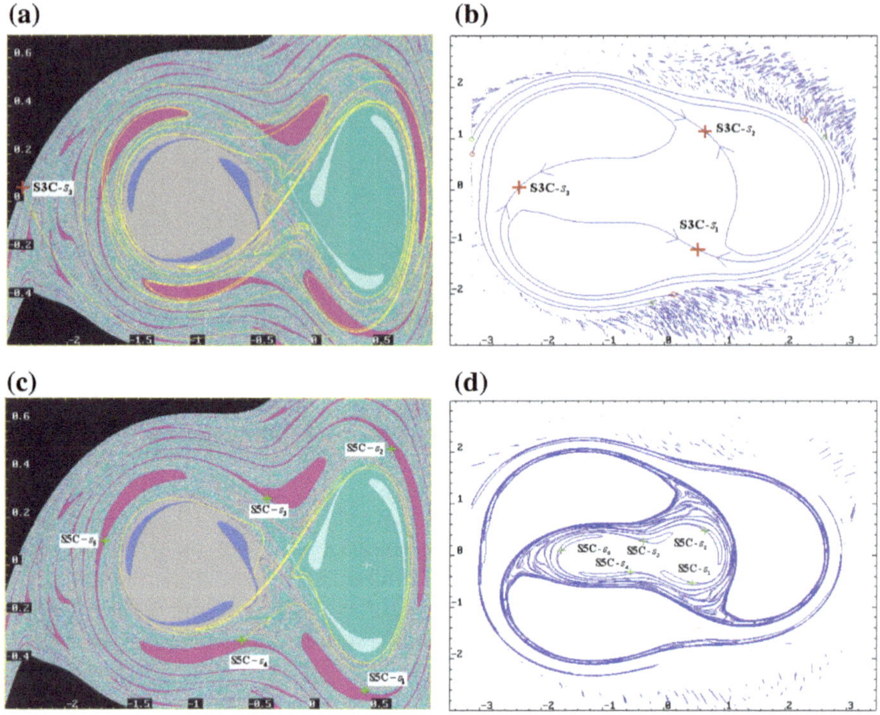

Fig. 6.3 **a** The unstable manifold (denoted by the *yellow curve*) of the period-3 saddle ($S3C_{-s_1}$) on the basin boundary intersecting six basins (P1L, P1R, P3L, P3R, P5 and P7); **b** The stable manifold of the S3C; **c** The unstable manifold (denoted by the *yellow curve*) of the period-5 saddle ($S5C_{-s_4}$) on the basin boundary intersecting five basins (P1L, P1R, P3L, P3R and P7); **d** The stable manifold of the S5C

Figure 6.3a shows a saddle $S3C_{-s_3}$ (marked by red '+') on the basin boundary and one branch of the unstable manifold (marked by yellow curves) of the saddle $S3C_{-s_3}$. It shows that the unstable manifold intersects six basins (all basins in the region B: $P1R$, $P1L$, $P3R$, $P3L$, $P5$ and $P7$). Similarly, the unstable manifold of $S3C_{-s_1}$ and the unstable manifold of $S3C_{-s_2}$ intersect each of the six basins in the region B. In fact, only a piece of what is shown is needed to show that the unstable manifold intersects six basins. Figure 6.3b shows the stable manifold of $S3C$. Theorem 6.1 shows that every point on the stable manifold of $S3C$ is a Wada point (in fact, every point on the closure of the stable manifold is a Wada point). Therefore, the boundaries of the regions B have Wada property. In the region B, it is easy to visualize from an intuitive point of view that the purple basin of $P5$ has Wada basin boundaries. We will verify the Wada basin boundaries with some rigor. We obtain a period-5 saddle-hyperbolic orbit $S5C$ on the basin boundary which is accessible from the purple basin of $P5$. Figure 6.3c shows a saddle $S5C_{-s_4}$ (marked by red '+') on the basin

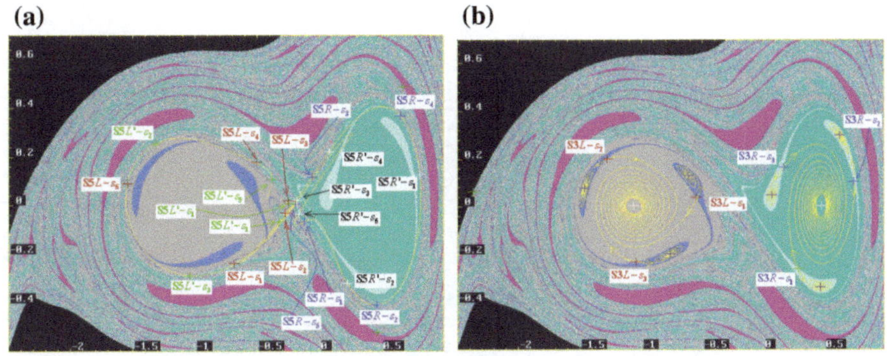

Fig. 6.4 **a** Two period-5 saddles (S5L and S5L′) that are accessible from the basin of P3L. Two period-5 saddles (S5R and S5R′) that are accessible from the basin of P3R. The unstable manifold (denoted by the *yellow curve*) of $S5L_{-s_4}$ intersects four basins (P1L, P1R, P3L and P3R); **b** The figure shows the unstable manifold (denoted by the *yellow curve*) of the period-3 saddles (S3L) on the basin boundary that is accessible from the basin of P3L. This manifold intersects only two basins (P3L and P1L) that is accessible from the basin of P3L. The unstable manifold (denoted by the *yellow curve*) of the period-3 saddles (S3R) on the basin boundary that is accessible from the basin of P3R. This manifold intersects only two basins (P3R and P1R) that is accessible from the basin of P3R

boundary and one branch of the unstable manifold (marked by yellow curves) of the saddle $S5C_{-s_4}$. It shows that the unstable manifold of the only accessible periodic orbit intersects five basins ($P1R$, $P1L$, $P3R$, $P3L$, and $P7$). Figure 6.3d shows the stable manifold of $S5C$. Applying Theorem 6.1 we find that the basin boundary of $P5$ is a Wada basin boundary. In other words, all the points on the stable manifold are Wada points. The whole basin boundaries of $P5$ are the stable manifold of $S5C$. We can also observe that more Wada basin boundaries are created by a saddle-node bifurcation as the parameter f_0 is increased to some critical values.

The light grey basin of $P1L$ and the dark green basin of $P1R$ have partially Wada basin boundaries. We obtain two period-5 saddle-hyperbolic orbits ($S5L$ and $S5L'$) on the basin boundary which is accessible from the grey basin of $P1L$. Figure 6.4a shows that one branch of the unstable manifold (marked by yellow curves) of the saddles $S5L'_{-s_5}$ intersects four basins. We find that all accessible period-5 saddles are on the manifold. We also obtain another a period-3 saddle-hyperbolic orbits ($S3L$) on the basin boundary which is accessible from the gray basin of $P1L$. Figure 6.4b shows that the unstable manifolds (marked by yellow curves) of the saddles $S3L$ intersects only two basins. Applying Theorem 6.2 we find that the grey basin of $P1L$ has partially Wada basin boundaries. Similarly, the dark green basin boundaries of $P1R$ are partially Wada basin boundaries (see Fig. 6.4a, b). All saddle-hyperbolic orbits are calculated by a Quasi-Newton method (Table 6.6). In the region C, the basin boundaries are more complex. We can observe that most of the basins have partially Wada basin boundaries. For example, we obtain two period-6 saddle-hyperbolic orbits ($S6A1$ and $S6A2$ shown in Table 6.7 and Fig. 6.5a) which are accessible from

Table 6.6 The accessible periodic saddles in Region B

Saddles	s_1	s_2	s_3	s_4	s_5
S3C	(0.5556, −1.1427)	(0.6678, 1.1502)	(−2.3686, 0.0632)		
S3L	(−0.5548, 0.0396)	(−1.2763, 0.2017)	(−1.0438, −0.2304)		
S3R	(0.2862, −0.2589)	(0.7462, 0.1059)	(0.1644, 0.1656)		
S5C	(0.4671, −0.5221)	(0.6735, 0.4862)	(−0.3382, 0.2787)	(−0.5451, −0.3110)	(−1.6722, 0.1004)
S5L	(−0.6891, −0.2395)	(−0.2733, −0.0317)	(−0.2457, 0.0197)	(−0.5078, 0.1797)	(−1.5399, 0.0906)
S5R	(−0.1234, −0.0653)	(0.4623, −0.4068)	(−0.0580, 0.1187)	(0.6530, 0.3677)	(−0.1880, 0.0128)
S5L′	(−0.3235, −0.0663)	(−1.3119, 0.2567)	(−1.0359, −0.2873)	(−0.2439, 0.0056)	(−0.3819, 0.1086)
S5R′	(0.8090, 0.1246)	(0.2270, −0.3213)	(−0.1869, 0.0176)	(09.0860,0.2226)	(−0.1539, −0.0343)

Table 6.7 The accessible periodic saddles in Region C

Saddles	s_1	s_2	s_3	s_4	s_5	s_6
S6A1	(0.9778, 2.3635)	(0.7678, 2.2855)	(0.9178, 2.2201)	(1.2338, 2.1670)	(1.4578, 2.1895)	(1.3438, 2.3065)
S6A2	(0.4365, 2.4611)	(0.1809, 2.0649)	(1.3811, 2.0341)	(1.7950, 2.1611)	Out of the Fig. 6.5a	Out of the Fig. 6.5a

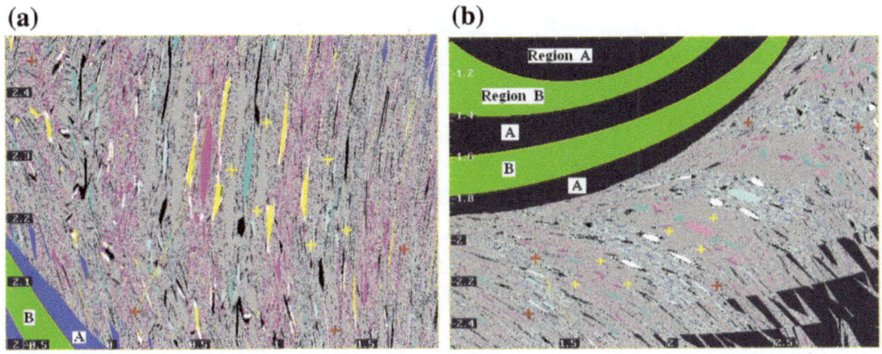

Fig. 6.5 **a** Some saddles S6A1 marked by the *yellow* '+' and S6A2 marked by the *red* '+' in the region C; **b** Some saddles S6A1′ marked by the *yellow* '+' and S6A2′ marked by the *red* '+' in the region D

the light grey basin of $P6A1$. Their unstable manifolds cross more than three basins. This yields a complex interwoven structure of the basins. Another interesting fact is that the basin boundaries of other attractors (in the regions A and region B) spread over the basin boundaries of these attractors (in the region C). It is difficult to observe because the basins of other attractors cover only 0.7 %. It may be explained by the stable manifold of $P3C$ (Fig. 6.3b) and the stable manifold of $P5$ (Fig. 6.3d). In the region D, the basin boundaries are also complex. We can also observe that most of the basins have partially Wada basin boundaries. We obtain two period-6 saddle-hyperbolic orbits ($S6A1'$ and $S6A1'$ shown in Table 6.8 and Fig. 6.5b) which are accessible from the light grey basin of $P6A1'$. Their unstable manifolds cross more than three basins. This yields a complex interwoven structure of the basins. The interesting fact is that the basin boundaries of other attractors (in regions D and B) spread over the basin boundaries of these attractors (in region D). It is difficult to observe because the basins of other attractors cover only 0.6 %.

An interesting observation is that the basins (in regions E and F) seem to have Wada property which appears to spread over most of the state space. This yields a complex interwoven structure of the basins. Extremely small changes in the initial conditions are sufficient to shift a point from one basin to another one (at least 20 basins of the coexisting attractors can be alternative). Two representative regions are shown in Fig. 6.6a ($[-3, 3] \times [5, 10]$) denoted by $E1$) and Fig. 6.6b

Table 6.8 The accessible periodic saddles in Region D

Saddles	s_1	s_2	s_3	s_4	s_5	s_6
S6A1′	(2.2247, −1.8664)	(2.1687, −2.0144)	(2.0147, −1.9354)	(1.7007, −2.0734)	(1.8307, −2.1874)	(1.5867, −2.1814)
S6A2′	(1.3677, −2.3258)	(2.8987, −1.4355)	(1.4012, −2.0577)	(2.3867, −1.4095)	(2.2407, −2.1955)	Out of the Fig. 6.5b

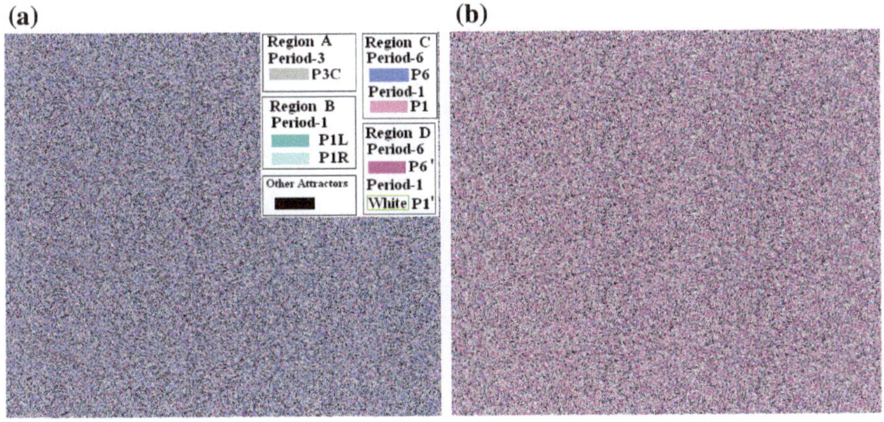

Fig. 6.6 **a** The generated basins in the E region ($[-3, 3] \times [5, 10]$) and **b** The generated basins in the F region ($[-3, 3] \times [-5, -10]$)

(($[-3, 3] \times [-10, -5]$) denoted by $F1$). In order to visualize the plot clearly, we give the generalized basins by eight colors. In region A, there is only one attractor $P3C$ and its basin is shown in light grey. In region B, all the attractors are divided into three groups: the left period-1 attractor $P1L$, the right period-1 attractor $P1R$ and other attractors ($P3L$, $P3R$, $P5$ and $P7$). Their basins are shown in dark green, light green and black, respectively. In region C, the eight attractors are divided into two groups: period-6 attractors ($P6$) and period-1 attractors ($P1$). Their basins are color coded by light blue and light red, respectively. In region D, the eight attractors are divided into two groups: period-6 attractors ($P6'$) and period-1 attractors ($P1'$). Their basins are color coded by dark red and white, respectively. If there are other attractors out of these regions (A, B, C and D), their basins are also coded by black. These eight basins are intermingled in the sense that every initial condition can be perturbed, with positive probability, into a different general attractor's basin by an arbitrarily small random change in the location of the initial condition. The light grey basin of $P3C$ spreads over most of regions $E1$ and $F1$ at the same probability level. The probability tends to limiting values of 22.1 % for the attractor $P3C$. The basins of $P1L$ and $P1R$ also spread over most of regions $E1$ and $F1$ at the same probability level (the probability is approximately 4.1 %). In the region $E1$, the probabilities tend to limiting values of 41.7 % for the attractor $P6$ and 20.7 % for the attractor $P6'$. While in the region $F1$, the probabilities tend to limiting values of 20.5 % for the attractor $P6$ and 41.6 % for the attractor $P6'$. Table 6.9 gives the size of the basins of all the attractors in Fig. 6.2a ($[-50, 50] \times [-50, 50]$). It turns out that most of the periodic attractors have low periods (<10); stable higher period orbits are found rarely. These are likely to have the similar properties, but it is difficult to confirm.

Basin boundaries with fractal and Wada property lead to final state sensitivity. Final state sensitivity can be quantified by the uncertainty exponent ζ, which was firstly introduced by Grebogi et al. [23]. The uncertainty exponent ζ is defined as

Table 6.9 The basin area of the attractors in the region $([-50, 50] \times [-50, 50])$, S is the area of this region

Attractors	Basin area	Attractors	Basin area	Attractors	Basin area	Attractors	Basin area
P3C	22.7 %	P1L	2.6 % S	P6A1	19.3 % S	P6A1′	19.1 % S
		P1R	2.6 % S	P6A2	1.2 % S	P6A2′	1.8 % S
		P3L	0.3 % S	P6A3	1.2 % S	P6A3′	2.4 % S
		P3R	0.3 % S	P6A4	1.1 % S	P6A4′	1.0 % S
		P5	0.7 % S	P1A1	4.2 % S	P1A1′	5.2 % S
		P7	0.07 % S	P1A2	1.2 % S	P1A2′	1.2 % S
				P1A3	1.2 % S	P1A3′	1.0 % S
				P1A4	1.2 % S	P1A4′	3.4 % S

follows. Choose randomly an initial condition x_0. Define $x_0' = x_0 + \delta$, where δ is a small perturbation. Determine whether the asymptotic dynamics of the system using these two initial conditions is qualitatively different. For a given perturbation δ, a fraction of uncertain initial conditions $f(\delta)$ can be computed by randomly choosing many initial conditions and determining if they are uncertain. The uncertainty exponent ζ is determined by fitting the scaling law $f(\delta) \sim \delta\zeta$. The uncertainty dimension $\mathcal{D}$ of the fractal set embedded in the initial conditions is obtained from $\mathcal{D} = N - \zeta$, where N is the dimension of the phase space. Therefore the uncertainty dimension is defined in the range $\mathcal{D} \in [N - 1, N]$. The fractal dimension is a common way to measure the complexity of a fractal basin boundary. If it is an integer number, then the boundary is said to be non-fractal (or smooth), while if it is not an integer, the boundary is fractal. The uncertain exponent is estimated from the slope in the $logf(\delta) - log\delta$ plot as shown in Fig. 6.7. Figure 6.7a shows $log_{10}f(\delta)$ versus $log_{10}\delta$, where the initial conditions are drawn on the line defined by $x' = -0.2$ in Fig. 6.2c in the region B. The uncertainty exponent is estimated to be $\zeta \approx 0.23$ and $d_s \approx 0.77$. Figure 6.7b shows $log_{10}f(\delta)$ versus $log_{10}\delta$, where the initial conditions are drawn on the line defined by $x' = 2.3$ in Fig. 6.2d in the region B. The uncertainty exponent is estimated to $\zeta \approx 0.12$ and $d_s \approx 0.88$. Figure 6.7c shows $log_{10}f(\delta)$ versus $log_{10}\delta$, where the initial conditions are drawn on the line defined by $x' = 8$ (or $x' = -8$) in Fig. 6.6a (or 6.6b). The uncertainty exponent is estimated to be $\zeta \approx 0.007$ (or $\zeta \approx 0.005$) and $d_s \approx 0.993$ ($d_s \approx 0.995$), i.e., the basin boundaries is nearly two dimensional and it indicates a highly fractal structure in the region E and in the region F. This large uncertain dimension can be regarded as an additional indication of the high complexity of the basin structure. The basin boundaries are so intertwined that serious difficulties in prediction arise if there are small uncertainties in fixing the initial condition. The capability of the SD oscillator to make precise predictions was questioned with the discovery of the Wada properties.

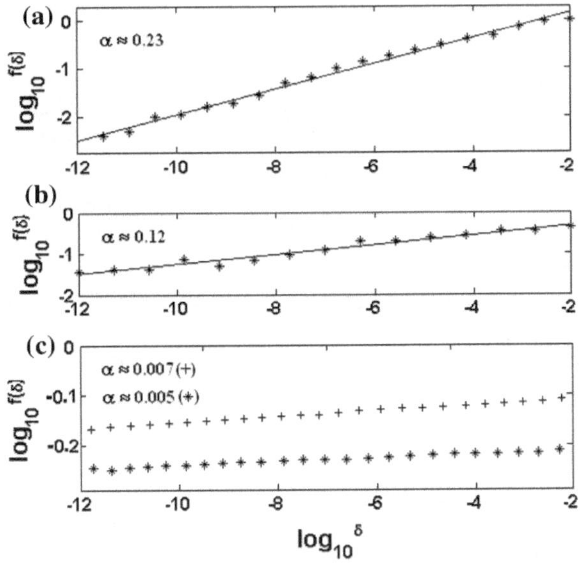

Fig. 6.7 **a** A linear fit yields the following uncertainty exponent for the region B: $\alpha \approx 0.23$ [$\log_{10} f(\delta)$ versus $\log_{10} \delta$ for the basin boundaries between the *gray* and other basins]; **b** A linear fit yields the following uncertainty exponent for the region C: $\alpha \approx 0.12$ [$\log_{10} f(\delta)$ versus $\log_{10} \delta$ for the basin boundaries between the *gray* and other basins]; **c** Two linear fits yield the following uncertainty exponent for the region E: $\alpha \approx 0.007$ and for the region F: $\alpha \approx 0.005$ [$\log_{10} f(\delta)$ versus $\log_{10} \delta$ versus for the basin boundaries between the *gray* and other basins]

6.5 Wada Basin Boundaries

Since the Wada property might have very huge consequences on the applicability of a system, the present subsection just begins the discussion about engineering applications, which will be the topic of subsequent chapters. An important application of the SD oscillator is in the dynamics of arch bridges, arch beams and other public arch structures. One of the main mechanical consequences of Wada basin boundaries is the existence of a high number of oscillatory patterns and the difficulty of predicting to which of them a given initial condition leads. We show here that some Wada basin boundaries can be extremely complex because the simple oscillator possesses more than 20 coexisting low-period periodic motions in some physical circumstances. This type of topological complexity of basin boundaries provides insight into the qualitative unpredictability and uncertain dynamics for the mechanical oscillator. The implications of these results for mechanical systems are cautionary. Usually, the vibrating systems have one parameter combination that results in two distinct oscillatory patterns. It may be of interest for an engineer to predict which of the two will exist for a given initial condition. Stipulation of alternative possibilities based on simple basin structure could be misleading because this type of Wada basin boundaries has not been reported in mechanical systems. Especially in the region E and in the region F (see Fig. 6.2a), the engineers cannot predict which of the 23 patterns will exist due to a major transformation of the basins themselves, implying unpredictability over a broad range of initial conditions.

The important point is that the extent of final state sensitivity here is different from the typical Wada basin boundaries. When the previous cautionary tales emphasized the impossibility of making quantitative predictions, the existence of Wada properties suggests something more troublesome [5]. Therefore, the existence of the Wada properties should be considered when building a mechanical device. From the point of view of practical applications, many mechanical devices governed by nonlinear motion equations may exhibit irregular changes in their motion for a given parameter set. It is important to be aware of the possibility of such sudden jumps to what could be "dangerous" states in the design of structures, and thus to avoid them. For the shallow arch model, the deformed shape can also undergo a sudden jump called snap-through buckling [24–26] which is often undesirable. The results here may provide a Wada basin dynamics perspective of this phenomenon. In the real world or in the outside environment, any system is subjected to random external disturbances described as noise. The macroscopic mechanical systems also experience internal noise from thermal vibrations of individual molecules. The dynamics of a shallow arch oscillator is extremely sensitive to initial conditions and very small perturbations of the initial state can influence the final state. In this case, the system is extremely sensitive to noise and noise may cause a hopping process between various attractors. There exists a competition between the attractiveness toward regular motion in the neighborhood of an attractor and the jumping among the different attractors. In reality, the noise can kick the orbit out of the open neighborhood of the attractor into the Wada basin boundaries. The trajectory spends a certain amount of time and switches to the same attractor or another attractor. Once this process keeps repeating, the system (Fig. 6.1) may undergo some sudden jumps and lead to snap-through buckling phenomenon.

The evidence that the model (with $f_0 = 1.0$ as a paradigm) possesses 23 coexisting low-period periodic attractors for other fixed parameters (β, α and ω) (and even varying parameter f_0 we can also observe more than 20 coexisting low-period periodic attractors (e.g., $0.7 \leq f_0 \leq 1.0$)) shows how a simple mechanical model can generate qualitative unpredictability. The results here suggest that although a mechanical system is perched on the boundary between two alternative states, it may in fact wind up in a third (or fourth, even twentieth) state, resulting in unpredictably so. The complexity of the basins is due to the appearance of different periodic saddles in basin boundaries. In particular, the Wada basin boundaries appear to spread over most of the state space in regions E and F. In regions E and F, all the basins are closely interwoven that perturbations of the magnitude round-off error are sufficient to move an initial condition from one basin to another.

6.6 Summary

In this chapter the complicated Poincaré maps have been used to examine the SD oscillator with viscous damping subjected to a sinusoidal variation of the central load. The system may possess more than 20 coexisted low-period periodic attractors for a given set of parameters. The large number of stable orbits yields a complex structure of closely interwoven basins of attraction, whose boundaries fill almost the whole state space. We describe the Wada basin boundaries and partially Wada basin boundaries by the mathematically rigorous method for some basins. The consequence is the existence of a large number of possible final states which is difficult to be predicted. Perturbations in such a system render uncertain type even the qualitative fate of a given initial state. The complexity of the oscillator depends on the strong nonlinearity in the form of irrational type differing from the polynomial system like Duffing equation.

References

1. Holmes, P., & Marsden, J. E. (2014). Bifurcations of dynamical systems and nonlinear oscillations in engineering systems. *Dynamics and Design Conference: D & D* (pp. 561–582).
2. Mcdonald, S. W., Grebogi, C., Ott, E., & Yorke, J. A. (1985). Fractal basin boundaries. *Physica D-Nonlinear Phenomena, 17*(2), 125–153.
3. Moon, F. C., & Li, G. X. (1985). Fractal basin boundaries and homoclinic orbits for periodic motions in a two-well potential. *Physics Review Letters, 55*(14), 1439–1442.
4. Kennedy, J., & Yorke, J. A. (1991). Basin of wada. *Physica D-Nonlinear Phenomena, 51*(1–3), 213–225.
5. Aguirre, J., Viana, R. L., & Sanjuán, M. A. F. (2009). Fractal structures in nonlinear dynamics. *Review of Modern Physics, 81*, 333–386.
6. Aguirre, J., & Sanjuán, M. A. F. (2002). Unpredictable behavior in the duffing oscillator: Wada basins. *Physica D-Nonlinear Phenomena, 171*(1–2), 41–51.
7. Zhang, Y., & Luo, G. (2012). Unpredictability of the wada property in the parameter plane. *Physics Letters A, 376*(45), 3060–3066.
8. Viana, R. L., & Sanjuán, M. A. F. (2011). Fractal structures in nonlinear plasma physics. *Philosophical Transactions of the Royal Society A, 369*(1935), 371–395.
9. Bellido, F., & Ramirez-Malo, J. B. (2006). Periodic and chaotic dynamics of a sliding driven oscillator with dry friction. *International Journal of Non-Linear Mechanics, 41*(6), 860–871.
10. Vandermeer, J. (2004). Wada basins and qualitative unpredictability in ecological models: A graphical interpretation. *Ecological Modelling, 176*(1–2), 65–74.
11. Seoane, J. M., & Sanjuán, M. A. F. (2013). New developments in classical chaotic scattering. *Reports on Progress in Physics, 76*(1), 016001.
12. Feudel, U., Grebogi, C., Hunt, B. R., & Yorke, J. A. (1996). Map with more than 100 coexisting low-period periodic attractors. *Physics Review E, 54*(1), 71–81.
13. Breban, R., Nusse, H. E., & Ott, E. (2003). Lack of predictability in dynamical systems with drift: Scaling of indeterminate saddle-node bifurcations. *Physics Letters A, 319*(1), 79–84.
14. Aguirre, J., Vallejo, J. C., & Sanjuán, M. A. F. (2012). Wada basins and unpredictability in hamiltonian and dissipative systems. *International Journal of Modern Physics B, 17*(17), 22–24.
15. Feudel, U., Witt, A., & Lai, Y. C. (1998). Basin bifurcation in quasiperiodically forced systems. *Physics Review E, 58*(58), 3060–3066.
16. Zhang, Y. (2013). Strange nonchaotic attractors with wada basins. *Physica D-Nonlinear Phenomena, 259*(3), 26–36.
17. Zhang, Y. (2013). Wada basins of strange nonchaotic attractors in a quasiperiodically forced system. *Physics Letters A, 337*(18), 1269–1273.
18. Nusse, H. E., Ott, E., & Yorke, J. A. (1995). Saddle-node bifurcations on fractal basin boundaries. *Physical Review Letters, 75*(13), 2482–2485.
19. Guckenheimer, J., & Holmes, P. (1983). *Nonlinear oscillation, dynamical system and bifurcation of vector fields*. New York: Springer.
20. Nusse, H. E., & Yorke, J. A. (1989). A procedure for finding numerical trajectories on chaotic saddles. *Physica D-Nonlinear Phenomena, 36*(1–2), 137–156.

21. You, Z., Kostelich, E., & Yorke, J. A. (1991). Calculating stable and unstable manifolds. *Internatioal Journal of Bifurcation and Chaos*, *1*(3), 605–624.
22. Krauskopf, B., & Osinga, H. (1998). Growing 1d and quasi 2d unstable manifolds of maps. *Journal of Computational Physics*, *146*(1), 404–419.
23. Grebogi, C., McDonald, S. W., Ott, E., & Yorke, J. A. (1983). Final state sensitivity: An obstruction to predictability. *Physical Letters A*, *99*(9), 415–418.
24. Pippard, A. B. (1990). The elastic arch and its modes of instability source. *European Journal of Physics*, *11*(6), 359–365.
25. Mallon, N. J., Feya, R. H. B., Nijmeijera, H., & Zhang, G. Q. (2006). Dynamic buckling of a shallow arch under shock loading considering the effects of the arch shape. *International Journal of Nonlinear Mechanics*, *41*(9), 1065–1075.
26. Chen, J., & Lin, J. (2004). Dynamic snap-through of a shallow arch under a moving point load. *Journal of Vibration and Acoustics ASME*, *126*(4), 514–519.

Chapter 7
Piecewise Linear Approach

This chapter and the following ones introduce a sequence of methodologies for analytical and semi-analytical computations of the SD oscillator whose nonlinearity is of irrational type. In this chapter, we introduce a triple linear procedure depending on the property of the nonlinearity of the SD oscillator of the irrational type. And then in the following chapters we will introduce an extended averaging method in Chap. 8, irrational elliptic functions method in Chap. 9 and the so-called cell-mapping method in Chap. 10. These methodologies will be valid for both smooth and discontinuous cases. All the methods which will be presented in the subsequent chapters aim at overcoming or removing the barriers of the conventional methodologies. For example an analytical analysis based upon the polynomial theory is invalid for this irrational nonlinearity and the Runge–Kutta direct integration method is no more applicable for the discontinuous case.[1]

7.1 Introduction

We will introduce a triple linear technique for the SD oscillator whose nonlinearity is of irrational type. This procedure is valid for both smooth and discontinuous stages. The triple linearization process is proposed $0 \leq \alpha < 1$ when the system with the double well behaviour based on original oscillator with the same equilibria and the same derivatives. This approximation is valid for the limit case of $\alpha = 0$, at which the triple linear system becomes the original discontinuous case, that is the boundary value $x_0 \to 0$, as $\alpha \to 0$. This procedure is also valid for $\alpha \geq 1$, which is treated as the case of $x_0 = 0$. For the perturbed system, chaotic attractors are presented for both the

[1]This chapter relies a common work with Professors Marian Wiercigroch, Ekaterina Pavlovskaia, Celso Grebogi, J. Michael T. Thompson, Center for Applied Dynamics Research, Aberdeen University, King's College, Aberdeen AB24 3UE, Scotland, UK.

Q. Cao and A. Léger, *A Smooth and Discontinuous Oscillator*, Springer Tracts in Mechanical Engineering, DOI 10.1007/978-3-662-53094-8_7

original oscillator and for the triple linear approximation with a excellent agreement for the topological structures and Lyapunov exponents. Other methodologies such as the multiple scale procedure, numerical-analytical method, bifurcation analysis for discontinuous dynamics have been proposed in the literature (see e.g. [1–3]).

7.2 Unperturbed Dynamics

In this section, the triple linear approximation is proposed for the SD oscillator, and the unperturbed dynamics or the triple linear system is discussed to show the good agreement with the original system.

7.2.1 Triple Linear Approach

Recall the dimensionless SD oscillator, the mechanical model being still associated with Fig. 2.1 and, as discussed in the previous chapters, the unperturbed system is rewritten here

$$x'' + x\left(1 - \frac{1}{\sqrt{x^2+\alpha^2}}\right) = 0, \tag{7.1}$$

and in the discontinuous case

$$x'' + x(1 - \text{Sign}(x)) = 0, \quad \text{Sign}(x) = \begin{cases} 1, & x > 0 \\ 0, & x = 0 \\ -1, & x < 0 \end{cases} \tag{7.2}$$

where $\omega_0^2 = 1$ is assumed in this chapter and the equilibria of the system (7.1) have been investigated in previous chapters.

The Hamiltonian function obtained from Eq. (7.1), is the following

$$H(x, y) = \frac{1}{2}x^2 + \frac{1}{2}y^2 - \sqrt{x^2+\alpha^2} + \alpha, \tag{7.3}$$

from which the possible types of trajectories can be classified and analyzed for different values of the Hamiltonian function $H(x, y) = E$.

The corresponding phase portraits have been represented in Fig. 2.3 of Chap. 2, which is similar to that of the double-well Duffing oscillator [4]. For the discontinuous case, i.e. $\alpha = 0$, the orbits for $E > 0$ are comprised between two large segments of circles with their center located at $(-1, 0)$ and $(1, 0)$ connected at $x = 0$. The case $E < 0$ involves two families of circles. When $\alpha = 0$, the unperturbed system (7.1) reduces to system (7.2), whose Jacobian at $(0, 0)$ does not exist. This case is the most

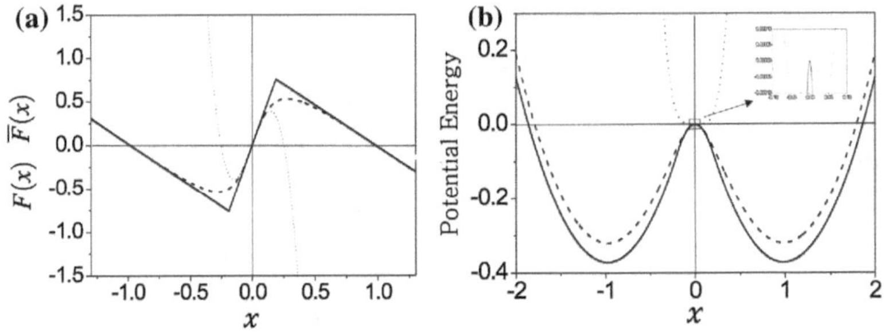

Fig. 7.1 Restoring forces in (**a**) and potential wells in (**b**) for $\alpha = 0.2$: The *solid curve* is for the nonlinear restoring force $F(x)$ in (7.17), the *dashed curve* denotes the piecewise linear force $\bar{F}(x)$ in (7.6) and *dotted curves* presents the Duffing approximation force

interesting as the discontinuous system (7.2) exhibits two circles centered at $(\pm 1, 0)$, connecting at the singularity point $(0, 0)$, and forming a special singular homoclinic-like orbit. The structure around the point indicates a saddle-like behaviour. This point $(0, 0)$ is an isolated singularity without eigenvalues or eigenvectors. The two circles excluding the point $(0, 0)$ are not the manifolds of the singularity, but the flow along these circles approaches the point $(0, 0)$ at a rate of $dy/dx \to \pm\infty$ as $x \to 0$ and it will be trapped by the singularity.

Now the piecewise linear approach is introduced so that it will be possible to go on in a theoretical investigation of the Hamiltonian, the homoclinic orbit and the periodic orbits of the approximate system. In the smooth case, the restoring force $F(x) = -x(1 - \frac{1}{\sqrt{x^2 + \alpha^2}})$, shown in Fig. 7.1a by solid curve, is expanded into a Taylor series in the neighborhood of $x = 0$ as,

$$F(x) = \frac{1}{\alpha}(1 - \alpha)x + \sum_{n=1}^{\infty} \frac{P_{2n}(0)}{\alpha^{2n+1}} x^{2n+1}, \tag{7.4}$$

where $P_n(x)$ are the Legendre polynomials defined by

$$P_n(x) = \frac{1}{2^n n!} \left[\frac{d^n}{dx^n} \left(x^2 - 1\right)^n \right], \quad n = 1, 2, \ldots,$$

the first values of which are

$$P_2(0) = -\frac{1}{2}, \quad P_4(0) = \frac{3}{8}, \quad P_6(0) = -\frac{5}{16}, \ldots.$$

It is worth noticing that the truncation of the Taylor expansion, Eq. (7.4), after the cubic term and substituting the truncation into Eq. (7.17) gives exactly the Duffing oscillator, $F_d(x) = \frac{1-\alpha}{\alpha} x - \frac{x^3}{2\alpha^3}$, as shown in the dotted curve in Fig. 7.1a. Of course this truncation would fail for the case $\alpha = 0$. To approximate the function $F(x)$, a trilinear function is introduced. It is shown in Fig. 7.1a and built in the following way. The middle section of this trilinear curve, i.e. $|x| \leq x_0$ is obtained by taking the first term in the Taylor expansion Eq. (7.4). The other two sections of the trilinear function for $|x| > x_0$ are straight lines which are tangential to the function $F(x)$ at the equilibrium points $(\pm\sqrt{1-\alpha^2}, 0)$. To fit these lines, the derivative of the function $F(x)$

$$F_x(x) = -\left(1 - \frac{1}{\sqrt{x^2+\alpha^2}}\right) - \frac{x^2}{\sqrt{\left(x^2+\alpha^2\right)^3}},$$

is evaluated at $x = \pm\sqrt{1-\alpha^2}$ as

$$F_x(\pm\sqrt{1-\alpha^2}) = -(1-\alpha^2). \tag{7.5}$$

Thus the trilinear approximation of the restoring force can be written as

$$\overline{F}(x) = \begin{cases} \omega_1^2 x, & |x| \leq x_0, \\ -\omega_2^2(x - \text{sign}(x)\omega_2), & |x| > x_0, \end{cases} \tag{7.6}$$

where $\omega_1^2 = (1-\alpha)/\alpha$, $\omega_2^2 = (1-\alpha^2)$ and x_0 is the boundary point for the trilinear function which is obtained by letting

$$(1-\alpha)x_0/\alpha = -(1-\alpha^2)(x_0 - \sqrt{1-\alpha^2}),$$

which gives

$$x_0 = \frac{\alpha\,(1+\alpha)\,\sqrt{1-\alpha^2}}{1+\alpha+\alpha^2}.$$

This piecewise linearized force, $\bar{F}(x)$, is plotted in Fig. 7.1a. The corresponding potentials are shown in Fig. 7.1b: the dashed, dotted and the solid curve are for the nonlinear, Duffing approach and the piecewise linear approach. Figure 7.1a, b, exhibit the kind of approximations and the agreement between the triple linear approach and the original nonlinear force. The cubic approximation of the SD oscillator is not satisfied with the comparison for both the forces and the potentials, even they are equivalent qualitatively for the smooth case, as seen in Fig. 7.1b.

The fact that $x_0 \to 0$ for α tending to zero, indicates that the triple linear approximation is valid for the case $\alpha = 0$, which is of interest in understanding the dynamics. This limit case leads to the discontinuous oscillator. We will use this piecewise linearized approach to get a better understanding of the dynamics. The detailed discussion is in the following sections.

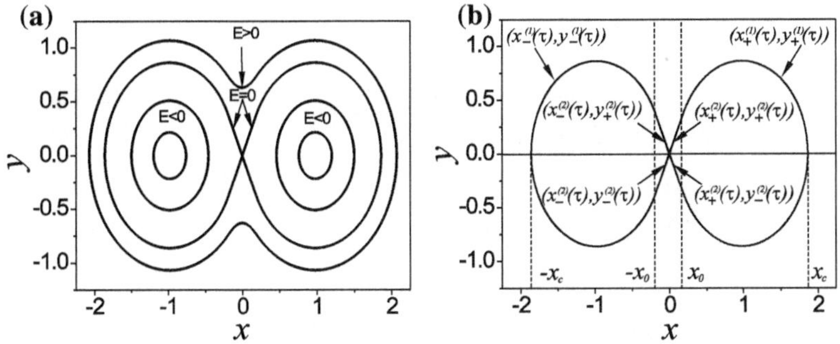

Fig. 7.2 **a** Phase portrait for the Hamiltonian function (7.8) of the piecewise linear system (7.7) for $\alpha = 0.2$, **b** the details of the homoclinic orbit of system (7.7)

7.2.2 Hamiltonian Function

The piecewise linearized system, obtained from system (7.6), is written as

$$\begin{cases} x' = y, \\ y' = \begin{cases} \omega_1^2 x, & |x| \le x_0, \\ -\omega_2^2 \left(x - \operatorname{sgn}(x)\omega_2\right), & |x| > x_0. \end{cases} \end{cases} \tag{7.7}$$

In the following part of this chapter $0 \le \alpha < 1$ is always assumed, as a matter of fact, $\alpha = 0$ implies $x_0 = 0$. So that it is reasonable to write the first part of the second equation of system (7.7) as $y' = 0$ for the case of $\alpha = 0$ when $x = 0$. Without loss of generality, $\alpha > 0$ is kept from now onwards in this chapter when $\alpha = 0$ is not mentioned explicitly.

The Hamilton function of Eq. (7.7) is the following,

$$H(x, y) = \frac{1}{2}y^2 + \begin{cases} -\frac{1}{2}\omega_1^2 x^2, & |x| \le x_0, \\ -\frac{1}{2}\omega_1^2 x_0^2 + \frac{1}{2}\omega_2^2\left[(x - \operatorname{sgn}(x)\omega_2)^2 - (x_0 - \omega_2)^2\right], & |x| > x_0, \end{cases} \tag{7.8}$$

The phase portrait plotted via the Hamiltonian function (7.8) for the piecewise linear system (7.7) is shown in Fig. 7.2a for $\alpha = 0.2$.

7.2.3 Homoclinic Orbit

The homoclinic orbit of system (7.7) obtained by taking $H(x, y) = 0$, intersects the x-axis at x_c, so that $H(x_c, 0) = 0$, i.e.

$$x_c = \omega_2 + \sqrt{\frac{\omega_1^2 x_0^2 + \omega_2^2 (x_0 - \omega_2)^2}{\omega_2^2}} = \sqrt{1-\alpha^2} + \sqrt{\frac{(1-\alpha^2)}{1+\alpha+\alpha^2}}.$$

This homoclinic orbit can be divided into several parts separated by vertical lines $x = \pm x_0$ into two ellipse segments, marked as $\left(x_+^{(1)}(\tau), y_+^{(1)}(\tau)\right)$ and $\left(x_-^{(1)}(\tau), y_-^{(1)}(\tau)\right)$, and four segments, marked as $\left(x_+^{(2)}(\tau), y_+^{(2)}(\tau)\right)$, $\left(x_+^{(2)}(\tau), y_-^{(2)}(\tau)\right)$, $\left(x_-^{(2)}(\tau), y_-^{(2)}(\tau)\right)$ and $\left(x_-^{(2)}(\tau), y_+^{(2)}(\tau)\right)$ meeting at the point $(0, 0)$ as $\tau \to \infty$, as shown in Fig. 7.2b. The ellipse segments of the homoclinic orbit for $|x| > x_0$ have been obtained in analytical form as parametric function of variable τ

$$\left(x_\pm^{(1)}(\tau), y_\pm^{(1)}(\tau)\right) = \left(\pm\sqrt{1-\alpha^2} \pm \frac{R}{\omega_2}\cos\omega_2\tau, \mp R\sin\omega_2\tau\right), \ \tau \in (-\tau_0, \tau_0), \tag{7.9}$$

where

$$R = \sqrt{\omega_1^2 x_0^2 + \omega_2^2 (x_0 - \sqrt{1-\alpha^2})^2},$$

$$\tau_0 = \frac{1}{\omega_2}\arccos\left(\frac{\omega_2\left(x_0 - \sqrt{1-\alpha^2}\right)}{R}\right),$$

and the segments of the homoclinic orbit in the regime $|x| \le x_0$ have been found similarly as

$$\left(x_\pm^{(2)}(\tau), y_\pm^{(2)}(\tau)\right) = \left(\pm x_0 e^{\pm\omega_1(\tau\pm\tau_0)}, \mp\omega_1 x_0 e^{\pm\omega_1(\tau\pm\tau_0)}\right), \ \tau \in (-\infty, -\tau_0] \bigcup [\tau_0, \infty). \tag{7.10}$$

The tangents to these line segments are

$$\lim_{\tau\to\pm\infty} \frac{dy}{dx} = \pm\omega_1 = \pm\sqrt{\frac{1-\alpha}{\alpha}}, \tag{7.11}$$

which tend to infinity as $\alpha \to 0$. In the limit case of $\alpha = 0$, it can be seen that $R = 1$ from (7.6) and $\omega_2 = 1$ from (7.6). Equations (7.9) and (7.10) lead to:

$$\Gamma_0 = \{(x_\pm(\tau), y_\pm(\tau)), \tau \in (-\pi, \pi\} \bigcup \{(0,0)\}, \tag{7.12}$$

where $(x_\pm(\tau), y_\pm(\tau)) = (\pm 1 \pm \cos\tau, \mp \sin\tau)$. This is the solution of the special homoclinic-like orbit of the discontinuous system, which is shown in Fig. 3.2b marked as $E = 0$. This analytical solution can also be obtained by considering both (7.9) and (7.10) for $\alpha \to 0$. This trajectory is then the limit case of the homoclinic orbit of the piecewise linear system (7.8). From this point of view the point (0, 0) is a saddle-like equilibrium and the corresponding trajectory is homoclinic-like orbit.

7.2.4 Periodic Orbits

In this subsection, periodic solutions are presented for the piecewise linear system. At the equilibrium points $\left(\pm\sqrt{1-\alpha^2}, 0\right)$, the Hamiltonian constant $2E$ is equal to $2H(\pm\sqrt{1-\alpha^2}, 0) = -\omega_1^2 x_0^2 - \omega_2^2\left(x_0 - \sqrt{1-\alpha^2}\right)^2$. If $0 > 2E > -\omega_1^2 x_0^2 - \omega_2^2\left(x_0 - \sqrt{1-\alpha^2}\right)^2$, two families of periodic orbits exist inside the homoclinic orbit surrounding the center equilibrium points. The parts of the orbits determined by $|x| > x_0$ can be written as

$$\left(x_\pm^{(E,1)}(\tau), y_\pm^{(E,1)}(\tau)\right) = \left(\pm\sqrt{1-\alpha^2} \pm \frac{R_E}{\omega_2}\cos\omega_2\tau, \mp R_E \sin\omega_2\tau\right), \tau \in [-\tau_E, \tau_E], \tag{7.13}$$

where

$$R_E = \sqrt{2E + \omega_1^2 x_0^2 + \omega_2^2\left(x_0 - \sqrt{1-\alpha^2}\right)^2},$$

$$\tau_E = \begin{cases} \dfrac{\pi}{\omega_2}, & R_E \le \sqrt{1-\alpha^2} - x_0, \\[2ex] \dfrac{1}{\omega_2}\arccos\left(\dfrac{\omega_2\left(x_0 - \sqrt{1-\alpha^2}\right)}{R_E}\right), & R_E > \sqrt{1-\alpha^2} - x_0. \end{cases}$$

In the same way, the parts of the orbits corresponding to $|x| \le x_0$ are:

$$\left(x_\pm^{(E,2)}(\tau), y_\pm^{(E,2)}(\tau)\right) = \left(\pm\frac{x_0\cosh\omega_1(\tau \pm \frac{\pi}{\omega_2})}{\cosh\omega_1(\tau_E \pm \frac{\pi}{\omega_2})}, \pm\frac{x_0\omega_1\sinh\omega_1(\tau \pm \frac{\pi}{\omega_2})}{\cosh\omega_1(\tau_E \pm \frac{\pi}{\omega_2})}\right),$$
$$\tau \in \left[-\frac{\pi}{\omega_1}, -\tau_E\right] \bigcup \left[\tau_E, \frac{\pi}{\omega_1}\right]. \tag{7.14}$$

For $E > 0$, a family of periodic orbits exists outside the homoclinic orbit surrounding all the singularities. For $|x| > x_0$, these parts of the orbits have been obtained as

$$\begin{gathered}\left(x_{\pm}^{(E,1)}(\tau), y_{\pm}^{(E,1)}(\tau)\right) = \left(\pm\sqrt{1-\alpha^2} \pm \frac{R_E}{\omega_2}\cos\omega_2\tau, \mp R_E \sin\omega_2\tau\right), \\ \tau \in [0, \tau_E] \bigcup \left[\tfrac{\pi}{\omega_1} + \tau_E, \tfrac{2\pi}{\omega_1}\right],\end{gathered} \tag{7.15}$$

where

$$R_E = \sqrt{2E + \omega_1^2 x_0^2 + \omega_2^2\left(x_0 - \sqrt{1-\alpha^2}\right)^2},$$

$$\tau_E = \frac{1}{\omega_2}\arccos\left(\frac{x_0 - \sqrt{1-\alpha^2}}{R_E}\right).$$

Furthermore, for $|x| \le x_0$, these orbits can be obtained and written as

$$\begin{gathered}\left(x_{\pm}^{(E,2)}(\tau), y_{\pm}^{(E,2)}(\tau)\right) = \left(\pm\frac{x_0 \sinh\omega_1(\tau \pm \frac{\pi}{\omega_2})}{\sinh\omega_1(\tau_E \pm \frac{\pi}{\omega_2})}, \pm\frac{x_0\omega_1 \cosh\omega_1(\tau \pm \frac{\pi}{\omega_2})}{\sinh\omega_1(\tau_E \pm \frac{\pi}{\omega_2})}\right), \\ \tau \in \left[\tau_E, \tau_E + \tfrac{\pi}{\omega_1}\right].\end{gathered} \tag{7.16}$$

7.3 Perturbed System

The attention is now drawn to the perturbed triple linear system obtained above. A chaotic behaviour may occur and the Melnikov method will be used to detect its boundary when the parameter varies. Semi-analytical method is used to investigate the nonlinear dynamics of the perturbed system.

7.3.1 Melnikov Method

The following perturbed SD oscillator (7.1) with linear viscous damping and external forcing excitation

$$x'' + 2\xi x' + x\left(1 - \frac{1}{\sqrt{x^2+\alpha^2}}\right) = f_0\cos\omega\tau \tag{7.17}$$

can be written in the following form by using the triple linearized system (7.7).

$$\begin{cases} x' = y, \\ y' = -2\xi\dot{x} + \overline{F}(x) + f_0\cos\omega\tau, \end{cases} \tag{7.18}$$

where $\overline{F}(x)$ is precisely the triple linearized force given in (7.6). Next we introduce the following notations in system (7.18)

$$\overline{f}\left(\overline{X}\right) = \begin{pmatrix} y \\ \overline{F}(x) \end{pmatrix}, \quad \overline{g}\left(\overline{X}, \tau\right) = \begin{pmatrix} 0 \\ -2\xi y + f_0\cos\omega\tau \end{pmatrix},$$

and for the homoclinic orbit, we have

$$\text{Trace}\left(D\overline{f}\right) = \text{Trace}\begin{pmatrix} 0 & 1 \\ \overline{\omega}^2 & 0 \end{pmatrix} = 0,$$

with

$$\overline{X} = \begin{pmatrix} x \\ y \end{pmatrix}, \quad \overline{\omega}^2 = \begin{cases} \omega_1^2, & |x| \le x_0, \\ \omega_2^2, & |x| > x_0, \end{cases}$$

and

$$\overline{f}\left(\overline{x}_\pm(\tau)\right) \wedge \overline{g}\left(\overline{x}_\pm(\tau), \tau + \tau_1\right) = \left[-2\xi\left(y_\pm(\tau)\right)^2 + y_\pm(\tau) f_0\cos\omega(\tau + \tau_1)\right], \tag{7.19}$$

where the operator $\wedge$ is defined as

$$\mathbf{a} \wedge \mathbf{b} = a_1 b_2, -a_2 b_1 \tag{7.20}$$

for any $\mathbf{a} = (a_1, a_2)^T$ and $\mathbf{b} = (b_1, b_2)^T$. The corresponding Melnikov function of Eq. (7.18), (see e.g. [4–6]), is then given by

$$\begin{aligned} M_\pm(\tau_1) &= \int_{-\infty}^{\infty}\left[\overline{f}\left(\overline{x}_\pm(\tau)\right) \wedge \overline{g}\left(\overline{x}_\pm(\tau), \tau + \tau_1\right)\right] d\tau \\ &= \int_{-\infty}^{\infty}\left[-2\xi\left(y_\pm(\tau)\right)^2 + y_\pm(\tau) f_0\cos\omega(\tau + \tau_1)\right] d\tau. \end{aligned} \tag{7.21}$$

Omitting the tedious computations, Melnikov function is obtained as the following,

$$M_\pm(\tau_1) = 2\xi\Omega(\omega_1, \omega_2, x_0, R) \pm f_0\Psi(\omega, \omega_1, \omega_2, \tau_0, R)\sin\omega\tau_1, \tag{7.22}$$

It can be seen that the equation

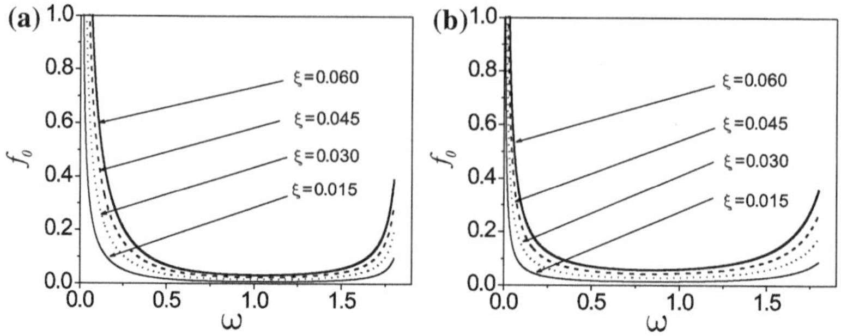

Fig. 7.3 Melnikov-detected chaotic boundary for system (7.18). The *thin solid*, *dotted*, *dashed* and the *thick solid curves* mark the Melnikov-detected chaotic boundaries for $\xi = 0.015, 0.030, 0.045$ and 0.060, **a** for $\alpha = 0.2$ and **b** for $\alpha = 0$, respectively

$$M_{\pm}(\tau_1) = 0 \tag{7.23}$$

has simple zero for τ_1 if and only if the following inequality holds,

$$f_0 > \frac{\Omega(\omega_1, \omega_2, x_0, R)}{\Psi(\omega, \omega_1, x_0, \tau_0, R)} 2\xi = \xi \overline{R}(\omega, \omega_1, \omega_2, \tau_0, x_0, R). \tag{7.24}$$

Melnikov-detected chaotic boundary for system (7.17) are obtained for $\alpha = 0.2$, shown in Fig. 7.3a, and $\alpha = 0$, shown in Fig. 7.3b, for different choices of parameter ξ for $\xi = 0.015, 0.030, 0.045$ and 0.060, (a) for $\alpha = 0.2$ and (b) for $\alpha = 0$,. In the area above the detected boundaries the chaotic motions might be detected, i.e., in the area of $|f_0|/|\xi| > |\overline{R}(\omega, \omega_1, \omega_2, \tau_0, x_0, R)|$.

7.3.2 Semi-analytical Analysis

A semi-analytical analysis, see [7], can now be given for the calculation of the solutions of Eq. (7.18) of the triple linearized system in the Melnikov-detected chaotic region. The Lyapunov exponents and the corresponding dimension of the attractors are now displayed. The analytical solution for the perturbed system (7.18) is expressed as follows.

If $|x| \le x_0$, the solution is

$$\begin{cases} x(\tau) = C_1 e^{\alpha_1(\tau-\tau_0)} + C_2 e^{\alpha_2(\tau-\tau_0)} + x_m(\tau), \\ y(\tau) = \alpha_1 C_1 e^{\alpha_1(\tau-\tau_0)} + \alpha_2 C_2 e^{\alpha_2(\tau-\tau_0)} + y_m(\tau), \end{cases} \tag{7.25}$$

where $\alpha_{1,2} = -\xi \pm \sqrt{\omega^2 + \xi^2}$, and

$$x_m(\tau) = \frac{-(\omega_1^2 + \omega^2) f_0 \cos(\omega\tau) + 2 f_0 \xi \omega \sin(\omega\tau)}{4\xi^2\omega^2 + (\omega_1^2 + \omega^2)^2},$$
$$y_m(\tau) = \frac{\omega(\omega_1^2 + \omega^2) f_0 \sin(\omega\tau) + 2 f_0 \xi \omega^2 \cos(\omega\tau)}{4\xi^2\omega^2 + (\omega_1^2 + \omega^2)^2},$$

and with

$$C_1 = \alpha_2 \left(x(\tau_0) - x_m(\tau_0)\right) - \left(y(\tau_0) - y_m(\tau_0)\right),$$
$$C_2 = -\alpha_1 \left(x(\tau_0) - x_m(\tau_0)\right) + \left(y(\tau_0) - y_m(\tau_0)\right).$$

When $|x| > x_0$, the solution is

$$\begin{cases} x(\tau) = x_0(\tau) + x_g(\tau), \\ y(\tau) = y_0(\tau) + y_g(\tau), \end{cases} \tag{7.26}$$

where

$$x_0(\tau) = \frac{(\omega_2^2 - \omega^2) f_0 \cos(\omega\tau) + 2 f_0 \xi \omega \sin(\omega\tau)}{4\xi^2\omega^2 + (\omega_2^2 - \omega^2)^2}$$
$$y_0(\tau) = \frac{-\omega(\omega_2^2 - \omega^2) f_0 \sin(\omega\tau) + 2 f_0 \xi \omega^2 \cos(\omega\tau)}{4\xi^2\omega^2 + (\omega_2^2 - \omega^2)^2}$$

with

$$x_g(\tau) = e^{-\xi(\tau-\tau_0)} \left(C_1 \cos\omega_d(\tau - \tau_0) + C_2 \sin\omega_d(\tau - \tau_0)\right) + \text{sign}(x(\tau_0))\omega_2,$$
$$y_g(\tau) = e^{-\xi(\tau-\tau_0))} \left((\omega_d C_2 - \xi C_1)\cos\omega_d(\tau - \tau_0) - (\omega_d C_1 + \xi C_2)\sin\omega_d(\tau - \tau_0)\right),$$

The constants C_1 and C_2 are obtained due to initial conditions.

$$C_1 = x(\tau_0) - x_0(\tau_0) - \text{sign}(x(\tau_0))\omega_2,$$
$$C_2 = \frac{1}{\omega_d}\left[y(\tau_0) - y_0(\tau_0)\right] + \frac{\xi}{\omega_d}\left[x(\tau_0) - x_0(\tau_0) - \text{sign}(x(\tau_0))\omega_2\right].$$

so that the solution finally reads:

$$\begin{aligned} x(\tau) = x_0(\tau) + \frac{e^{-\xi(\tau-\tau_0)}}{\omega_d} &\left[(y(\tau_0) - y_0(\tau_0) + \xi\,(x(\tau_0) - x_0(\tau_0)\right. \\ &-\text{sign}(x(\tau_0))\omega_2))\sin\omega_d(\tau - \tau_0) + \omega_d\,(x(\tau_0) - x_0(\tau_0) \\ &\left.-\text{sign}(x(\tau_0))\omega_2)\cos\omega_d(\tau - \tau_0)\right] + \text{sign}(x(\tau_0))\omega_2, \end{aligned}$$

$$\begin{aligned} y(\tau) = y_0(\tau) + \frac{e^{-\xi(\tau-\tau_0)}}{\omega_d} &\{\omega_d\,(y(t_0) - y_0(t_0))\cos\omega_d(\tau - \tau_0) \\ &-\left[\left(\xi^2 + \omega_d^2\right)(x(\tau_0) - x_0(\tau_0) - \text{sign}(x(t_0))\omega_2)\right. \\ &+\xi\,(y(\tau_0) - y_0(\tau_0))]\sin\omega_d(\tau - \tau_0)\}. \end{aligned}$$

The phase space is divided into three regimes $x < -x_0$, $-x_0 \le x \le x_0$, and $x > x_0$. The constants above depend on the initial conditions and also on the transient points when the trajectory crosses the boundaries $x = \pm x_0$.

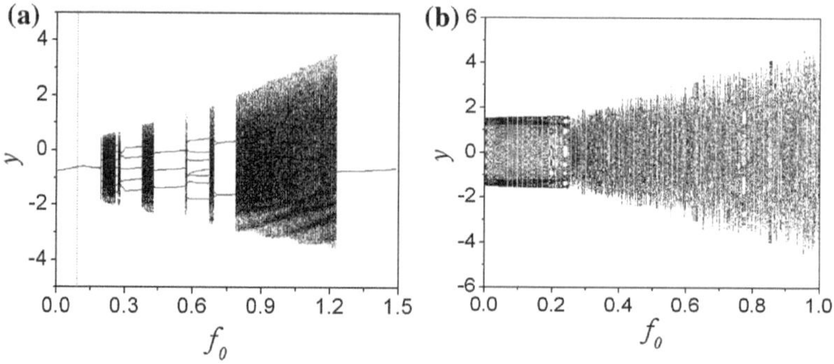

Fig. 7.4 Bifurcation diagrams for y versus the external forcing amplitude f_0 for $\alpha = 0.2, \omega = 1.05$ for system (7.18); **a** for $\xi = 0.015$, **b** for $\xi = 0$

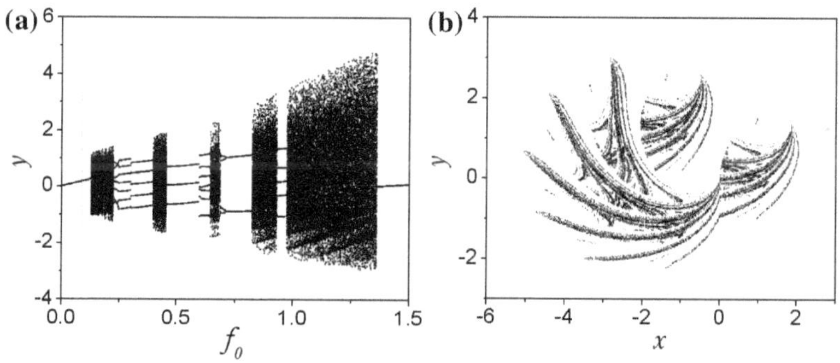

Fig. 7.5 **a** Bifurcation diagrams for f_0 versus y for $\omega = 1.05, \xi = 0.015$ and $\alpha = 0$ for system (7.18), **b** the chaotic attractor for $f_0 = 0.83, \omega = 1.05, \xi = 0.015$ and $\alpha = 0$ with the Lyapunov exponents $(\lambda_1, \lambda_2) = (-0.07851, 0.04735)$ and Lyapunov dimension $D_L = 1.60321$

7.3.3 Bifurcation Diagrams and Dynamical Analysis

Now numerical results are displayed to show the dynamical behaviour of the triple linearized system. For both plots, the same initial conditions $(x_0, y_0) = (1, 0)$ are taken and 5000 cycles are taken as the pre-iterates to eliminate the transients. For the Poincaré section, 200 cycles are recorded for each parameter value.

The bifurcation diagrams are calculated for $\alpha = 0.2$, $\omega = 1.05$ and $\xi = 0.015$ in Fig. 7.4a, $\alpha = 0.2$, $\omega = 1.05$ and $\xi = 0$ Fig. 7.4b, and $\alpha = 0.0$, $\omega = 1.05$ and $\xi = 0.015$ Fig. 7.5a. All these bifurcation diagrams show the global complexity for f_0 changing from zero, crossing the Melnikov boundary, the dashed vertical lines in Figs. 7.4a and 7.5a, and entering the Melnikov detected region, while Fig. 7.5a shows that the Melnikov chaotic boundary is $f_0 = 0$ as shown in inequality (7.24) for $\xi = 0$. This also means that the driven system without damping behaves more complicated and may exhibit chaos for any small perturbation f_0 positive.

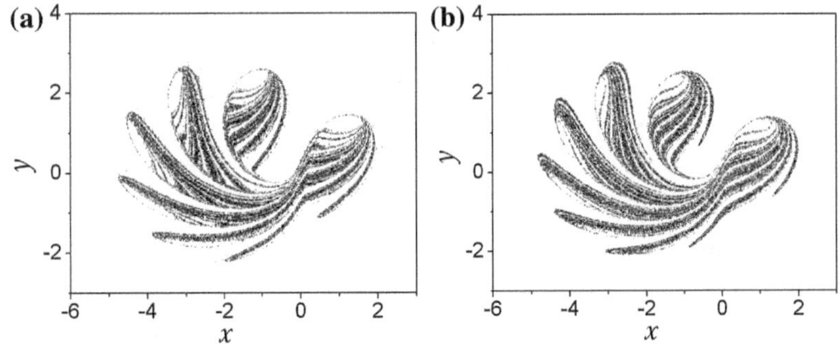

Fig. 7.6 Chaotic attractors for triple linear system at $\alpha = 0.2$, $\omega = 1.05$, $\xi = 0.015$ and $f_0 = 0.83$; **a** For triple linear system (7.18) with the Lyapunov exponents $(\lambda_1, \lambda_2) = (-0.11382, 0.08578)$ and Lyapunov dimension $D_L = 1.75363$, **b** for system (7.17) with the Lyapunov exponents $(\lambda_1, \lambda_2) = (-0.11716, 0.08887)$ and Lyapunov dimension $D_L = 1.77585$

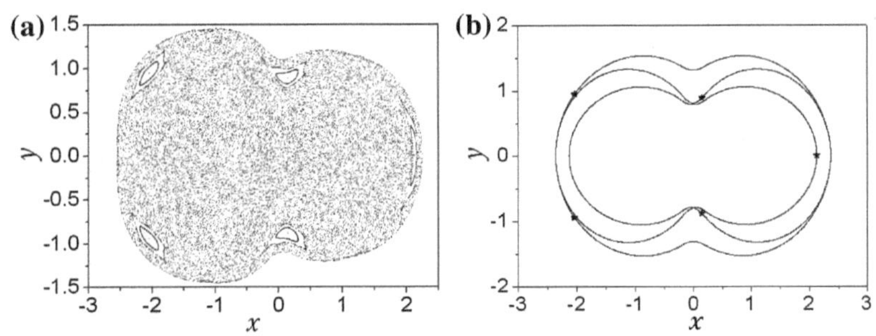

Fig. 7.7 Chaotic sea with islands and periodic solution for a small perturbation in system (7.18) for $f_0 = 0.2$, $\omega = 1.05$, $\alpha = 0.2$ and $\xi = 0.0$; **a** the chaotic sea with quasi-periodic islands and periodic solution of period 5, **b** the corresponding trajectory for the resonance solution, the stars mark the Poincaé section

Chaotic attractors are given in Fig. 7.5b for $f_0 = 0.83$, $\alpha = 0$, $\omega = 1.05$ and $\xi = 0.015$. The Lyapunov exponents and the fractal dimension are calculated as shown in the caption.

The chaotic attractors for both the smooth and the triple linearized systems for the same parameters $f_0 = 0.83$, $\alpha = 0.2$, $\omega = 1.05$ and $\xi = 0.015$ are shown in Fig. 7.6a for system (7.17) and Fig. 7.6b system (7.18) respectively. The comparison can be made to show the adequacy of the triple linear approximation to the original nonlinear system not only in the attractor topological structure, [8, 9], but also the digital characteristics for Lyapunov exponents and the fractal dimension as shown in the corresponding captions.

Chaotic sea with islands and periodic solution for a small perturbation in system (7.18) for $f_0 = 0.2$, $\omega = 1.05$, $\alpha = 0.2$ and $\xi = 0.0$ is presented in Fig. 7.7a and the corresponding trajectory with Poincaré section for the resonance solution is also shown in Fig. 7.7b.

7.4 Summary

This chapter introduced a triple linear approach for the SD oscillator with the nonlinearity of irrational type, which is valid for both smooth and discontinuous stages. The triple linearization process is proposed for $0 \leq \alpha < 1$ for the system with the double well behaviour based on the original oscillator with the same equilibria and the same derivatives. This approximation is valid for the limit case of $\alpha = 0$, at which the triple linear system becomes the original discontinuous case, that is the boundary value $x_0 \to 0$, as $\alpha \to 0$. Chaotic attractors are presented for both the original oscillator and for the triple linear approximation with a excellent agreement for the topological structures and Lyapunov exponents.

References

1. Lai, S. K., & Xiang, Y. (2010). Application of a generalized senator-bapat perturbation technique to nonlinear dynamical systems with an irrational restoring force. *Computers and Mathematics with Applications*, *60*(7), 2078–2086.
2. Santhosh, B., Padmanabhan, C., & Narayanan, S. (2014). Numeric-analytic solutions of the smooth and discontinuous oscillator. *International Journal of Mechanical Sciences*, *84*, 102–119.
3. Chen, H. B., & Xie, J. H. (2016). Harmonic and subharmonic solutions of the sd oscillator. *Nonlinear Dynamics*, 1–10.
4. Guckenheimer, J., & Holmes, P. (1983). *Nonlinear oscillation, dynamical system and bifurcation of vector fields*. New York: Springer.
5. Melnikov, V. K. (1963). On the stability of the center for time period perturbations. *Transactions of the Moscow Mathematical Society*, *12*, 3–52.
6. Chen, Y. S., & Leung, A. Y. T. (1998). *Bifurcation and chaos in engineering*. London: springer.
7. Pavlovska, E., & Wiercigroch, M. (2004). Analytical drift reconstruction for visco-elastic impact oscillators operating in periodic and chaotic regimes. *Chaos, Solitons and Fractals*, *19*(1), 151–161.
8. Thompson, J. M. T., & Ghaffari, R. (1983). Chaotic dynamics of an impact oscillator. *Physical Review A*, *27*(3), 1741–1743.
9. Thompson, J. M. T., & Stewart, H. B. (2002). *Nonlinear dynamics and chaos* (2nd ed.). Chichester: Wiley.

Chapter 8
Extended Averaging Method

This chapter aims at studying the periodic solutions and the Hopf bifurcations of the SD oscillator using the so-called averaging method. This will be done in the case where the system has a viscous damping and an external harmonic excitation. A four dimensional averaging method is introduced by using the complete Jacobian elliptic integrals, directly to obtain the perturbed primary responses which bifurcate from both the hyperbolic saddle point and the non-hyperbolic centers of the unperturbed system. The stability of these periodic solutions is obtained by examining the four dimensional averaged equation using Lyapunov method. Essentially the results presented here are valid either for the smooth or for the discontinuous cases.[1]

8.1 Introduction

Much attention has been paid to the periodic solutions of nonlinear dynamical systems in the past decades [1–3]. Simultaneously, methodologies have been developed to investigate the periodic solutions of nonlinear systems, such as averaging method [4], matrix measure and fixed point theorem [5, 6], as well. As far as we know, these conventional methods are always limited by a small parameter, ε, which means that they are applicable to weakly nonlinear systems, while strongly nonlinear systems [7–10] are often concerned by engineering applications. Usually, the nonlinearities which occur in practice are studied after a truncation of a Taylor's expansion which change the pattern into a polynomial dynamical system. For example, the simple pen-

[1] This chapter relies on a common work with Dr. Zhixin Li, Center for Nonlinear Dynamics Research, School of Astronautics, Harbin Institute of Technology, Harbin 150001 Chian, and Professor Ruilan Tian, Center for Nonlinear Dynamics Research, Shijiazhang Tiedao University, Shijiazhuang 050043 China.

Q. Cao and A. Léger, *A Smooth and Discontinuous Oscillator*, Springer Tracts in Mechanical Engineering, DOI 10.1007/978-3-662-53094-8_8

dulum [11, 12] is frequently studied using Taylor expansion method, which requires the assumption that the angle θ is small.

An extended averaging method is developed to the SD oscillator which is applicable for both smooth and discontinuous dynamics to carry out the analytical analysis for the perturbed periodic behaviours. The results presented here show that the centres may shift around under the external excitations for the case $0 \leq \alpha < 0$. It is worth noticing that the results presented here in this chapter using the extended four dimensional averaging method answer to why the averaging theorem spectacularly fails to describe the stability of Duffing system expressed in [4] near the saddle point $(0, 0)$.

8.2 The Averaging Procedure

It is assumed that the oscillator is subjected to a viscous damping and an external harmonic excitation for both smooth and discontinuous cases in the following sections. We aim at investigating the periodic responses of the oscillator. The dimensionless equation of the perturbed oscillator, reads

$$\ddot{x} + \delta\dot{x} + \omega_0^2 x\left(1 - \frac{1}{\sqrt{x^2 + \alpha^2}}\right) = f\cos\omega t. \tag{8.1}$$

Without loss of generality, let us set $\tau = \omega t$, then system (8.1) reads

$$\omega^2\ddot{x} + \omega\delta\dot{x} + \omega_0^2 x\left(1 - \frac{1}{\sqrt{x^2 + \alpha^2}}\right) = f\cos\tau. \tag{8.2}$$

where the variables are defined as mentioned in the previous chapters. The periodic solutions of the system will now be examined in a neighborhood of each of the equilibrium states.

8.2.1 *Periodic Solutions Near* $(0, 0)$

Let us assume that the periodic solutions with slowly varying amplitude and phase are of the form

$$\begin{cases} x = a\cos(\tau + \theta), \\ \dot{x} = -a\sin(\tau + \theta), \end{cases} \tag{8.3}$$

which is substituted into Eq. (8.2). This substitution yields

$$\begin{cases} \dot{a}\cos(\tau + \theta) = a\dot{\theta}\sin(\tau + \theta), \\ -\omega^2[\dot{a}\sin(\tau + \theta) + a\dot{\theta}\cos(\tau + \theta)] = H(a, \theta, \tau), \end{cases} \tag{8.4}$$

where $H(a, \theta, \tau) = \delta\omega a \sin(\tau + \theta) + (\omega^2 - \omega_0^2)a \cos(\tau + \theta) + \dfrac{\omega_0^2 a \cos(\tau + \theta)}{\sqrt{a^2 \cos^2(\tau + \theta) + \alpha^2}} + f \cos \tau$. The above derivation gives the following system for the slowly varying amplitude and phase equations:

$$\begin{cases} \dot{a} = -\dfrac{1}{\omega^2} H(a, \theta, \tau) \sin(\tau + \theta), \\ \dot{\theta} = -\dfrac{1}{a\omega^2} H(a, \theta, \tau) \cos(\tau + \theta). \end{cases} \tag{8.5}$$

By averaging Eq. (8.5) over a period $[0, 2\pi]$, the averaged equation can be written as

$$\begin{cases} \dot{a} = -\dfrac{1}{2\omega^2} (\delta\omega a + f \sin \theta), \\ \dot{\theta} = -\dfrac{1}{2a\omega^2} \left(aG(a, \alpha) + a(\omega^2 - \omega_0^2) + f \cos \theta\right), \end{cases} \tag{8.6}$$

which, neglecting the phase variable, leads to the amplitude equation

$$\left[G(a, \alpha) + (\omega^2 - \omega_0^2)\right]^2 + \delta^2\omega^2 = \frac{f^2}{a^2}, \tag{8.7}$$

where

$$G(a, \alpha) = \frac{4\omega_0^2}{\pi a^2} \sqrt{a^2 + \alpha^2} E\left(\sqrt{\frac{a^2}{a^2 + \alpha^2}}\right) - \frac{4\omega_0^2 \alpha^2}{\pi a^2 \sqrt{a^2 + \alpha^2}} K\left(\sqrt{\frac{a^2}{a^2 + \alpha^2}}\right),$$

$K(k)$ and $E(k)$ [13, 14] are the complete Jacobian elliptic integrals of the first and second kind, with $0 < k < 1$, which are defined as

$$K(k) = \int_0^{\pi/2} \frac{1}{\sqrt{1 - k^2 \sin^2 \theta}} d\theta, \; E(k) = \int_0^{\pi/2} \sqrt{1 - k^2 \sin^2 \theta} d\theta.$$

It has been proved that $\lim_{\alpha \to 0} G(a, \alpha) = \dfrac{4\omega^2}{\pi a}$, that is the limit of Eq. (8.7) for $\alpha \to 0$, can be obtained as follows

$$\left[\frac{4\omega_0^2}{\pi a} + (\omega^2 - \omega_0^2)\right]^2 + \delta^2\omega^2 = \frac{f^2}{a^2}. \tag{8.8}$$

This could have been also obtained directly from the system (8.2) for the case $\alpha = 0$. From the above analysis, it is clear that the Eq. (8.7) is valid for both cases, smooth ($\alpha > 0$) and discontinuous ($\alpha = 0$).

8.2.2 Periodic Solution Near $(\sqrt{1-\alpha^2}, 0)$

Let us now introduce the existence of periodic solutions around the other equilibrium states. An extension of the conventional averaging method to a generalized four dimensional is needed, which will be applicable to the non-hyperbolic centers. As in the case of the equilibrium point (0, 0), slowly varying amplitude and phase are introduced, but now with a center shift c [15]:

$$\begin{cases} x = a\cos(\tau+\theta) + c, \\ \dot{x} = -a\sin(\tau+\theta) + \dot{c}. \end{cases} \tag{8.9}$$

Then the averaged equations for the equilibrium $(\sqrt{1-\alpha^2}, 0)$ take the form:

$$\begin{cases} \omega^2\ddot{c} + \delta\omega\dot{c} = -\omega_0^2\left(c - Q_1(a,c)\right), \\ \dot{a} = -\dfrac{1}{2\omega^2}(\delta\omega a + f\sin\theta), \\ \dot{\theta} = -\dfrac{1}{2a\omega^2}\left[a(\omega^2-\omega_0^2) + \omega_0^2 Q_2(a,c) + f\cos\theta\right], \end{cases} \tag{8.10}$$

where

$$Q_1(a,c) = \frac{1}{2\pi}\int_0^{2\pi} \frac{a\cos(\tau+\theta)+c}{\sqrt{(a\cos(\tau+\theta)+c)^2+\alpha^2}}\,d(\tau+\theta),$$

$$Q_2(a,c) = \frac{1}{\pi}\int_0^{2\pi} \frac{[a\cos(\tau+\theta)+c]\cos(\tau+\theta)}{\sqrt{(a\cos(\tau+\theta)+c)^2+\alpha^2}}\,d(\tau+\theta).$$

By letting $\ddot{c}, \dot{c} = 0$, $\dot{a} = 0$, $\dot{\theta} = 0$, the generalized averaging follows

$$\begin{cases} c = \dfrac{1}{2\pi}\displaystyle\int_0^{2\pi} \frac{a\cos(\tau+\theta)+c}{\sqrt{(a\cos(\tau+\theta)+c)^2+\alpha^2}}\,d(\tau+\theta), \\ \left[\omega^2 - \omega_0^2 + \dfrac{\omega_0^2}{a}Q_2(a,c)\right]^2 + \delta^2\omega^2 = \dfrac{f^2}{a^2}, \end{cases} \tag{8.11}$$

in which the first line is an implicit function of c, and the second is the response equation.

In the same way as for the equilibrium (0, 0), the limitation of Eq. (8.11) when $\alpha \to 0$ is

$$(\omega^2 - \omega_0^2)^2 + \delta^2\omega^2 = \frac{f^2}{a^2}, \tag{8.12}$$

which can be as previously obtained directly from system (8.2) when $\alpha = 0$; the response of equation (8.11) is valid for both the smooth and discontinuous cases.

8.3 Stability Analysis

Let's consider the stability of the periodic solutions using the eigenvalues of the linearized four dimensional averaged equation. Numerical calculations are carried out to show the efficiency of the theoretical analysis, which is undertaken for both the cases near the saddle (0, 0) and near the pair of centers $(\sqrt{1-\alpha^2}, 0)$.

Case I: The stability near (0, 0)

Letting $c = 0$, $t = \tau/\omega$ and $\dot{c} = d$, then it follows from Eq. (8.10):

$$\begin{cases} \dot{a} = -\dfrac{1}{2\omega}(\delta\omega a + f\sin\theta), \\ \dot{\theta} = -\dfrac{1}{2a\omega}\left[a(\omega^2 - \omega_0^2) + aG(a, \alpha) + f\cos\theta\right], \\ \dot{c} = d, \\ \dot{d} = \omega_0^2(\tilde{Q}_1 - 1)c - \delta d, \end{cases} \tag{8.13}$$

where

$$\tilde{Q}_1 = \frac{1}{\alpha}\sum_{n=0}^{+\infty}(-1)^n\left(\frac{(2n-1)!!}{(2n)!!}\right)^2(2n+1)\left(\frac{a}{\alpha}\right)^{2n}.$$

Let $(a_s, \theta_s, 0, 0)$ be an equilibrium of Eq. (8.13). The variational equation is obtained by choosing $a = a_s + \xi$, $\theta = \theta_s + \eta$, $c = c'$, and $d = d'$ and is written as

$$\dot{\mathbf{X}} = \mathbf{A}\mathbf{X}, \tag{8.14}$$

where

$$\mathbf{X} = (\xi, \eta, c', d')', \mathbf{A} = \begin{pmatrix} \mathbf{A}_1 & 0 \\ 0 & \mathbf{A}_2 \end{pmatrix}, \mathbf{A}_1 = \begin{pmatrix} -\dfrac{1}{2\omega}\left(\dfrac{\partial\Phi}{\partial a}\right)_s & \dfrac{1}{2\omega}\Psi_s \\ -\dfrac{1}{2\omega a_s}\left(\dfrac{\partial\Psi}{\partial a}\right)_s & -\dfrac{1}{2\omega a_s}\Phi_s \end{pmatrix},$$

$$\mathbf{A}_2 = \begin{pmatrix} 0 & 1 \\ \omega_0^2(\tilde{Q}_1 - 1) & -\delta \end{pmatrix}, \qquad \Phi = \delta\omega a, \Psi = aG(a, \alpha) + a(\omega^2 - \omega_0^2).$$

and the symbol $()_s$ denotes the value of the function at equilibrium point. $\mathbf{A}$ is a block diagonal matrix, which enables to undertake the stability analysis from matrices $\mathbf{A}_1$ and $\mathbf{A}_2$.

The characteristic equation of $\mathbf{A}_1$ follows from

$$\begin{vmatrix} \lambda + \frac{1}{2\omega}\left(\frac{\partial \Phi}{\partial a}\right)_s & -\frac{1}{2\omega}\Psi_s \\ \frac{1}{2\omega a_s}\left(\frac{\partial \Psi}{\partial a}\right)_s & \lambda + \frac{1}{2\omega a_s}\Phi_s \end{vmatrix} = 0, \tag{8.15}$$

which leads to

$$\lambda^2 + \delta\lambda + a_5 = 0. \tag{8.16}$$

Equation (8.16) implies that (a_s, θ_s) is asymptotically stable for $a_5 > 0$, where

$$a_5 = \frac{1}{4\omega^2 a_s}[\Phi\left(\frac{\partial \Phi}{\partial a}\right)_s + \Psi\left(\frac{\partial \Psi}{\partial a}\right)_s] = \frac{1}{8\omega^2 a_s}\left(\frac{\partial W}{\partial a}\right)_s,$$

$$W(a) = \Phi(a)^2 + \Psi(a)^2 - f^2,$$

which implies

$$\left(\frac{\partial W}{\partial a}\right)_s > 0. \tag{8.17}$$

Now the characteristic equation of matrix $\mathbf{A}_2$ reads

$$\begin{vmatrix} \lambda & -1 \\ \omega_0(1 - \tilde{Q}_1) & \lambda + \delta \end{vmatrix} = 0, \tag{8.18}$$

leading to

$$\lambda^2 + \delta\lambda + \omega_0(1 - \tilde{Q}_1) = 0, \tag{8.19}$$

which implies that $(0, 0)$ is asymptotically stable for $1 - \tilde{Q}_1 > 0$. From the above analysis, we conclude that the equilibrium $(\alpha_s, \theta_s, 0, 0)$ is asymptotically stable, if and only if the two following conditions are satisfied

$$\begin{cases} \left(\frac{\partial W}{\partial a}\right)_s > 0, \\ 1 - \tilde{Q}_1 > 0. \end{cases} \tag{8.20}$$

It is clear that condition (8.20) answers the question why the averaging theorem fails for the medium strength of external forcing for the Duffing system, see details in [4, 16, 17]. This question has been proposed over thirty years, the reason is that the stability analysis was done for (a_s, θ_s) without considering the stability of (c, d) at $(0, 0)$.

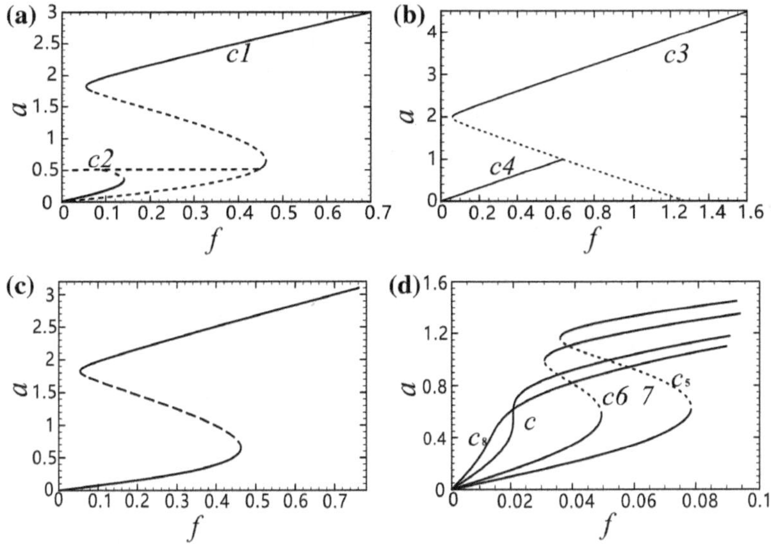

Fig. 8.1 Response curves of the system (8.2), $\delta = 0.05$, $\omega = 0.6$, $\omega_0 = 1$: **a** $\alpha = 0.5$ for both Eqs. (8.7)(c_1) and (8.11)(c_2); **b** $\alpha = 0$ for both Eqs. (8.8)(c_3) and (8.12)(c_4); **c** $\alpha = 1$ for Eq. (8.7) and **d** for (c_5)–(c_8) for $\alpha = 1.2$, 1.3, 1.438 and 1.5, respectively

The response curve of Eq. (8.7) near (0, 0) for $\alpha = 0.5$ is plotted in Fig. 8.1a, marked as c_1. The stable and unstable branches are drawn with solid and dashed curves respectively.

Case II: The stability near $(\pm\sqrt{1-\alpha^2}, 0)$.

Setting $t = \tau/\omega$ and $\dot{c} = d$, Eq. (8.10) becomes

$$\begin{cases} \dot{a} = -\dfrac{1}{2\omega}(\delta\omega a + f\sin\theta), \\ \dot{\theta} = -\dfrac{1}{2a\omega}\left[a(\omega^2-\omega_0^2) + \omega_0^2 Q_2 + f\cos\theta\right], \\ \dot{c} = d, \\ \dot{d} = -\omega_0^2(c - Q_1) - \delta d. \end{cases} \tag{8.21}$$

Let $(a_s, \theta_s, c_s, 0)$ be an equilibrium of Eq. (8.21), where

$$c_s = \frac{1}{2\pi}\int_0^{2\pi} \frac{a_s\cos(\tau+\theta) + c_s}{\sqrt{(a_s\cos(\tau+\theta)+c_s)^2+\alpha^2}} d(\tau+\theta).$$

The linearization of Eq. (8.21) is obtained with $a = a_s + \xi$, $\theta = \theta_s + \eta$, $c = c_s + \gamma$, and $d = \beta$, which leads to

$$\dot{\mathbf{Y}} = \mathbf{B}\mathbf{Y}, \tag{8.22}$$

where $\mathbf{Y} = (\xi, \eta, \gamma, \beta)'$, $\mathbf{B} = \begin{pmatrix} b_{11} & b_{12} & 0 & 0 \\ b_{21} & b_{22} & b_{23} & 0 \\ 0 & 0 & 0 & 1 \\ b_{41} & 0 & b_{43} & -\delta \end{pmatrix}$, and

$$b_{11} = b_{22} = -\frac{\delta}{2}, \qquad b_{12} = \frac{1}{2\omega}\left[a_s(\omega^2 - \omega_0^2) + \omega_0^2 Q_2(a_s, c_s)\right],$$
$$b_{21} = -\frac{1}{2\omega a_s}\left[\omega^2 - \omega_0^2 + \omega_0^2\left(\frac{\partial Q_2}{\partial a}\right)_s\right], \quad b_{23} = -\frac{\omega_0^2}{2\omega a_s}\left(\frac{\partial Q_2}{\partial c}\right)_s,$$
$$b_{41} = \omega_0^2\left(\frac{\partial Q_1}{\partial a}\right)_s, \qquad b_{43} = -\omega_0^2 + \omega_0^2\left(\frac{\partial Q_1}{\partial c}\right)_s.$$

The characteristic equation of matrix $\mathbf{B}$ reads

$$\lambda^4 + k_1\lambda^3 + k_2\lambda^2 + k_3\lambda + k_4 = 0, \tag{8.23}$$

where

$$\begin{aligned}
k_1 &= 2\delta, \\
k_2 &= b_{11}b_{22} - b_{12}b_{21} - \delta(b_{11} + b_{22}) - b_{43} \\
&= \frac{9\delta^2}{4} + \omega_0^2 - \omega_0^2\left(\frac{\partial Q_1}{\partial c}\right)_s \\
&\quad + \frac{1}{4\omega^2 a_s}\left[a_s(\omega^2 - \omega_0^2) + \omega_0^2 Q_{2_s}\right]\left[\omega^2 - \omega_0^2 + \omega_0^2\left(\frac{\partial Q_2}{\partial a}\right)_s\right], \\
k_3 &= \delta(b_{11}b_{22} - b_{12}b_{21}) + b_{43}(b_{11} + b_{22}) \\
&= \frac{\delta}{4\omega^2 a_s}\left[a_s(\omega^2 - \omega_0^2) + \omega_0^2 Q_{2_s}\right]\left[\omega^2 - \omega_0^2 + \omega_0^2\left(\frac{\partial Q_2}{\partial a}\right)_s\right] \\
&\quad + \frac{\delta^3}{4} - \delta\left[-\omega_0^2 + \omega_0^2\left(\frac{\partial Q_1}{\partial c}\right)_s\right], \\
k_4 &= b_{43}(b_{12}b_{21} - b_{11}b_{22}) - b_{12}b_{23}b_{41} \\
&= \frac{\omega_0^4}{4\omega^2 a_s}\left[a_s(\omega^2 - \omega_0^2) + \omega_0^2 Q_{2_s}\right]\left(\frac{\partial Q_2}{\partial c}\right)_s\left(\frac{\partial Q_1}{\partial a}\right)_s \\
&\quad + \left(\frac{\delta^2}{4} + \frac{1}{4\omega^2 a_s}\left[a_s(\omega^2 - \omega_0^2) + \omega_0^2 Q_{2_s}\right]\left[\omega^2 - \omega_0^2 + \omega_0^2\left(\frac{\partial Q_2}{\partial a}\right)_s\right]\right) \\
&\quad \times\left[-\omega_0^2\left(\frac{\partial Q_1}{\partial c}\right)_s + \omega_0^2\right].
\end{aligned}$$

The stability of equilibrium $(a_s, \theta_s, c_s, 0)$ is determined by examining the signs of the real parts of the eigenvalues of $\mathbf{B}$. According to Hurwitz criteria on [18], since $H_1 = 2\delta > 0$, $(a_s, \theta_s, c_s, 0)$ is asymptotically stable if and only if

$$\begin{cases} H_2 = \begin{vmatrix} k_1 & k_3 \\ 1 & k_2 \end{vmatrix} = k_1k_2 - k_3 > 0, \\ H_3 = \begin{vmatrix} k_1 & k_3 & 0 \\ 1 & k_2 & k_4 \\ 0 & k_1 & k_3, \end{vmatrix} = k_1k_2k_3 - k_3^2 - k_1^2k_4 > 0, \\ H_4 = \begin{vmatrix} k_1 & k_3 & 0 & 0 \\ 1 & k_2 & k_4 & 0 \\ 0 & k_1 & k_3 & 0 \\ 0 & 1 & k_2 & k_4, \end{vmatrix} = k_1k_2k_3k_4 - k_3^2k_4 - k_1^2k_4^2 > 0. \end{cases} \tag{8.24}$$

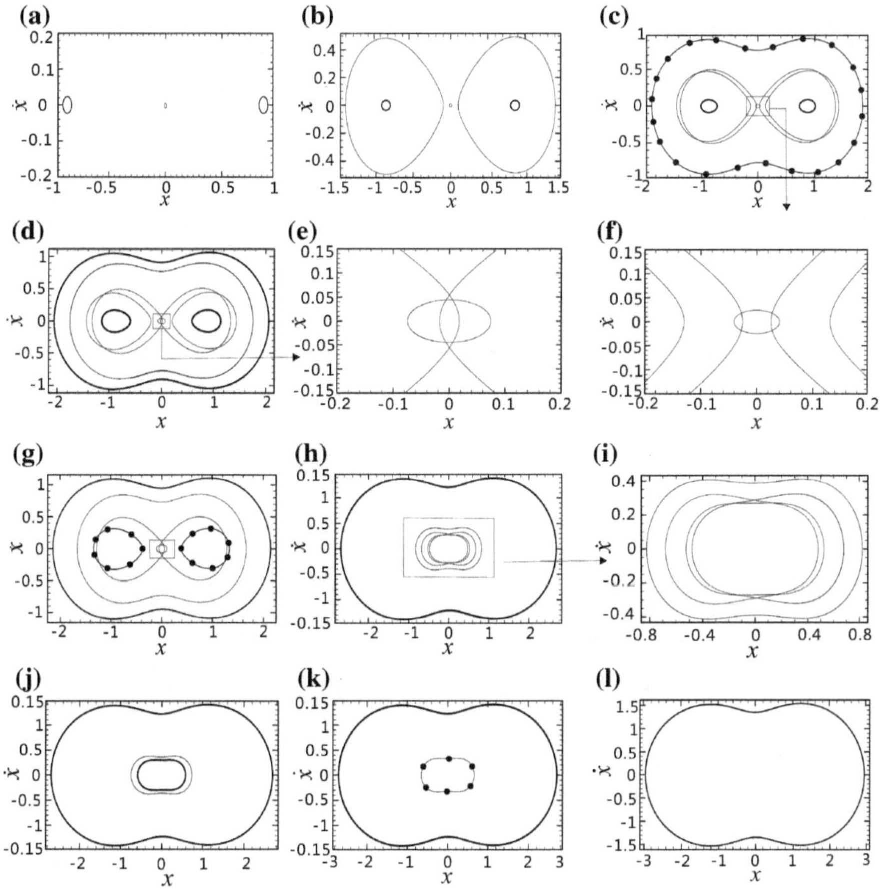

Fig. 8.2 Periodic solutions computed for $\alpha = 0.5$, $\delta = 0.05$, $\omega = 0.6$, $\omega_0 = 1$. **a** for $f = 0.015$; **b** for $f = 0.023$; **c** for $f = 0.055$; **d** for $f = 0.1$; **e**, **f** and **i** present a blow-up of the central part of (**d**), (**c**) and (**h**), respectively; **g** for $f = 0.1405$, the structure of the boxed area is similar as that described in (**e**); **h** for $f = 0.44$; **j** for $f = 0.45$; **k** for $f = 0.454$ and **l** for $f = 0.6$

Figure 8.1a also plots the response curves for $\alpha = 0.5$ near the pair of the centers for Eq. (8.13) marked as c_2. The stable branches are marked with solid and the unstable ones with dashed lines.

Both responses for $\alpha = 0$ near (0, 0) for Eq. (8.8) and near the pair of centers for Eq. (8.12), marked as c_3 and c_4 respectively, are plotted in Fig. 8.1b. Figure 8.1c shows the responses of Eq. (8.7) near (0, 0) for $\alpha = 1$.

The response curves of (8.7) for different values of parameter α are plotted in Fig. 8.1d, where the stable and unstable branches are plotted with solid and dashed. When α increases continuously, changes occur in the response curves, for example for $\alpha = 1.438$, the hysteresis phenomenon appears.

The periodic solutions of system (8.2) for $\alpha = 0.5, \alpha = 0$ and $\alpha = 1$ are shown in Figs. 8.2, 8.3 and 8.4 respectively. They show a good agreement with the theoretical analysis, and so do the stability of the solutions. Figure 8.2e, f, i present a blow-up of the central part of Fig. 8.2d, c, h respectively. The structure of the boxed area in

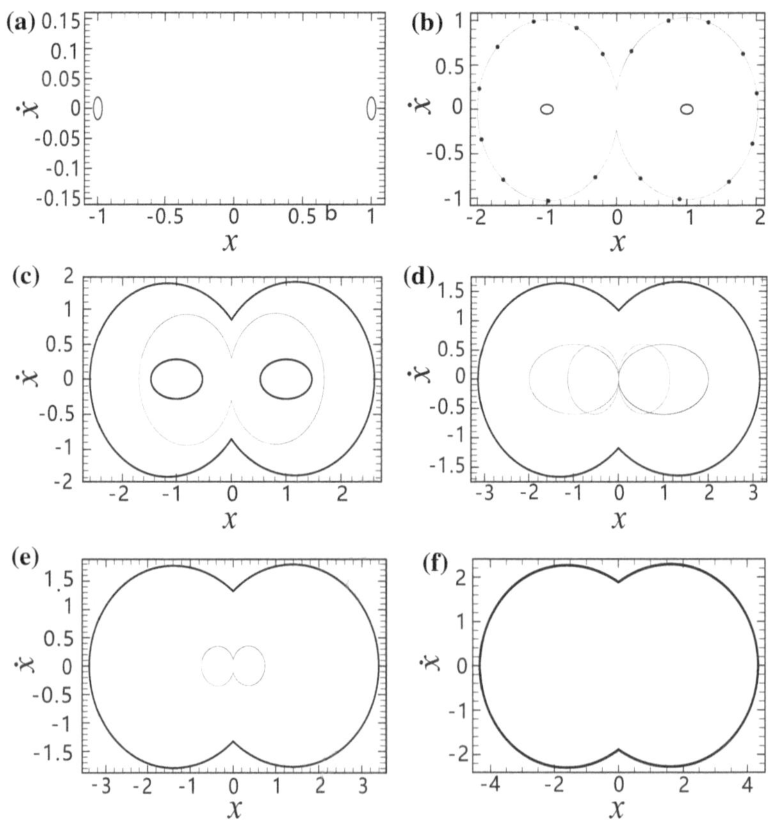

Fig. 8.3 Periodic solutions computed for $\alpha = 0, \delta = 0.05, \omega = 0.6, \omega_0 = 1$: **a–f** for $f = 0.02, 0.057, 0.3, 0.64, 0.8$ and 1.4, respectively

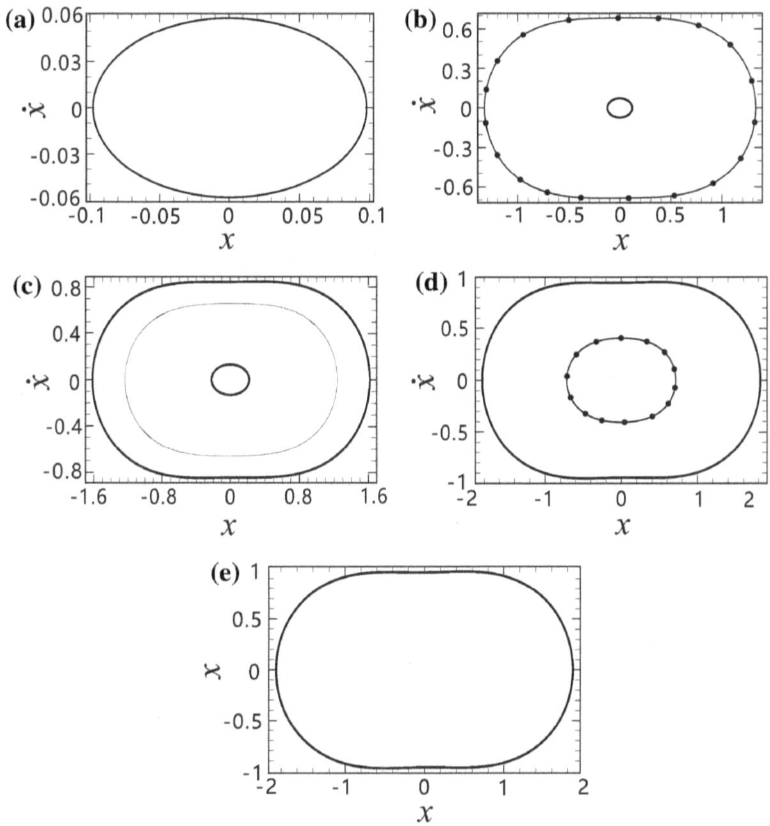

Fig. 8.4 Periodic solutions computed for $\alpha = 1$, $\delta = 0.05$, $\omega = 0.6$, $\omega_0 = 1$. **a**–**e** for $f = 0.02, 0.043, 0.075, 0.154$ and 0.17 respectively

Fig. 8.2g is similar as that described in Fig. 8.2e. The thick, dotted and thin lines represent respectively stable, semi-stable and unstable periodic solutions.

The stability properties of equilibria of the averaged equation can be shown by employing the following transformation.

$$\begin{cases} a = \sqrt{u^2 + v^2}, \\ \theta = -\arctan \dfrac{u}{v}. \end{cases} \tag{8.25}$$

By setting $t = \tau/\omega$, system (8.6) takes the averaged form:

$$\begin{cases} \dot{u} = \dfrac{1}{2\omega}\left[-(\omega^2 - \omega_0^2)v - \delta\omega u - \dfrac{4\omega_0^2 v}{\pi(u^2+v^2)}h(u,v,\alpha)\right], \\ \dot{v} = \dfrac{1}{2\omega}\left[(\omega^2 - \omega_0^2)u - \delta\omega v + \dfrac{4\omega_0^2 u}{\pi(u^2+v^2)}h(u,v,\alpha) + f\right], \end{cases} \tag{8.26}$$

where

$$h(u, v, \alpha) = \sqrt{u^2 + v^2 + \alpha^2} E\left(\sqrt{\frac{u^2 + v^2}{u^2 + v^2 + \alpha^2}}\right) - \frac{\alpha^2}{\sqrt{u^2 + v^2 + \alpha^2}} K\left(\sqrt{\frac{u^2 + v^2}{u^2 + v^2 + \alpha^2}}\right).$$

Figure 8.5 shows the phase portraits of the averaged system (8.26) for $\alpha = 1$. Using the averaging theorem, it is found that the equilibria of the averaged system (8.26) correspond to the periodic solutions of the system (8.2).

8.4 Hopf Bifurcations

Hopf bifurcation [19, 20] consists in the occurrence of a family of limit cycles bifurcating from an initially stable (or non-hyperbolic) equilibrium.

The point $(x, \dot{x}) = (0, 0)$ is a trivial equilibrium state for the SD oscillator without forcing. We study the occurrence of Hopf bifurcation near this point for the forced oscillator with a nonlinear damping that is for equation

$$\ddot{x} + (\xi + \gamma x^2)\dot{x} + x\left(1 - \frac{1}{\sqrt{x^2 + \alpha^2}}\right) = F \cos \Omega t. \tag{8.27}$$

where Ω may take arbitrary positive integer values while ξ and γ are arbitrary real numbers. Let us introduce new variable a and θ and $x = a \cos(t + \theta), \dot{x} = -a \sin(t + \theta)$, it follows

$$\begin{cases} \dot{a} = -f(a, \theta, t) \sin(t + \theta), \\ a\dot{\theta} = -f(a, \theta, t) \cos(t + \theta), \end{cases} \tag{8.28}$$

where

$$f(a, \theta, t) = a[\xi + \gamma a^2 \cos^2(t + \theta)] \sin(t + \theta) + \frac{a \cos(t + \theta)}{\sqrt{a^2 \cos^2(t + \theta) + \alpha^2}} + F \cos \Omega t. \tag{8.29}$$

Averaging Eq. (8.28) over a period $[0, 2\pi]$, it follows that

$$\begin{cases} \dot{a} = -\dfrac{1}{2\pi} \displaystyle\int_0^{2\pi} f(a, \theta, t) \sin(t + \theta) d(t + \theta), \\ a\dot{\theta} = -\dfrac{1}{2\pi} \displaystyle\int_0^{2\pi} f(a, \theta, t) \cos(t + \theta) d(t + \theta). \end{cases} \tag{8.30}$$

Equation (8.30) is now averaged for $\alpha \geq 0$ both the continuous and the discontinuous cases.

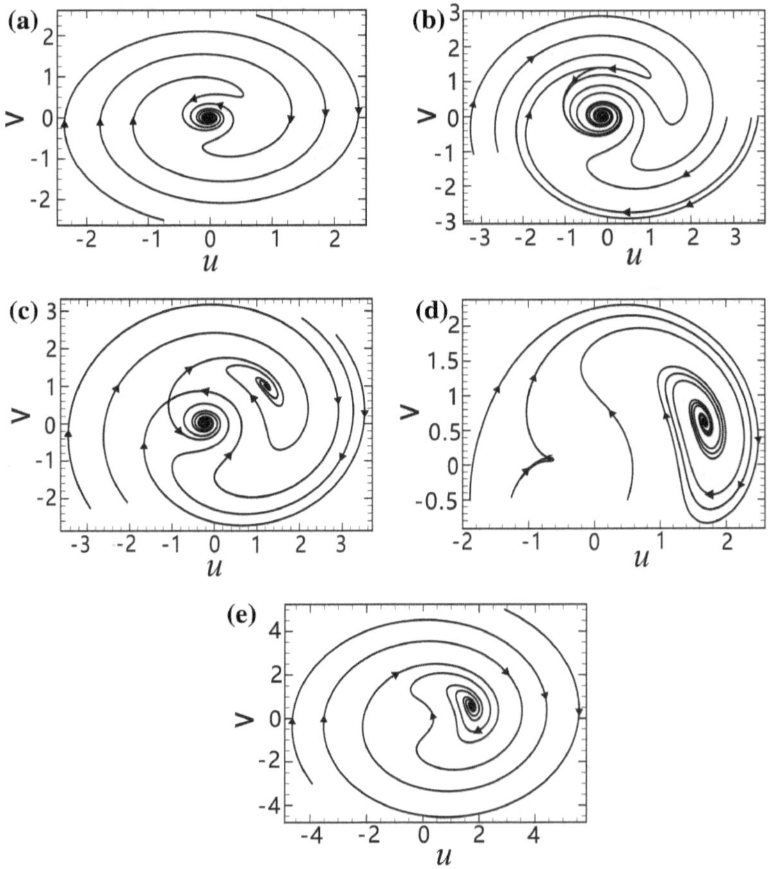

Fig. 8.5 Phase portraits of the averaged Eq. (8.26) for $\alpha = 1$, $\delta = 0.05$, $\omega = 0.6$, $\omega_0 = 1$. **a–e** for $f = 0.02, 0.043, 0.075, 0.154$ and 0.17 respectively

Let us first consider $\forall \alpha \geq 0$,

$$SDE[k] = \begin{cases} kE\left[\sqrt{\frac{a^2}{k^2}}\right], & |k| > a, \\ \lim\limits_{|k|\to a}\left(kE\left[\sqrt{\frac{a^2}{k^2}}\right]\right) = a, & |k| = a, \end{cases} \tag{8.31}$$

and

$$SDK[k] = \begin{cases} \frac{k^2 - a^2}{k}K\left[\sqrt{\frac{a^2}{k^2}}\right], & |k| > a, \\ \lim\limits_{|k|\to a}\left(\frac{k^2 - a^2}{k}K\left[\sqrt{\frac{a^2}{k^2}}\right]\right) = 0, & |k| = a, \end{cases} \tag{8.32}$$

where $E[m]$ and $K[m]$ denote respectively the complete elliptic integrals of the first and of the second kind, respectively.

Then it follows from system (8.30).

$$\begin{aligned}\int_0^{2\pi} \frac{a^2\cos^2(t+\theta)}{\sqrt{a^2\cos^2(t+\theta)+\alpha^2}}d(t+\theta) &= \int_{-\pi}^{\pi} \frac{a^2\cos y\sin y}{\sqrt{a^2\cos^2 y+\alpha^2}}dy \\ &= 4\left(SDE\left[\sqrt{a^2+\alpha^2}\right]-SDK\left[\sqrt{a^2+\alpha^2}\right]\right).\end{aligned} \tag{8.33}$$

The following identities are easily obtained

$$\int_0^{2\pi} \cos(t+\theta)\sin(t+\theta)d(t+\theta) = \int_0^{2\pi} \cos^3(t+\theta)\sin(t+\theta)d(t+\theta) = 0, \tag{8.34}$$

$$\int_0^{2\pi} \sin^2(t+\theta)d(t+\theta) = \int_0^{2\pi} \cos^2(t+\theta)d(t+\theta) = \pi, \tag{8.35}$$

$$\int_0^{2\pi} \cos^2(t+\theta)\sin^2(t+\theta)d(t+\theta) = \frac{\pi}{4}, \tag{8.36}$$

$$\int_0^{2\pi} \cos\Omega t\sin(t+\theta)d(t+\theta) = \frac{1}{2}\int_0^{2\pi} \sin((1-\Omega)t+\theta)d(t+\theta), \tag{8.37}$$

and

$$\int_0^{2\pi} \cos\Omega t\cos(t+\theta)d(t+\theta) = \frac{1}{2}\int_0^{2\pi} \cos((1-\Omega)t+\theta)d(t+\theta). \tag{8.38}$$

Due to these identities, Eqs. (8.29), (8.33) and (8.38) can be inserted into (8.30) which reads, after simplifications

$$\begin{cases} \dot{a} = -\dfrac{1}{2}\left(a\xi + \dfrac{a^3\gamma}{4} + \dfrac{F}{2\pi}\displaystyle\int_0^{2\pi} \sin((1-\Omega)t+\theta)d(t+\theta)\right), \\ a^2\dot{\theta} = -\dfrac{2}{\pi}\left(SDE\left[\sqrt{a^2+\alpha^2}\right]-SDK\left[\sqrt{a^2+\alpha^2}\right]\right) \\ \qquad -\dfrac{aF}{4\pi}\displaystyle\int_0^{2\pi} \cos((1-\Omega)t+\theta)d(t+\theta). \end{cases} \tag{8.39}$$

Note that Ω takes arbitrary positive integer values, which leads to the equation given by

$$\frac{F}{2\pi}\int_0^{2\pi} \sin((1-\Omega)t+\theta)d(t+\theta) = \begin{cases} 0, & \Omega \neq 1, \\ F\sin\theta, & \Omega = 1, \end{cases} \tag{8.40}$$

and

$$\frac{aF}{2\pi}\int_0^{2\pi}\cos((1-\Omega)t+\theta)d(t+\theta)=\begin{cases}0, & \Omega\neq 1,\\ aF\cos\theta, & \Omega=1.\end{cases} \tag{8.41}$$

Consider the case of $\Omega \neq 1$. Obviously, Eq. (8.39) is expressed as the normal form (which could be classically called Poincaré-Birckoff normal form)

$$\begin{cases}\dot{a}=-\frac{1}{2}\left(a\xi+\frac{a^3\gamma}{4}\right),\\ a^2\dot{\theta}=-\frac{2}{\pi}\left(SDE\left[\sqrt{a^2+\alpha^2}\right]-SDK\left[\sqrt{a^2+\alpha^2}\right]\right),\end{cases} \tag{8.42}$$

which leads to the equation given by

$$\begin{cases}a\xi+\frac{a^3\gamma}{4}=0,\\ SDE\left[\sqrt{a^2+\alpha^2}\right]-SDK\left[\sqrt{a^2+\alpha^2}\right]=0,\end{cases} \tag{8.43}$$

Hence, when $\alpha \geq 0$, a periodic solution may exit near $(x,\dot{x})=(0,0)$ as ξ varies from $\xi<0$ to $\xi>0$. It is noticed that when $\alpha=0$ system (8.27) is discontinuous at (0, 0) which is named the saddle-like singularity (see Chap. 2 for details); when $0<\alpha<1$, it is a saddle point; when $\alpha=1$, (0, 0) is a non-hyperbolic degenerate center point; and when $\alpha>1$, the equilibrium (0, 0) of the system is a center. From Hopf bifurcation theorem [21, 22], only if $\alpha>1$, Hopf bifurcation of Eq. (8.27) near $(x,\dot{x})=(0,0)$ may occur as ξ varies from $\xi<0$ to $\xi>0$. Furthermore, the stability of limit cycle bifurcated from the Hopf bifurcation point $(x,\dot{x})=(0,0)$ of Eq. (8.27) can be analyzed and it follows that,

1. For $\gamma<0$, subcritical Hopf bifurcation occurs as ξ varies from $\xi<0$ to $\xi>0$, while the trivial solution changes from an unstable focus to an asymptotically stable focus from which an unstable periodic solution bifurcates.
2. For $\gamma>0$ supercritical Hopf bifurcation occurs as ξ varies from $\xi>0$ to $\xi<0$, while the trivial solution changes from an asymptotically stable focus to an unstable focus from which an asymptotically stable limit cycle bifurcates (Fig. 8.6).

Considering the case of $\Omega=1$ and substituting Eqs. (8.40–8.41) into Eq. (8.39) yields the normal form

$$\begin{cases}\dot{a}=-\frac{1}{2}\left(a\xi+\frac{a^3\gamma}{4}+F\sin\theta\right),\\ \dot{\theta}=-\frac{2}{\pi a^2}\left[SDE\left[\sqrt{a^2+\alpha^2}\right]-SDK\left[\sqrt{a^2+\alpha^2}\right]\right]\\ \quad -\frac{1}{2a}F\cos\theta,\end{cases} \tag{8.44}$$

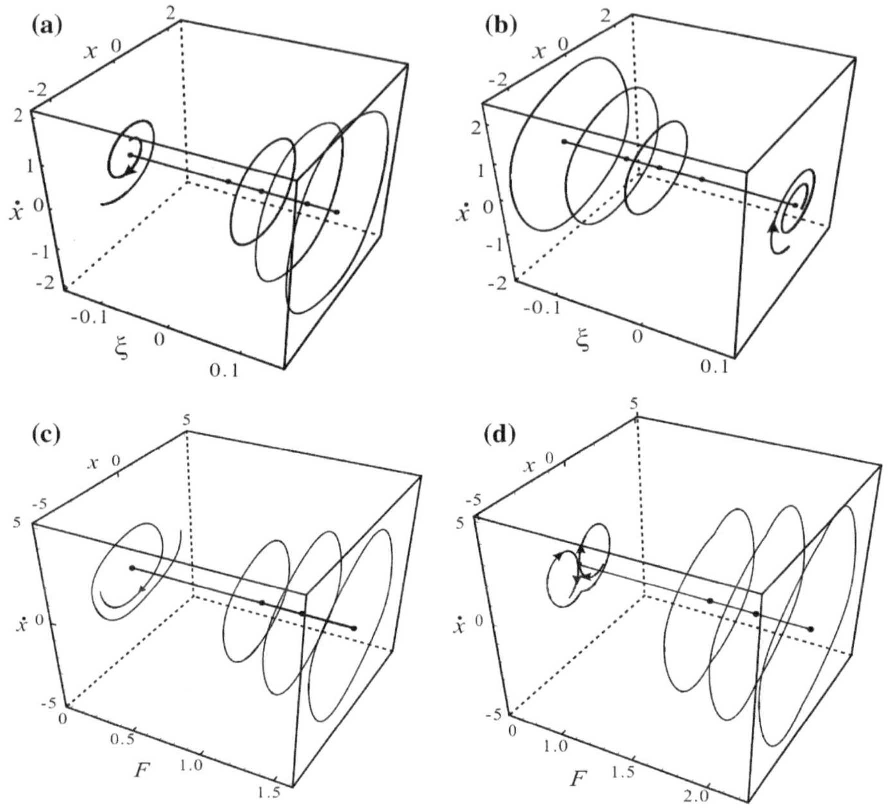

Fig. 8.6 The phase portraits of system (8.27) near $(x, \dot{x}) = (0, 0)$. **a** Subcritical Hopf bifurcation for $\Omega = 2, \alpha = 2, \gamma = -0.1, F = 0.001$; **b** supercritical Hopf bifurcation for $\Omega = 2, \alpha = 2, \gamma = 0.1, F = 0.001$; **c** supercritical Hopf bifurcation for $\Omega = 1, \alpha = 2, \gamma = 0.1, \xi = -0.1$ and **d** non-Hopf periodic solutions for $\Omega = 1, \alpha = 0, \gamma = 0.1, \xi = -0.1$

which leads to the equation given by

$$\frac{16}{\pi^2 a^2}\left(SDE\left[\sqrt{a^2+\alpha^2}\right] - SDK\left[\sqrt{a^2+\alpha^2}\right]\right)^2 + \left(a\xi + \frac{a^3\gamma}{4}\right)^2 = F^2. \tag{8.45}$$

Especially, when $\alpha = 0$, Eq. (8.45) yields

$$\frac{16}{\pi^2} + \left(a\xi + \frac{a^3\gamma}{4}\right)^2 = F^2. \tag{8.46}$$

So, whenever $\alpha \geq 0$, from Eqs. (8.45), (8.46), we know that Hopf bifurcation of Eq. (8.27) near $(x, \dot{x}) = (0, 0)$ could not appear as ξ and γ vary. But, when $\alpha > 1$,

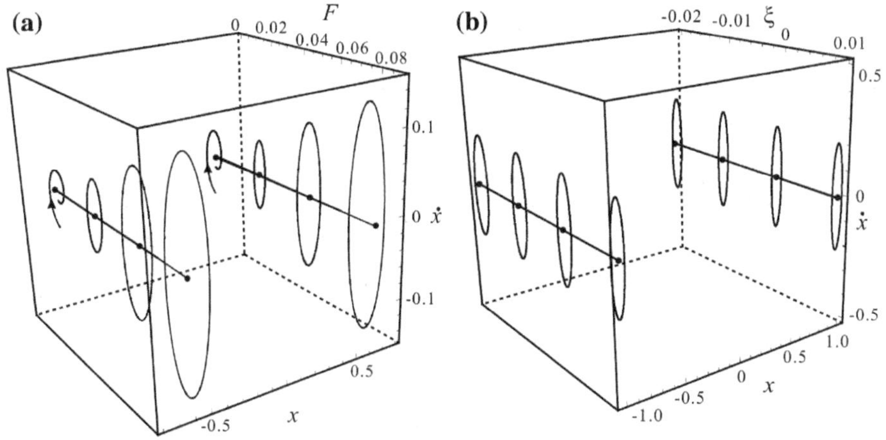

Fig. 8.7 The phase portraits of system (8.27) near $(x, \dot{x}) = (\pm\sqrt{1-\alpha^2}, 0)$. **a** Supercritical Hopf bifurcation for $\Omega = 1, \alpha = 0.8, \gamma = 0.8$ and $\xi = -0.06$; **b** non-Hopf periodic solutions for $\Omega = 4, \alpha = 0, \gamma = 0.7, F = 0.8$

Hopf bifurcation of Eq. (8.27) near $(x, \dot{x}) = (0, 0)$ may occur as F varies. By the averaging theorem (see [16]), the stability of periodic solutions can be obtained using the stability first approximation.

Numerical simulations are carried out for system (8.27) near $(0, 0)$ and $(\pm\sqrt{1-\alpha^2}, 0)$, respectively. Figure 8.4a presents subcritical Hopf bifurcation for $\Omega = 2, \alpha = 2, \gamma = -0.1$ and $F = 0.001$. Figure 8.4b displays supercritical Hopf bifurcation for $\Omega = 2, \alpha = 2, \gamma = 0.1$ and $F = 0.001$. Figure 8.4c shows supercritical Hopf bifurcation for $\Omega = 1, \alpha = 2, \gamma = 0.1$ and $\xi = -0.1$. While Fig. 8.4d gives the non-Hopf periodic solutions near $(0, 0)$ for $\Omega = 1, \alpha = 0, \gamma = 0.1$ and $\xi = -0.1$.

Figure 8.7a plots the Hopf bifurcations near $(x, \dot{x}) = (\pm\sqrt{1-\alpha^2}, 0)$ for $\Omega = 1, \alpha = 0.8, \gamma = 0.8$ and $\xi = -0.06$ and Fig. 8.7b plots the periodic solutions for $\Omega = 4, \alpha = 0, \gamma = 0.7, F = 0.8$. They indicate a good agreement between the analytical prediction and the numerical simulation results, which laid a theoretical foundation for parameter design, stable operation and fault diagnosis of a real system.

8.5 Summary

A four dimensional averaging method is introduced to the SD oscillator for both smooth and discontinuous stages to obtain the periodic solutions without any truncation by constructing the complete Jacobian elliptic integrals of the first and the second kinds. The stability of the system has also been analyzed by using Lyapunov criteria, which answers the question why the conventional averaging theorem fails to describe the stability of Duffing system near the saddle point. Meanwhile, the Hopf bifurcations behaviour of the oscillator for both smooth and discontinuous stages are demonstrated directly. Although periodic solutions exist near the trivial equilibrium as parameters vary for $\alpha \geq 0$, Hopf bifurcation occurs only for $\alpha \geq 1$ when the system presents both nonlinear damping and the external excitation.

References

1. Jane, C. (1960). An upper bound for the number of periodic solutions of a perturbed system. *Journal of Mathematical Analysis and Applications*, *1*(3–4), 334–341.
2. Mawhin, J. (1971). Periodic solutions of nonlinear functional differential equations. *Journal of Differential Equations*, *10*(10), 240–261.
3. James, R. W. (1979). The existence of periodic solutions for nonlinearly perturbed conservative systems. *Nonlinear Analysis: Theory, Methods, Applications*, *3*(5), 697–705.
4. Holmes, P. (1979). A nonlinear oscillator with a strange attractor. *Philosophical Transactions of the Royal Society of London: Mathematics and Physics Sciences A*, *292*(1394), 419–448.
5. Shao, G. J., & Xu, Z. (1995). Asymptotic method for primary resonance of a strongly nonlinear vibration system with many degrees of freedom. *Acta Mechanica Sinica*, *11*(5), 577–586.
6. Roy, R. V. (1994). Averaging method for strongly nonlinear oscillators with periodic excitations. *International Journal of Non-Linear Mechanics*, *29*(5), 737–753.
7. Mahmoud, G. M. (1993). On the generalized averaging method of a class of strongly nonlinear forced oscillators. *Physica A*, *199*(1), 87–95.
8. Cardone, A., Ferro, M., Ixaru, L. G., & Paternoster, B. (2010). A family of exponential fitting direct quadrature methods for volterra integral equations. In *Icnaam: International Conference of Numeri* (pp. 2204–2207).
9. Chen, S. H., Yang, X. M., & Cheung, Y. K. (1999). Periodic solutions of strongly quadratic non-linear oscillators by the elliptic lindstedt-poincaré method. *Journal of Sound and Vibration*, *227*(5), 1109–1118.
10. Lakrad, F., & Belhaq, M. (2002). Periodic solutions of strongly nonlinear oscillators by the multiple scales method. *Journal of Sound and Vibration*, *258*(4), 677–700.
11. Mousa, A. A., Sayed, M., Eldesoky, I. M., & Zhang, W. (2014). Nonlinear stability analysis of a composite laminated piezoelectric rectangular plate with multi-parametric and external excitations. *International Journal of Dynamics and Control*, *2*(4), 494–508.
12. Gamal, M. (1997). Periodic solutions of strongly nonlinear mathieu oscillators. *Journal of Non-linear Mechanics*, *32*(6), 1177–1185.
13. Byrd, P. F. (1954). *Handbook of elliptic integrals for engineers and physicists*. Berlin: Springer.
14. Lavrentiev, M. A., & Shabbat, B. V. (1973). *Methods in the theory of functions in a complex variable*. Moscow: Nauka.
15. Han, M. A., & Gu, S. S. (2001). *Theory and method of nonlinear system*. China: Science Press.
16. Guckenheimer, J., & Holmes, P. (1983). *Nonlinear oscillation, dynamical system and bifurcation of vector fields*. New York: Springer.
17. Holmes, P., & Marsden, J. E. (2014). Bifurcations of dynamical systems and nonlinear oscillations in engineering systems. *Dynamics and Design Conference: D & D* (pp. 561–582).
18. Ma, Z. E., & Zhou, Y. C. (2001). *Qualitative and stability method of ordinary differential equation*. Beijing: Science Press of China.
19. Hopf, E. (1942). Abzweigung einer periodischen Lösung von einer stationären Lösung. *Berich. Sächs. Akad. Wiss., Leipzig, Math. Phys. Kl*, *94*(19), 15–25.
20. Hopf, E. (1943). Abzweigung einer periodischer Lösung von einer stationären Lösung eines Differentialsystems. *Ber. Verh. Sächs. Akad. Wiss. Leipzig. Math-Nat. K1*, *95*, 3–22.
21. Wiggins, S. (1990). *Introduction to applied nonlinear dynamical systems and chaos*. New York: Springer.
22. Liu, W. M. (1994). Criterion of hopf bifurcations without using eigenvalues. *Journal of Mathematical Analysis and Applications*, *182*(1), 250–256.

Chapter 9
Elliptic and Hyperbolic Functions

In this chapter, we go on into the methods for obtaining the analytical solutions of the SD oscillator. A series of irrational elliptic functions and hyperbolic functions is proposed for the unperturbed oscillator to provide the analytical solutions for both the smooth and discontinuous cases with periodic solutions and the homoclinic ones which could not be expressed using classical tools, the traditional methodologies being applicable only for rational or polynomial systems. It is found that the solutions of the discontinuous case can also be given by letting $\alpha \to 0$. With the help of the defined elliptic functions and the hyperbolic functions for the periodic and homoclinic orbits, the chaotic behaviours of the perturbed system can be detected analytically.[1]

9.1 Introduction

The specificity of the SD oscillator is the irrational restoring force. This leads to barriers for the conventional methods to investigate the dynamical behaviour directly. Two kinds of irrational elliptic functions and one kind of hyperbolic function are introduced in the real domain to formulate the analytical solutions of the system. As preliminary mathematical results, the properties of these functions are obtained among which differentiability, periodicity and parity. Having these functions, the chaotic thresholds of the damped and forced oscillator can be depicted using subharmonic orbits and Melnikov method.

Using these functions, this chapter will give an answer to the open problem proposed in [1] for irrational integrals. We propose a series of irrational elliptic functions [2, 3] and hyperbolic functions defined in the real domain, which give the analytical

[1]This chapter relies on a common work with Dr. Dan Wang and Professor Yushu Chen, Center for Nonlinear Dynamics Research, School of Astronautics, Harbin Institute of Technology, Harbin 150001 China.

Q. Cao and A. Léger, *A Smooth and Discontinuous Oscillator*, Springer Tracts in Mechanical Engineering, DOI 10.1007/978-3-662-53094-8_9

periodic solutions and homoclinic solutions of the irrational system. We discuss the chaotic threshold of the oscillator by Melnikov methods with the hyperbolic functions which we introduced. Meanwhile, the basic properties of the irrational elliptic functions and hyperbolic functions are obtained as a part of applied mathematics theory.

Recall the unperturbed SD oscillator in the dimensionless form:

$$x'' + x(1 - \frac{1}{\sqrt{x^2 + \alpha^2}}) = 0. \tag{9.1}$$

It is irrational in the smooth case i.e. when $\alpha > 0$ and discontinuous when $\alpha = 0$.

As described in above chapters, a pitchfork bifurcation occurs for the equilibria of the system at $\alpha = 1$: a pair of centers $(\pm 1, 0)$ and a saddle-like $(0, 0)$ co-exist for $\alpha = 0$, a pair of centers $(\pm\sqrt{1-\alpha^2}, 0)$ and a saddle $(0, 0)$ for $0 < \alpha < 1$ and a unique center $(0, 0)$ for $\alpha \geq 1$.

We now aim at making a step toward the construction of analytical solutions of the problem and use them to discuss its qualitative behaviour. Hamiltonian function of (9.1) has already been used in the previous chapters; it reads as

$$H = U(x) + \frac{1}{2}y^2, \tag{9.2}$$

where $\dot{x} = y$, $U(x) = \frac{1}{2}x^2 - \sqrt{x^2 + \alpha^2} + \frac{1+\alpha^2}{2}$.

The phase portraits are plotted by the different values of α as shown in Fig. 9.1: Fig. 9.1 plots the corresponding phase portraits in (a)–(d) again for parameters $\alpha = 0.0, 0.4, 1.0$ and $\alpha = 2.0$ for the discontinuous dynamics, double well potential, bifurcation or degenerate and single well behaviours, respectively. The corresponding irrational elliptic functions or the hyperbolic functions will be constructed in the following sections, the irrational elliptic functions of the first kind for $0 \leq \alpha < 1$ and of the second type for $\alpha \geq 1$, respectively.

9.2 Irrational Elliptic Functions of the First Kind

9.2.1 Definitions

Introducing a notation $k^2/2 = H$, it follows

$$H = \frac{1}{2}k^2 = \frac{1}{2}y^2 + U(x), \tag{9.3}$$

from which the previous classification of the trajectories is rewritten. Denoting by x_0 the maximal intersection point of the orbits with the x-axis, it follows that $-2\sqrt{1-\alpha} \leqslant x_0 \leqslant 2\sqrt{1-\alpha}$ for $k \leq 1 - \alpha$ and the orbits can be classified according to the values of k as shown in Table 9.1, where k is determined by x_0.

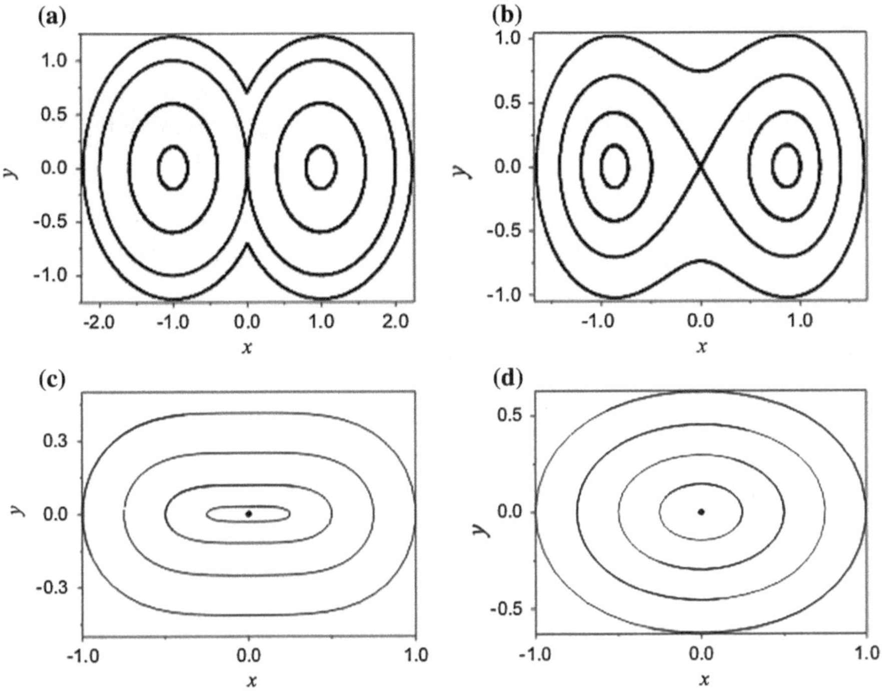

Fig. 9.1 Phase portraits: **a** for $\alpha = 0.0$ with a pair of centers and a saddle, **b** for $\alpha = 0.4$ with a pair of centers and a saddle-like equilibrium and **c** for $\alpha = 1.0$ and **d** for $\alpha = 2.0$, respectively

Table 9.1 Unperturbed Orbits classified by parameter k

Orbits	Centers	Small periodic orbits	Homoclinic orbits	Large periodic orbits
k	$k = 0$	$0 < k < 1 - \alpha$	$k = 1 - \alpha$	$k > 1 - \alpha$
$2H$	$2H = 0$	$0 < 2H < (1 - \alpha)^2$	$2H = (1 - \alpha)^2$	$2H > (1 - \alpha)^2$

The time τ from $(x_0, 0)$ to (x, y) along the lower branch of the periodic orbit is obtained by the following integral

$$\tau = -\int_{x_0}^{x} \frac{1}{\sqrt{k^2 - \left(\sqrt{t^2 + \alpha^2} - 1\right)^2}} dt. \tag{9.4}$$

Let us introduce

$$x = \pm\sqrt{(k \cdot \cos\varphi + 1)^2 - \alpha^2}, \tag{9.5}$$

where $\varphi \in [0, \pi]$, $k \in [0, 1 - \alpha)$, it follows that the time τ given by Eq. (9.4) is rewritten as

$$\tau = \int_0^{\varphi} \frac{1}{\sqrt{1 - \frac{\alpha^2}{(1+k\cdot\cos\eta)^2}}} d\eta, \tag{9.6}$$

and let us denote

$$I(k, \alpha) = \int_0^{\pi} \frac{1}{\sqrt{1 - \frac{\alpha^2}{(1+k\cdot\cos\eta)^2}}} d\eta. \tag{9.7}$$

The variable $\varphi = am\tau$ which has been defined by Eq. (9.5) is called the angular of τ and the irrational elliptic functions of the first kind are defined as follows

$$sd(\tau, k, \alpha)\sin\varphi = \sin(am\tau), \tag{9.8}$$

$$ad(\tau, k, \alpha)\cos\varphi = \cos(am\tau), \tag{9.9}$$

$$hd(\tau, k, \alpha) = \sqrt{(1 + k \cdot \cos(am\tau))^2 - \alpha^2}. \tag{9.10}$$

9.2.2 Properties

The properties of the irrational elliptic functions of the first kind defined above are listed below using the simpler notations $hd\tau$, $sd\tau$, $ad\tau$ instead of $hd(\tau, k, \alpha)$, $sd(\tau, k, \alpha)$, $ad(\tau, k, \alpha)$ without confusion.

1. **Identities**

$$sd^2\tau + ad^2\tau = 1,$$
$$hd^2\tau + \alpha^2 = (1 + k \cdot ad\tau)^2.$$

2. **Parities**

   From integral (9.6) and the above definitions, one immediately get that angular φ is an odd function of time variable τ, $sd\tau$ is an odd function of τ and that $hd\tau$ and $ad\tau$ are even functions of τ, then

$$sd(-\tau) = -sd\tau,$$
$$hd(-\tau) = hd\tau,$$
$$ad(-\tau) = ad\tau.$$

3. **Differentiability**
From the integral (9.6) and above definitions, the angular φ and the irrational elliptic functions of first kind $sd\tau, hd\tau, ad\tau$ can be differentiated with respect to τ, which gives

$$\begin{aligned}
\frac{\partial\varphi}{\partial\tau} &= \frac{hd\tau}{1+k\cdot ad\tau},\\
\frac{\partial sd\tau}{\partial\tau} &= \frac{ad\tau\cdot hd\tau}{1+k\cdot ad\tau},\\
\frac{\partial ad\tau}{\partial\tau} &= -\frac{sd\tau\cdot hd\tau}{1+k\cdot\sqrt{1-sd^2\tau}},\\
\frac{\partial hd\tau}{\partial\tau} &= -k\cdot sd\tau.
\end{aligned}$$

4. **Periodicity**
The periodicity of the irrational functions can be obtained and written as follows by considering the periodicity of the integrand of (9.6).

$$\begin{aligned}
\varphi &= am\tau,\\
\tau &\to \tau+2I(k,\alpha),\\
am(\tau+2I(k,\alpha)) &= am\tau+2\pi,\\
am(-\tau+2I(k,\alpha)) &= -am\tau+2\pi,
\end{aligned}$$

then

$$\begin{aligned}
sd(\tau+2I(k,\alpha)) &= sd\tau,\\
sd(-\tau+2I(k,\alpha)) &= -sd\tau,\\
ad(\tau+2I(k,\alpha)) &= ad\tau\\
ad(-\tau+2I(k,\alpha)) &= -ad\tau,
\end{aligned}$$

which leads to the result that $2I(k,\alpha)$ is of the period of $sd\tau$ and $ad\tau$ and it follows that $2I(k,\alpha)$ is also of the period of $hd\tau$, that is

$$\begin{aligned}
hd(\tau+2I(k,\alpha)) &= hd\tau,\\
hd(-\tau+2I(k,\alpha)) &= -hd\tau.
\end{aligned}$$

The graphs of the elliptic functions are plotted in Fig. 9.2, for $sd(t)$ (a), $ad(t)$ (b) and $hd(t)$ (c), for particular values of the parameters Special values of the irrational elliptic functions of the first kinds are listed in Table 9.2.

Let us now write the Taylor expansion of the integrand of Eq. (9.7).

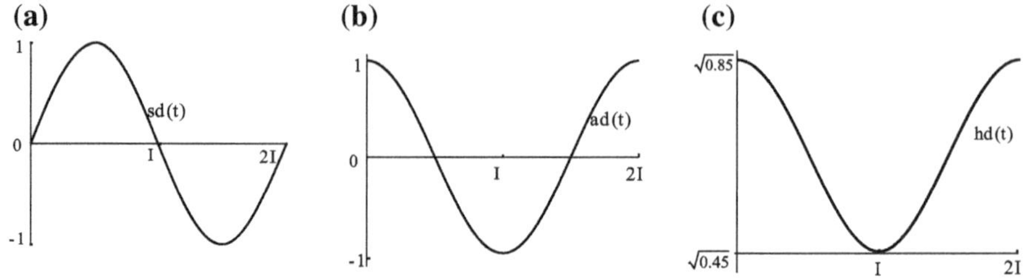

Fig. 9.2 Graphs for the elliptic functions of the first kinds: **a** for $sd(t)$, **b** for $ad(t)$ and **c** for $hd(t)$ with $\alpha = 0.6$ and $k = 0.1$

Table 9.2 Particular values of the irrational elliptic functions of the first kind

$\varphi = 0, \tau = 0$	$\varphi = \pi, \tau = I(k, \alpha)$
$sd(0) = 0, ad(0) = 1$	$sd(I) = 0, ad(I) = -1$
$hd(0) = \sqrt{(1+k)^2 - \alpha^2}$	$hd(I) = \sqrt{(1-k)^2 - \alpha^2}$

$$\begin{aligned}\frac{1}{\sqrt{1 - \frac{\alpha^2}{(1+k\cdot\cos\varphi)^2}}} &= \frac{1}{\sqrt{1-\alpha^2}} - \frac{k\cdot\alpha^2}{(1-\alpha^2)^{3/2}}\cdot\cos\varphi \\ &+ \frac{3\cdot k^2\cdot\alpha^2}{2\cdot(1-\alpha^2)^{5/2}}\cdot\cos^2\varphi - \frac{5\cdot k^3}{2\cdot(1-\alpha^2)^{7/2}}\cdot\cos^3\varphi \\ &+ \frac{5\cdot k^4\cdot\alpha^2\cdot(4+3\cdot\alpha^2)}{8\cdot(1-\alpha^2)^{9/2}}\cdot\cos^4\varphi \\ &- \frac{3\cdot k^5\cdot\alpha^2\cdot(8+12\cdot\alpha^2+\alpha^4)}{8\cdot(1-\alpha^2)^{11/2}}\cdot\cos^5\varphi + \cdots \end{aligned} \tag{9.11}$$

which leads to the following expansion of the integral

$$I(k,\alpha) = \frac{3\cdot k^2\cdot\alpha^2\cdot\pi}{4\cdot(1-\alpha^2)^{5/2}} + \frac{\pi}{\sqrt{1-\alpha^2}} - \frac{15\cdot k^4\cdot\alpha^2\cdot(3\cdot\alpha^4+\alpha^2-4)\cdot\pi}{64\cdot(1-\alpha^2)^{1/2}} + \cdots \tag{9.12}$$

And Eq. (9.6) can be rewritten as

$$\tau = az + bz^3 + cz^5 + \ldots \tag{9.13}$$

where

$$a = \frac{1+k}{(1+2k+k^2-\alpha^2)^{1/2}},$$

$$b = \frac{k\alpha^2}{6(1+2k+k^2-\alpha^2)^{3/2}},$$

$$c = \frac{k\alpha^2(\alpha^2 + 8k^2 + 7k - 1)}{120(1 + 2k + k^2 - \alpha^2)^{5/2}}.$$

Then, the anti-series form of τ in terms of variable z in Eq. (9.13) can be written as

$$z = A\tau + B\tau^3 + E\tau^5 + \dots \tag{9.14}$$

where

$$A = \frac{1}{a},\ B = -\frac{b}{a^4},\ E = \frac{3b^2 - ac}{a^7}.$$

The series forms of the irrational elliptic functions are then given as follows

$$sd\tau = s_1\tau - s_3\tau^3 + s_5\tau^5 + \dots \tag{9.15}$$

where

$$s_1 = \frac{(1 + 2k + k^2 - \alpha^2)^{1/2}}{1 + k},$$

$$s_3 = \frac{(1 + 2k + k^2 - \alpha^2)^{1/2}}{3!(1 + k)^4} \cdot (1 + 3k + 3k^2 + k^3 - \alpha^2),$$

$$s_5 = \frac{(1 + 2k + k^2 - \alpha^2)^2}{5!(1 + k)^7} \cdot [15k^4 + 6k^5 + k^6 + (\alpha^2 - 1)^2 - \\ + 3k^2(5 + 4\alpha^2) + k^3(20 + 7\alpha^2) + k(6 + 3\alpha^2 - 9\alpha^4)].$$

$$ad\tau = 1 - a_2\tau^2 + a_4\tau^4 - a_6\tau^6 + \dots \tag{9.16}$$

where

$$a_2 = \frac{(1 + k)^2 - \alpha^2}{2!(1 + k)^2},$$

$$a_4 = \frac{(1 + k)^2 - \alpha^2}{4!(1 + k)^5} \cdot [1 + 3k^2 + k^3 - \alpha^2 + 3k(1 + \alpha^2)],$$

$$a_6 = \frac{1}{6!(1 + k)^8} \cdot [8k^7 + k^8 - (\alpha^2 - 1)^3 + 8k(\alpha^2 - 1)^2(1 + 3\alpha^2) - \\ - 8k^5(13\alpha^2 - 7) - k^6(31\alpha^2 - 28) + k^4(70 - 109\alpha^2 + 75\alpha^4) - \\ + 8k^3(7 - 2\alpha^2 + 13\alpha^4) + k^2(28 + 31\alpha^2 - 14\alpha^4 - 45\alpha^6)].$$

So that the periodic orbits is given by the following expansion, as a power series with respect to τ:

$$hd\tau = p_0 - p_2\tau^2 + p_4\tau^4 - p_6\tau^6 - \dots \tag{9.17}$$

where

$$p_0 = \sqrt{(1+k)^2 - \alpha^2},\ p_2 = \frac{k\sqrt{(1+k)^2 - \alpha^2}}{2!(1+k)},$$

$$p_4 = \frac{k\sqrt{(1+k)^2 - \alpha^2}}{4!(1+k)^4}(1 + 3k + 3k^2 + k^2 - \alpha^2),$$

$$p_6 = \frac{k\sqrt{(1+k)^2 - \alpha^2}}{6!(1+k)^7} \times \{15k^4 + 6k^5 + k^6 + (1-\alpha^2)^2 + 3k^2(5 + 4\alpha^2) + k^3(20 + 7\alpha^2) + k(6 + 3\alpha^2 - 9\alpha^4)\}.$$

9.2.3 The Analytical Expression of the Periodic Orbits

When $0 \leqslant k < 1 - \alpha$, the analytical expression of the periodic orbits, written in the explicit form of x as a function of τ. The first and the second derivatives of the solution x are obtained by the differential properties of irrational elliptic functions of the first kind, and read respectively

$$\frac{dx}{d\tau} = \mp k \cdot sd\tau, \tag{9.18}$$

$$\frac{d^2x}{d\tau^2} = \mp hd\tau \cdot (1 - \frac{1}{\sqrt{hd^2\tau + \alpha^2}}) = -x \cdot \left(1 - \frac{1}{\sqrt{x^2 + \alpha^2}}\right), \tag{9.19}$$

which leads to

$$x = \pm hd(\tau, k, \alpha).$$

This is the periodic solution of the inside homoclinic orbits of the oscillator. The period of the these orbits is given by $T_k = 2I(k, \alpha)$, and T_k increases monotonically with k with the following limits

$$\lim_{k\to 0} T_k = \frac{2\pi}{\sqrt{1-\alpha^2}},\ \lim_{k\to 1-\alpha} T_k = \infty.$$

Moreover, when $k \to 0$ we have

$$\lim_{k \to 0} I(k, \alpha) = \int_0^{\pi} \frac{1}{\sqrt{1-\alpha^2}} dx = \frac{\pi}{\sqrt{1-\alpha^2}}, \tag{9.20}$$

which is the limit case $\alpha = 0$ satisfying $I(k, \alpha) = \pi$ and $I(k, \alpha) = \infty$ at $\alpha = 1$.

9.3 The Hyperbolic Functions

9.3.1 *Definitions*

Let us come back to the Hamiltonian again. We have already defined $k^2/2 = y^2/2 + U(x)$ which represents the energy of the homoclinic orbits when $k = 1 - \alpha$. Let

$$\tau = \int_0^{\varphi} \frac{1}{\sqrt{1 - \frac{\alpha^2}{[1+(1-\alpha)\cdot\cos\eta]^2}}} d\eta. \tag{9.21}$$

Then the hyperbolic functions of the oscillator are defined by the limits

$$\varphi = tam\tau, \lim_{k \to 1-\alpha} sd(\tau, k, \alpha) \triangleq tsd(\tau, \alpha), \tag{9.22}$$

$$\lim_{k \to 1-\alpha} ad(\tau, k, \alpha) = \sqrt{1 - tsd^2(\tau, \alpha)} \triangleq tad(\tau, \alpha), \tag{9.23}$$

$$\lim_{k \to 1-\alpha} hd(\tau, k, \alpha) = \sqrt{[(1-\alpha)\cdot tad(\tau, \alpha) + 1]^2 - \alpha^2} \triangleq thd(\tau, \alpha), \tag{9.24}$$

where $0 \le \alpha < 1$ and $0 \le k < 1 - \alpha$. The graphs of the hyperbolic functions $thd(t)$ and $tsd(t)$ are shown in Fig. 9.3.

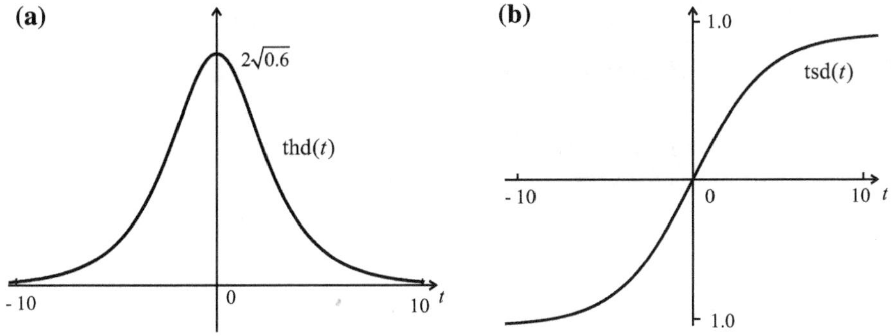

Fig. 9.3 Graphs of the hyperbolic function for $\alpha = 0.4$: **a** for $thd(t)$ and **b** for $tsd(t)$

9.3.2 Properties

In the same way as for the elliptic functions of the first kind, the properties of the hyperbolic functions are listed as the following.

1. **Identities**

$$tsd^2\tau + tad^2\tau = 1,$$
$$thd^2\tau + \alpha^2 = [1 + (1-\alpha)\cdot tad\tau]^2.$$

2. **Parity**
   It can be seen that $tsd\tau$ is an odd function of τ and $thd\tau, tad\tau$ are even functions of τ due to the odd property of φ, that is

$$tsd(-\tau) = -tsd\tau,$$
$$thd(-\tau) = thd\tau,$$
$$tad(-\tau) = tad\tau.$$

3. **Differentiability**
   The differentiation of the hyperbolic functions are obtained by the definitions and identities as follows

$$\frac{\partial\varphi}{\partial\tau} = \frac{thd\tau}{1 + (1-\alpha)\cdot tad\tau},$$
$$\frac{\partial tsd\tau}{\partial\tau} = \frac{tad\tau \cdot thd\tau}{1 + (1-\alpha)\cdot tad\tau},$$
$$\frac{\partial tad\tau}{\partial\tau} = -\frac{tsd\tau \cdot thd\tau}{1 + (1-\alpha)\cdot\sqrt{1 - tsd^2\tau}},$$
$$\frac{\partial thd\tau}{\partial\tau} = -(1-\alpha)\cdot tsd\tau.$$

9.3.3 Analytical Expressions of the Homoclinic Orbits

The application to the homoclinic orbits of the SD oscillator is immediate. When $k = 1 - \alpha$, the homoclinic orbits can be written

$$x = \pm thd(\tau, k, \alpha). \tag{9.25}$$

The first and second derivatives of the solution are obtained by the differential properties of the hyperbolic functions, namely

$$\frac{dx}{d\tau} = \mp(1-\alpha)\cdot tsd\tau, \tag{9.26}$$

$$\begin{aligned}\frac{d^2x}{d\tau^2} &= \mp thd\tau\cdot(1-\frac{1}{\sqrt{thd^2\tau+\alpha^2}})\\ &= -x\cdot\left(1-\frac{1}{\sqrt{x^2+\alpha^2}}\right),\end{aligned} \tag{9.27}$$

which leads to that the explicit analytical expression of $x = \pm thd(\tau, k, \alpha)$.

9.4 Irrational Elliptic Functions of the Second Kind

9.4.1 Definitions

We now study the case $\alpha \geq 1$. When, α is strictly larger than 1, (0, 0) is the unique center. The phase portrait is shown in Fig. 9.1c. Recall that the hamiltonian is given by

$$H = \frac{1}{2}k^2 = \frac{1}{2}y^2 + U(x) \geq \frac{(1-\alpha)^2}{2}, \tag{9.28}$$

where $k > \alpha - 1$ and $\alpha \geq 1$. The time τ from $(x_0, 0)$ to (x, y) along the lower branch of the periodic orbits is still given by the integral

$$\tau = -\int_{x_0}^{x} \frac{1}{\sqrt{k^2 - \left(\sqrt{t^2+\alpha^2}-1\right)^2}}dt, \tag{9.29}$$

where x_0 is the maximum intersect point of the orbits with x-axis. Let

$$x = \pm\sqrt{(k\cdot\cos\varphi + 1)^2 - \alpha^2}, \tag{9.30}$$

where $\varphi \in [0, \varphi_0]$ for $x \in [0, x_0]$ and $\varphi_0 = \arccos(\frac{\alpha-1}{k})$. Again the integral (9.29) can be simplified as

$$\tau = \int_0^{\varphi} \frac{1}{\sqrt{1-\frac{\alpha^2}{(1+k\cdot\cos\eta)^2}}}d\eta, \tag{9.31}$$

where the time variable τ is a function of k, φ, α.

Let us introduce a function $W(k,\alpha)$ by

$$W(k,\alpha)=\int_0^{\varphi_0}\frac{1}{\sqrt{1-\frac{\alpha^2}{(1+k\cdot\cos\varphi)^2}}}d\varphi, \tag{9.32}$$

and define

$$\varphi = Am\tau, \tag{9.33}$$

As previously, φ is defined as the angular of τ, and the irrational elliptic functions of the second kind are defined as

$$Sd(\tau,k,\alpha)\triangleq \sin\varphi=\sin(Am\tau), \tag{9.34}$$

$$Ad(\tau,k,\alpha)\triangleq \cos\varphi=\cos(Am\tau), \tag{9.35}$$

$$Hd(\tau,k,\alpha)\triangleq \sqrt{(1+k\cdot\cos(Am\tau))^2-\alpha^2}. \tag{9.36}$$

9.4.2 Properties

The properties of the irrational elliptic functions of the second kind defined above are listed below, using, as for elliptic functions of the first kind and for hyperbolic functions, the simplified notations: $Hd\tau$, $Sd\tau$, $Ad\tau$ instead of $Hd(\tau,k,\alpha)$, $Sd(\tau,k,\alpha)$, $Ad(\tau,k,\alpha)$, without confusion.

1. **Identities**

$$\begin{aligned} Sd^2\tau+Ad^2\tau&=1,\\ Hd^2\tau+\alpha^2&=(1+k\cdot Ad\tau)^2. \end{aligned}$$

2. **Parities**
It is easily seen that $Sd\tau$ is an odd function of τ and that $Hd\tau$ and $Ad\tau$ are even functions of τ due to the odd property of φ, which reads

$$\begin{aligned} Sd(-\tau)&=-Sd\tau,\\ Hd(-\tau)&=Hd\tau,\\ Ad(-\tau)&=Ad\tau. \end{aligned}$$

3. **Differentiability**
The derivatives of φ, $Sd\tau$, $Ad\tau$, $Hd\tau$ with respect to the time variable τ are readily obtained as previously

$$\frac{\partial \varphi}{\partial \tau} = \frac{Hd\tau}{1 + k \cdot Ad\tau},$$
$$\frac{\partial Sd\tau}{\partial \tau} = \frac{Ad\tau \cdot Hd\tau}{1 + k \cdot Ad\tau},$$
$$\frac{\partial Ad\tau}{\partial \tau} = -\frac{Sd\tau \cdot Hd\tau}{1 + k \cdot \sqrt{1 - Sd^2\tau}},$$
$$\frac{\partial Hd\tau}{\partial \tau} = -k \cdot Sd\tau.$$

4. **Periodicity**
   For the symmetry of the phase portraits, the period T of the solution is four times the time taken on the trajectory from $x = x_0$ to $x = 0$, which leads to the fact that the period of $Hd\tau$ is $4W(k, \alpha)$, that is

$$Hd(\tau + 4W(k, \alpha)) = Hd\tau,$$
$$Hd(-\tau + 4W(k, \alpha)) = -Hd\tau.$$

   The differential properties of the irrational elliptic functions of the second kind lead to the fact that the period of $Sd\tau$ is also $4W(k, \alpha)$ by the following calculation

$$\frac{\partial Hd(\tau + 4W(k.\alpha))}{\partial \tau} = -k \cdot Sd(\tau + 4W(k, \alpha)) = \frac{Hd\tau}{\partial \tau} = -k \cdot Sd\tau. \quad (9.37)$$

   that is

$$Sd(\tau + 4W(k, \alpha)) = Sd\tau,$$
$$Sd(-\tau + 4W(k, \alpha)) = -Sd\tau,$$

   and the period of $Ad\tau$ is also $4W(k, \alpha)$, that is

$$Ad\tau = \sqrt{1 - Sd^2\tau} = \sqrt{1 - Sd^2(\tau + 4W(k, \alpha))} = Ad(\tau + 4W(k, \alpha)), \quad (9.38)$$

$$Ad(\tau + 4W(k, \alpha)) = Ad\tau,$$
$$Ad(-\tau + 4W(k, \alpha)) = -Ad\tau.$$

Special useful values of the irrational elliptic functions of the second kind are listed in Table 9.3.

Table 9.3 Special values of the irrational elliptic functions of the second kind

$\varphi = 0, \tau = 0$	$\varphi = \pi, \tau = W(k, \alpha)$
$Sd(0) = 0, Ad(0) = 1$	$Hd(W) = 0$
$Hd(0) = \sqrt{(1+k)^2 - \alpha^2}$	$Ad(W) = \frac{\alpha-1}{k}, Sd(W) = \sqrt{1 - \frac{(\alpha-1)^2}{k^2}}$

9.4.3 Analytical Expressions of the Periodic Orbits

In the same way as when dealing with elliptic functions of the first kind and hyperbolic functions, the analytical expressions of the corresponding periodic orbits of the SD oscillator follow when $k > \alpha - 1$, and are given by

$$x = \pm Hd(\tau, k, \alpha). \tag{9.39}$$

As a matter of fact, the first and the second derivatives of the solution are obtained by the differential properties of the irrational elliptic functions of the second kind just listed above, namely

$$\frac{dx}{d\tau} = \mp k \cdot Sd\tau, \tag{9.40}$$

$$\begin{aligned} \frac{d^2x}{d\tau^2} &= \mp Hd\tau \cdot (1 - \frac{1}{\sqrt{Hd^2\tau + \alpha^2}}) \\ &= -x \cdot \left(1 - \frac{1}{\sqrt{x^2 + \alpha^2}}\right), \end{aligned} \tag{9.41}$$

so that $x = \pm Hd(\tau, k, \alpha)$ is explicitly the analytical solution of periodic orbits outside the center $(0, 0)$. Moreover, the period of the these orbits is given by

$$T_k^o = 4W(k, \alpha).$$

9.5 Chaotic Threshold

Let us come back to the perturbed SD oscillator involving a linear viscous damping and an external harmonic forcing:

$$x'' + 2\xi x' + x(1 - \frac{1}{\sqrt{x^2 + \alpha^2}}) = f_0 \cos(\omega\tau), \tag{9.42}$$

which is rewritten as

$$\begin{cases} x' = y \\ y' = -2\xi y - x(1 - \frac{1}{\sqrt{x^2+\alpha^2}}) + f_0 \cos(\omega\tau). \end{cases} \quad (9.43)$$

9.5.1 Melnikov's Method of the Homoclnic Orbits

The Melnikov function [2, 4] of system (9.43) is given by

$$\begin{aligned} M_{\pm}(\tau_1) &= \int_{-\infty}^{\infty} f(\mathbf{x}_{\pm}(\tau)) \wedge g(\mathbf{x}_{\pm}(\tau), \tau + \tau_1) d\tau \\ &= \int_{-\infty}^{\infty} -2\xi y_{\pm}(\tau)^2 + y_{\pm}(\tau) f_0 \cos[\omega(\tau + \tau_1)] d\tau \\ &= -2\xi \int_{-\infty}^{\infty} y_{\pm}(\tau)^2 d\tau + f_0 \int_{-\infty}^{\infty} y_{\pm}(\tau) \cos[\omega(\tau + \tau_1)] d\tau. \quad (9.44) \end{aligned}$$

Due to the symmetry of the system, the discussion can concentrate on $M_+(\tau_1)$.

When $0 \le \alpha < 1$, the displacement and the velocity on the homoclinic orbit are given by

$$x = thd(\tau), \qquad y = x' = -(1-\alpha)tsd(\tau) \quad (9.45)$$

Then, the Melnikov function of the system reads as follows

$$M_{\pm}(\tau_1) = -4\xi A \pm 2 f_0 \sin \omega\tau_1 B, \quad (9.46)$$

with

$$A = (1-\alpha)^2 \int_0^{\infty} tsd(\tau)^2 d\tau,$$

$$B = (1-\alpha) \int_0^{\infty} tsd(\tau) \sin \omega\tau d\tau = (1-\alpha)\pi i Res[tsd(\tau) \sin \omega\tau, iH(\alpha)],$$

where $Res[f, z]$ stands for the residue of f at point z.

It can be seen that $M_{\pm}(\tau_1) = 0$ has a simple zero for τ_1 if and only if the following inequality holds

$$|\frac{f_0}{\xi}| \ge 2\frac{|A|}{|B|}. \quad (9.47)$$

When the inequality is satisfied, the Poincaré map of Eq. (9.43) might be chaotic in the sense of Smale horseshoe, the intersection between the stable and unstable manifolds. The thresholds for the occurrence of chaos are plotted in Fig. 9.4a: the solid line for $\alpha = 0, k = 1$, the thin line for $\alpha = 0.4, k = 0.6$ and the dashed line for $\alpha = 0.8, k = 0.2$, the meaning of these curves being that chaos may occur if the parameters are above the corresponding curves.

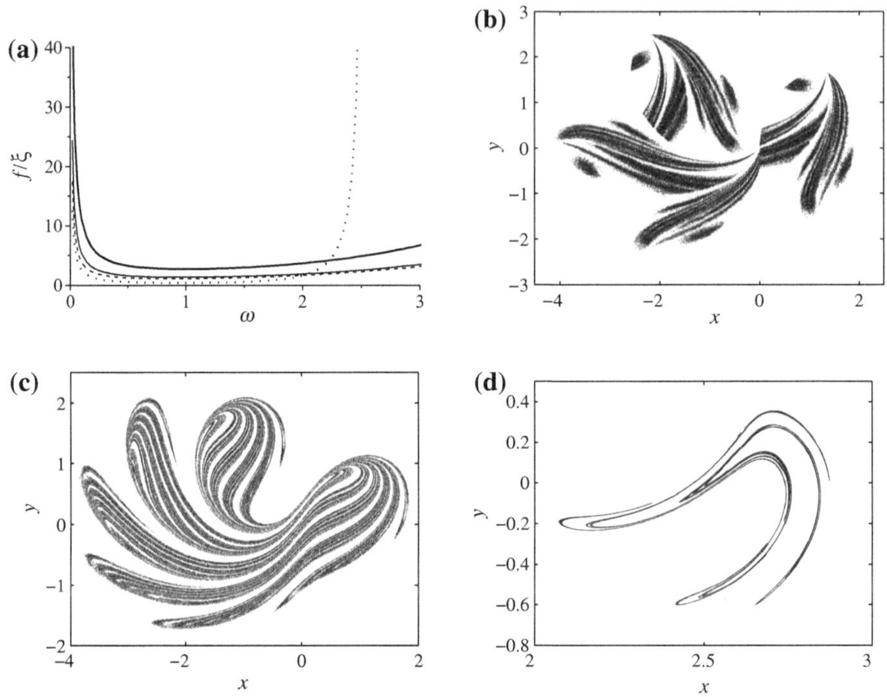

Fig. 9.4 Chaotic thresholds and the attractors: **a** Chaotic thresholds of the SD oscillator plotted with the *solid curve* for $\alpha = 0, k = 1$, the *thin* for $\alpha = 0.4, k = 0.6$, the *dashed* for $\alpha = 0.8, k = 0.2$ and the *dotted* for $\alpha = 1, k = 0.4$, respectively; **b** attractors for $\alpha = 0$, $\xi = 0.015$, $f_0 = 1.1$, $\omega = 1.3$; **c** for $\alpha = 0.4$, $\xi = 0.02$, $f_0 = 0.70$, $\omega = 1.06$ and **d** for $\alpha = 1$, $\xi = 0.015$, $f_0 = 0.80$, $\omega = 1.3$, respectively

9.5.2 Melnikov's Method of the Subharmonic Orbits

When $\alpha = 1$, the subharmonic Melnikov function of the subharmonic orbits is

$$M_1^+(\tau_1) = M_1^-(\tau_1) = \int_0^T -2\xi(y_k(\tau))^2 + y_k(\tau) f_0 \cos[\omega(\tau + \tau_1)]d\tau, \tag{9.48}$$

which can be rewritten as

$$M_1^+(\tau_1) = -8\xi P + 4 f_0 Q, \tag{9.49}$$

where

$$P = k^2 \int_0^{W(k,\alpha)} Sd(\tau)^2 d\tau,$$

and

$$Q = k \int_0^{W(k,\alpha)} Sd(\tau)sin(\omega\tau)d\tau.$$

It can be seen that $M_1^+(\tau_1) = 0$ has a simple zero for τ_1 if and only if the following inequality holds

$$|\frac{f_0}{\xi}| \geq 2\frac{|P|}{|Q|}. \quad (9.50)$$

When the inequality is satisfied, the subharmonic orbits may occur in the sense of Smale horseshoe map. The threshold boundary is plotted for $\alpha = 1, k = 0.4$ with the dotted curve in Fig. 9.4a.

9.5.3 *Numerical Experiments with Chaotic Attractors*

In this section, numerical analysis is carried out by using the fourth order Runge–Kutta method [5] to the system (9.42) to show the efficiency of the threshold given in (9.46). Figure 9.4 shows the corresponding attractors, Fig. 9.4b for $\alpha = 0, \xi = 0.015$, $f_0 = 1.1$, $\omega = 1.3$; Fig. 9.4c for $\alpha = 0.4$, $\xi = 0.02$, $f_0 = 0.70$, $\omega = 1.06$ and Fig. 9.4d for $\alpha = 1$, $\xi = 0.015$, $f_0 = 0.80$, $\omega = 1.3$. The parameter used in Fig. 9.4b–d are carefully chosen to meet the criterion given in (9.46) for $\alpha = 0$, $\alpha = 0.4$, and $\alpha = 1.0$, respectively, showing the reliability of the chaotic criterion obtained analytically depends on the introduced irrational hyperbolic function for the homoclinic orbits.

9.6 Summary

In this chapter, two types of irrational elliptic functions and a kind of irrational hyperbolic functions have been defined for the SD oscillator controlled by the geometrical parameter α, which overcome the limitation of the conventional mathematical tools due to the lack of analytical solutions for the system with irrational nonlinearity. All the solutions of the unperturbed SD oscillator have been formulated by the irrational elliptic functions and the homoclinic functions defined in this chapter. As one of the application, the chaotic threshold has been obtained using Melnikov functions based on the hyperbolic functions defined in this chapter, which is valid for $0 < \alpha < 1$. The method is also applicable for the homoclinic-like orbits and the degenerate case, which are the limits as $\alpha \rightarrow 0$ and $\alpha \rightarrow 1$, respectively.

References

1. Cao, Q. J., Wiercigroch, M., Pavlovskaia, E. E., Grebogi, C., & Thompson, J. M. T. (2008). Piecewise linear approach to an archetypal oscillator for smooth and discontinuous dynamics. *Philosophical Transactions of the Royal Society, A-Mathematical, Physical and Engineering Sciences, 366*(1865), 635–652.
2. Guckenheimer, J., & Holmes, P. (1983). *Nonlinear Oscillation, dynamical System and bifurcation of vector fields*. New York: Springer.
3. Whittaker, E. T., & Watson, G. N. (1952). *A Course of modern analysis*. Cambridge: Cambridge University Press.
4. Garcia-Margallo, J., & Bejarano, J. D. (1998). Melnikov's method for nonlinear oscillators with nonlinear excitations. *Journal of Sound and Vibration, 212*(2), 311–319.
5. Cartwright, J. H. E., & Piro, O. (1992). The dynamics of Runge-Kutta methods. *International Journal of Bifurcations Chaos, 2*(3), 427–449.

Chapter 10
Cell Mapping Method

By comparison with calculation methods built upon the specificity of the restoring force of the SD oscillator, which suggests either a triple linearization, a semi-analytical method, or an analytical method using elliptic functions, the cell mapping method is general for the investigation of any nonlinear dynamical system. It has nevertheless been used successfully in the case of the SD oscillator, and it seemed useful to present the corresponding results. The basic concepts of this method are close to algebra and graph theory, which requires, for the present book to be self-contained, to begin by several definitions and theoretical properties.[1]

10.1 Introduction

The cell mapping method [1–3] and the generalized cell mapping method [4–6] have previously been applied to well-known problems such as calculating the periodic solutions of van der Pol equation or investigating the qualitative properties of Duffing equation. Here the present chapter focusses on the behaviour of the SD oscillator, dealing in particular with the occurrence of crisis [7, 8] which will be defined as collisions between chaotic saddles and periodic solutions.

Firstly, the basic concepts of the cell mapping and the generalized cell mapping are explained with the relationship to the system. Then the basic properties of the classification of the state cells and the digraph procedure of the algorithm applied to the SD oscillator are given. Using this approach, an interior crisis is studied by means of the generalized cell mapping digraph method. Attractors, basins, basin boundaries and unstable solutions are obtained once through a global analysis at low computational cost. Moreover, the approach of a periodic attractor to a chaotic

[1]This chapter is due to the contribution by Professor Ling Hong, Mr. Jianhua Wang, Ms. Xiaoyan Zhang, Department of Engineering Mechanics, Xian Jiaotong University, Xian 710049 China.

Q. Cao and A. Léger, *A Smooth and Discontinuous Oscillator*, Springer Tracts in Mechanical Engineering, DOI 10.1007/978-3-662-53094-8_10

saddle in its basin attractor before an interior crisis, and the collision of the periodic attractor with the chaotic saddle when the crisis occurs, are explicitly shown.

10.2 Generalized Cell Mapping Method

10.2.1 Cell Mapping Method

As an introduction to cell mapping, the simplest case of the Simple Cell Mapping method reads as follows. Let a given dynamical system be described by an ordinary differential equation written formally

$$\mathscr{L}X = g(X) \text{ together with initial data,} \tag{10.1}$$

where $\mathscr{L}$ is some differential operator and X the state variable assumed to lie in some subset Ω of R^n. Assume for simplicity that Ω is a bounded parallelepipoid and let us divide this domain into parallelepipedic cells indexed by an integer j. If for example $\Omega \subset R^2$ with coordinates x_i, $i \in \{1, 2\}$, then Ω is defined by $[x_1^\ell, x_1^m] \times [x_2^\ell, x_2^m]$, where ℓ and m stand for both ends of an interval on each axis. Each interval is divided into M_i subintervals of length

$$h_i = \frac{x_i^m - x_i^\ell}{M_i},$$

so that the number of rectangular cells in Ω is $M = M_1 M_2$, which means that the integer M belongs to $\{1, 2, \dots M\}$.

Then, the fundamental statement of cell mapping is that the state of system (10.1) is given by a value of integer j and no longer by $X(t)$, which in other words means that X belongs to cell j at time t, or that the volume of cell j gives an approximation of $X(t)$.

Then the time interval of calculation is divided into time steps of length Δt indexed by n, and let the cell j be denoted by $\xi(n)$. Then the cell mapping procedure consists in defining a map $\mathscr{C}$, $\mathscr{C} : N \longrightarrow N$ such that:

$$\xi(n+1) = \mathscr{C}(\xi(n)).$$

The last step is the center point method which consists in:

- *Step 1.* calculating the center point of cell $\xi(n)$,
- *Step 2.* integrating Eq. (10.1) over Δt with the center point of cell $\xi(n)$ as initial data,
- *Step 3.* determining the index of the cell which contains the end point of this portion of trajectory.

Remark 10.1 *(i)* Let the index 0 denote the so-called sink, i.e. the open set R/Ω, then the map $\mathscr{C}$ satisfies $\mathscr{C}(0)) = 0$.

(ii) Assume there exists an integer k such that $\mathscr{C}^k(\xi) = \xi$, where $\mathscr{C}^k$ means that the map $\mathscr{C}$ has been iterated k times, then the trajectory is periodic with period k.

(iii) Since the number of cells M is finite by construction, either a periodic trajectory or the sink will be reached in a finite number of steps of the center point method.

(iv) The length h_i being given, the time interval Δt should be large enough for the end point not belonging to the same cell as the initial center point in general.

10.2.2 Generalized Cell Mapping

The generalized cell mapping method is based upon the fact that a given cell may have several images. This has nothing to do with a possible ill-posedness of problem (10.1), of course it is assumed to be well-posed, but it can be simply due to the approximation given by the size of the cell and that another initial data than the center point might lead to an end point in another cell. So that the state of the system is no longer given by an integer which is the index of a cell, but by a probability law. Let us now state this fact more rigorously.

The most direct and versatile way is the straightforward sampling method which in the present case consists in the following steps:

- *Step 1.* choosing L points distributed in a cell,
- *Step 2.* integrating Eq. (10.1) L times over Δt with all these L points as initial data,
- *Step 3.* determining the cells which contain the end points,
- *Step 4.* deducing the probability for a given end point to be in a cell.

Performing these four steps for all the indexes j leads to a so-called homogeneous finite Markov chain, so that the generalized cell mapping of the dynamical system is given by

$$\mathbf{p}(n+1) = \mathbf{P}\mathbf{p}(n) \quad \text{or} \quad \mathbf{p}(n) = \mathbf{P}^{\mathrm{n}}\mathbf{p}(0) \tag{10.2}$$

where the map $\mathscr{C}$ and the state ξ have been replaced as follows:

P denotes the one-step transition probability matrix, $\mathbf{p}^n$ denotes the n-step transition probability matrix, $\mathbf{p}(0)$ denotes the initial probability distribution vector, and $\mathbf{p}$ (n) denotes the n-step probability distribution vector. Let p_{ij} be the (i, j) element of **P**. It is called the one-step transition probability from cell j to cell i. Let p^n_{ij} be the (i, j)th element of $\mathbf{p}(n)$. It is called the n-step transition probability from cell j to cell i.

10.3 Basic Concepts and Classifications of State Cells

In order to prove the existence of two results of the self-cycling set and the persistent self-cycling set directly in cell state space, state cells are classified afresh on the basis of the classical theory of Markov chains [9].

Definition 10.1 (*Leads to*) If it is possible, through the mapping to go from cell j to cell i, we say cell j leads to cell i, symbolically $j \Rightarrow i$. Analytically, cell j leads to cell i if and only if there exists a positive integer m such that $p_{ij}^{(m)} > 0$. Similarly, if it is possible through one-step to go from cell j to cell i, we say that cell j leads to cell i in one step, symbolically $j \rightarrow i$.

Definition 10.2 (*Communicates with*) If cell i leads to cell j and cell j leads to cell i, then we say that cell i communicates with cell j or cell j communicates with cell i. This will be denoted by $i \Leftrightarrow j$.

Definition 10.3 A set of cells in which every member communicates with all the other members is a self-cycling set.

Definition 10.4 Assume Z stands for the set consisting of all state cells in the cell state space, N_+ stands for the set of all positive integers and C for a set of state cells, with $C \subset Z$. C is called a closed set if for every $j \in C$ and $i \notin C$ there exists $p_{ij} = 0$. Analytically, C is a closed set if and only if for every $j \in C$, $i \notin C$ and $n \in N_+$, there exists $p_{ij}^{(n)} = 0$.

Definition 10.5 If a self-cycling set is a closed set, it is called a persistent self-cycling set or simply a persistent set. A cell belonging to a persistent set is called a persistent cell.

Definition 10.6 If a persistent set contains only one cell j, then j is called an absorbing cell. For an absorbing cell j there exist $p_{jj} = 1$ and $p_{ij} = 0$ for every $i \neq j$.

Persistent sets represent possible long-term stable motions of the systems. In general, they are attractors. Therefore, they are undoubtedly the most significant entities of the generalized cell mapping of dynamical systems.

Definition 10.7 A self-cycling set that is not a closed set is called a transient self-cycling set. A cell belonging to a transient self-cycling set is called a transient self-cycling cell.

Transient self-cycling sets are usually associated with unstable fixed point and periodic solutions of the systems. For the overall analysis of the systems this is a significant result.

Definition 10.8 All the cells which are not persistent cells are called transient cells.

Definition 10.9 If a transient cell j leads to a persistent set Z_p, then Z_p is called a domicile of cell j.

Transient cells may be classified according to the number of domiciles they have.

Definition 10.10 Transient cells which have only one domicile are called single-domicile transient cells, and those which have more than one are called multiple-domicile transient cells.

The domiciles of the transient cells provide a very natural way of partitioning the transient cells into transient sets.

Definition 10.11 Let N_p be the number of persistent sets. These sets, as sets of cells, can be denoted by $Z_{pi},\ i = 1, 2 \ldots, N_p$. The set consisting of all the single-domicile transient cells having the ith persistent set as domicile is called the ith single-domicile transient set and is denoted by $B(i)$, where $i = 1, 2 \ldots, N_p$. It populates the domain of attraction of the ith persistent set. The set consisting of all multiple-domicile transient cells having i^{th} and j^{th} persistent sets as their only home is called the $(i, j)^{th}$ two-domicile transient set and is denoted by $B(i, j)$ where $i, j = 1, 2 \ldots, N_p$ and $i \neq j$. The area populated by this set $B(i, j)$ is called the boundary regions between the ith and j^{th} domains of attraction. Similarly, the $(i, j, k)^{th}$ three-domicile transient set $B(i, j, k)$, the $(i, j, k, l)^{th}$ four-domicile transient set $B(i, j, k, l)$, and so on for $i, j, k, l = 1, 2, \ldots, N_p$ and $i \neq j \neq k \neq l$.

For other properties such as the period, the limit probability distribution, the probability absorbing, and the expected absorption time, the reader can see [1].

10.4 Generalized Cell Mapping Digraph Method

Based on set theory, the correspondence between generalized cell mapping of dynamical systems and digraphs is theoretically justified. Thus, digraphs are directly introduced in dynamical systems. The availability of powerful algorithms from digraphs allows to determine the evolution of the global systems. The essential concepts and terminology of sets and digraphs, have been presented in [10, 11].

10.4.1 The Generalized Cell Mapping Systems and the Digraphs

Assume the number of all the state cells which are considered is equal to N. The set Z consisting of all the state cells is denoted by $\{Z = z_1, z_2, \ldots, z_j, \ldots, z_N\}$. Binary relationships are defined on the set Z. The relation 'leads to' and 'in one step' is defined on the set Z and is denoted by R_{a1}. Consider the diagram of relation R_{a1}. This is the weighted digraph and is denoted by $DR_{a1} = (V, A)$ where $V = Z$ and $Z = R_{a1}$. For the digraph DR_{a1}, if $z_j R_{a1} z_i$, there exists the arc (z_j, z_i) with the nonzero weight $p_{i,j}$. Such a digraph is called a one-step mapping digraph. The adjacency matrix of the one-step mapping digraph is denoted by $\mathbf{MR}_{a1} = [m_{ij}]$ and the one-step transition probability matrix $\mathbf{P} = [p_{ij}]$. The relationship between these two matrices is

$$m_{ij} = \begin{cases} 1 \ if & p_{ij} > 0, \\ 0 \ if & p_{ij} = 0. \end{cases} \tag{10.3}$$

The transitive closure of relation R_{a1} is denoted $t(R_{a1})$,

$$t(R_{a1}) = \bigcup_{1}^{N} R_{a1}^{i}. \tag{10.4}$$

The transitive closure set for each state cell z_i in Z is denoted by $\{z_i^+\},\ i = 1, \ldots, N\}$. Namely, $\{z_i^+\}$ is the set consisting of all elements z_i leads to and can be written as follows

$$\{z_i^+\} = \{z_j | z_i t(R_{a1}) z_j,\ z_i \in Z,\ z_j \in Z\}, \tag{10.5}$$

Or simply

$$\{z_i^+\} = \{z_j | z_i \Rightarrow z_j,\ z_i \in Z,\ z_j \in Z\}, \tag{10.6}$$

The relation *communicates with* is defined on Z and is denoted by R_c. The reflexive closure relation of R_c is denoted by $r(R_c)$. We have

$$r(R_c) = I_A \cup R_c. \tag{10.7}$$

It can be checked that the relation $r(R_c)$ is an equivalence relation. Thus $r(R_c)$ determines uniquely a partition on Z. This partition is denoted by $\pi(Z) = \{Z_\alpha\},\ \alpha = 1, \ldots, N_c$, N_c denotes the number of the blocks in $\pi(Z)$, and we denote by n_α the number of elements in each block Z_α. To obtain the method of the classification on Z, two results are necessary, given here without proof (see [12]).

Theorem 10.1 *Z_α is a self-cycling set if and only if $Z_\alpha \not\subseteq \{i\}^+$ for every $i \in Z_\alpha$.*

Theorem 10.2 *A self-cycling set Z_α is a closed set if and only if for every $i \in Z_\alpha$, there exists $Z_\alpha = \{i\}^+$.*

Based on Theorems 10.1 and 10.2, the following conclusions hold:

a.) Z_α is the single cell which consists of only one element (i.e. $n_\alpha = 1$), let $Z_\alpha = \{i\}$.
If $Z_\alpha = \{i\} = \{i\}^+$, cell i is an *absorbing cell.*
In view of the one-step mapping digraph, i is an absorbing cell if and only if it has a self-loop with weight one.
If $Z_\alpha = \{i\} \subset \{i\}^+$, cell i is a *transient self-cycling cell.*
On the one-step mapping digraph, i has a self-loop with weight less than one and its out-degree is more than one.
If $Z_\alpha = \{i\} \not\subseteq \{i\}^+$, namely $i \notin \{i\}^+$ (i does not lead to i), i is a *transient cell.*

b.) Z_α consists of more than one element (i.e. $n_\alpha > 1$).
Z_α is a *self-cycling set.*
On the one-step mapping digraph, the induced subdigraph $D[Z_\alpha]$ is a strongly connected component.

Definition 10.12 The digraph that results after contracting all the arcs in each strongly connected component of the one-step mapping digraph is called the condensed digraph of the one-step mapping digraph DR_{a1} and is denoted by D_c. The vertex in D_c which corresponds to the connected component strongly is called the condensed vertex of the component.

- If for every $i \in Z_\alpha$ there exists $Z_\alpha = \{i\}^+$, Z_α is a *persistent set.*
  On the condensed digraph, the out-degree of the condensed vertex of the component Z_α is zero.
- If for every $i \in Z_\alpha$ there exists $Z_\alpha \subset \{i\}^+$, Z_α is a *transient self-cycling set.*
  On the condensed digraph, the out-degree of the condensed vertex of Z_α is nonzero.

Following the previous analysis, we consider the set $\pi(Z) = \{Z_\alpha\}$ where $\alpha = 1, \ldots, N_c$. The representative element of Z_α is denoted by k_α, where $\alpha = 1, \ldots, N_c$. The set consisting of all the representative elements is denoted by K (i.e. $K = \{k_1, k_2, \ldots k_\alpha, \ldots k_{N_c}\}$). The relation *leads to* is defined on K and is denoted by R_α. R_α is irreflexive and transitive on K. R_α is a quasi order (i.e. strictly partial order) relationship. K is a quasi ordered set.

Definition 10.13 For a quasi-ordered set K, the process of extending quasi-ordering to total ordering is called a topological sorting. The total ordering is called a topological enumeration of K.

Carrying out the topological sorting for the set K, gives the overall transient evolution process and the ordered relationship of all the transient cells, and a classification of the transient cells as transient sets.

Topological sorting is a problem that fits naturally into the terminology of binary relationships and digraphs. There are many digraph algorithms for solving topological sorting problems.

10.4.2 Qualitative and Quantitative Properties

Global properties can be divided into two categories. The first category consists of qualitative or topological ones. In view of the two relationships *leads to* and *communicates with*, it follows that whether j leads to i only depends on whether there exists a direct path from i to j, but it is not related to the weight p_{ij} of the corresponding arc (i, j). Hence, the properties of the dynamical systems based on the relationships 'leads to' and 'communicates with' are called qualitative (or topological) properties. Namely, the qualitative properties are related to the locations of the nonzero elements of the transition probability matrix $\mathbf{P}$, but not to the value of nonzero elements.

Belonging to the second category are properties which are quantitative in nature, such as for example, the limiting probability distribution, the absorption probability and the expected absorption time. The significance of dividing the global properties into these two categories lies in the fact that certain information of the dynamical system under investigation which is needed for the determination of quantitative properties is not needed for studying the qualitative properties. This leads to a dichotomy in the method of treatment. Thus, in the analysis of the qualitative properties, only Boolean operations are used.

10.4.3 Construction of the Algorithm

After digraphs have been introduced in dynamical systems, very powerful algorithms in the field of graphs and digraphs can be adopted for the analysis of the systems. With respect to the building of concrete algorithms, different algorithms can be generated. The whole algorithm process is given as follows.

10.4.3.1 An Algorithm for Determining All the Self-cycling Sets

Warshall's algorithm [11, 13] by which a transitive closure matrix can be obtained directly from an adjacency matrix of a digraph, is already a fairly simple and common one. Its time complexity degree, considering the worst situation is $O(N^3)$ where N is the number of all the vertices of the digraph. Only $N \times N$ of computer memory units are occupied, and even in the process of operating, no extra memory units are used. Only Boolean operations are adopted. Thus Warshall's algorithm is absolutely accurate, reliable and time-saving, and superior to the known others. Because the Boolean matrix used for calculating the transitive closure is the one over the set $\{0, 1\}$, each element of the Boolean matrix can be stored by one bit of memory unit, namely, the vertices of the digraph are indicated by bit vectors, thus, the storage space is undoubtedly ten times less than in the previous case.

The concrete steps for determining all the self-cycling sets are as follows:

- *Step 1.* The one-step transition probability matrix $\mathbf{P} = [p_{ij}]$ is obtained by using the sampling method.
- *Step 2.* The adjacency matrix $\mathbf{MR}_{a1} = [m_{ij}]$ of the one-step mapping digraph is obtained by

$$m_{ij} = \begin{cases} 1 \ if \quad p_{ij} > 0, \\ 0 \ if \quad p_{ij} = 0. \end{cases} \tag{10.8}$$

- *Step 3.* The transitive closure matrix $\mathbf{TR}_{a1} = [t_{ij}]$ (i.e. the relationship matrix of $t(R_{a1})$) is calculated by using the standard Warshall subroutine, and the transitive closure set $\{i\}^+$ of every vertex i is recorded.

- *Step 4.* The relationship matrix $\mathbf{Q} = [q_[ij]]$ of the relationship 'communicates with' R_c is calculated by $\mathbf{TR}_{a1} \bigodot \mathbf{TR}_{a1}^T$, namely $q_{ij} = t_{ij}.t_{ji}$, where the superscript T denotes the transposed operation of the matrix, the symbol $\bigodot$ stands for the element product, the symbol '.' denotes the common product. Thus, it follows

$$q_{ij} = \begin{cases} 1 \ if \ i \Leftrightarrow j, \\ 0 \ if \text{ otherwise.} \end{cases} \tag{10.9}$$

- *Step 5.* Each set consisting of all the vertices that have the same raw vector (or the same column vector) on the matrix $\mathbf{Q}$ is a strongly connected component of the one-step mapping digraph, namely, a self-cycling set. In this way, all the self-cycling sets [14] can be easily recorded.

For computing the period of a self-cycling set, the self-cycling set can be regarded as a homogeneous finite irreducible Markov chain. Then, for the self-cycling set, we can similarly obtain its adjacency matrix $\mathbf{M}_{scs}$. Hence, the period of the self-cycling set is denoted by d_{scs} and can be described by a finite computation formula as follows:

$$d_{scs} = g.c.d\{i \mid Trace\mathbf{MQ}_{scs}^i = 1, \ 1 \leq i \leq n_\alpha\}. \tag{10.10}$$

where n is the number of all the cells in the a self-cycling set. $\mathbf{M}_{scs}^i$ is the ith power of Boolean $\mathbf{M}_{scs}$. $Trace\mathbf{M}_{scs}^i$ represents the traces of $\mathbf{M}_{scs}^i$, namely the Boolean sum $\vee$ of all diagonal elements of $\mathbf{M}_{scs}^i$ where the Boolean sum is $0 \vee 0 = 0$, $0 \vee 1 = 1$, $1 \vee 0 = 1$, $1 \vee 1 = 1$, $g.c.d$ represents the greatest common divisor.

The concrete steps for determining the period of a self-cycling set are the following:

- *Step 1.* Matrix $\mathbf{M}_{scs}$ of order $n_\alpha \times n_\alpha$ is the submatrix of the adjacency matrix $\mathbf{MR}_{a1}$ restricted to a self-cycling set. $\mathbf{M}_{scs}$ can be easily obtained by drawing out all the (i, j)th elements from $\mathbf{MR}_{a1}$ where i and j vertices both belong to the self-cycling set.
- *Step 2.* Period d_{scs} is given by an initial value, namely, $d_{scs} := 0$.
- *Step 3.* The k $(1 \leq k \leq n_\alpha)$ accessible step matrix $\mathbf{M}_{scs}^k$ can in turn be calculated by the recursive formula : $\mathbf{M}_{scs}^{k+1} = \mathbf{M}_{scs}^k \wedge \mathbf{M}_{scs}$ where $\wedge$ stands for the Boolean product i.e. $0 \wedge 0 = 0$, $0 \wedge 1 = 0$, $1 \wedge 0 = 0$, $1 \wedge 1 = 1$.
- *Step 4.* The middle value of the period d_{scs} is obtained by calculating the greatest common divisor of k and d_{scs} if $\mathbf{M}_{scs}^k 1$, or go to step 3 to calculate $\mathbf{M}_{scs}^{k+1}$ if $\mathbf{M}_{scs}^k = 0$.
- *Step 5.* If $d_{scs} = 1$, d_{scs} is already the required period of the self-cycling set and is recorded, the operating finishes. Otherwise, go to step 3 until $k = n_\alpha$. When $k = n_\alpha$, d_{scs} is the required period of the self-cycling set, and is recorded.

The time complexity degree of the above algorithm is $O(n_\alpha^3)$, the calculation is recursive and only Boolean operations are used. As a result, the algorithm is absolutely accurate and efficient. Self-cycling sets are further divided into persistent self-cycling sets and transient self-cycling sets on the basis of the conclusions a. and b. of Sect. 10.4.1. In concrete terms the steps are:

- *Step 1.* All the vertices that have the same transitive closure set are classified as one group on the matrix $\mathbf{TR}_{a1}$.
- *Step 2.* Z_α is a self-cycling set. If for every $i \in Z_\alpha$ there exists a $Z_\alpha = \{i\}^+$, Z_α is a *persistent self-cycling set.*
- *Step 3.* Otherwise, if for $i \in Z_\alpha$, there exists $Z_\alpha \subset \{i\}^+$, Z_α is a *transient self-cycling set.*

The transitive closure of a matrix condensed digraph D_c can be directly obtained by $\mathbf{TR}_{a1}$ and is denoted by $\mathbf{A}$. Naturally $\mathbf{A}$ is also the relationship matrix of a quasi-ordered set K, namely the adjacency matrix of the acyclic digraph representing the quasi ordered set K. A single algorithm can be used to obtain topological sorting. The time complexity degree of the algorithm is $O(N_c \times N_e)$ where N_c is the number of the vertices of the acyclic digraph, N_e is the number of arcs of the acyclic digraph. The algorithm of topological sorting is the following:

- *Step 1.* By means of the digraphic condensation method, the adjacency matrix $\mathbf{A}$ of the acyclic digraph of a quasi ordered set K can be directly obtained by merging all the vertices belonging to every self-cycling set on the matrix $\mathbf{TR}_{a1}$.
- *Step 2.* The vertex whose in-degree is zero is called a zero-in-degree vertex. Find all the zero-in-degree vertices and assign the same topological number to them. Here the topological number increases progressively in executing step 2 every time. When first executing step 2 there exists at least a zero-in-degree vertex because the quasi ordered set K is a finite set.
- *Step 3.* Delete all the zero-in-degree vertices found at step 2, and delete every arc whose initial vertex is a zero-in-degree vertex (let $e = (u, v)$, the u and v are called the ends of the arc e, u being the initial vertex and v the terminal vertex of e). Go to step 2 until every vertex of the acyclic diagraph is assigned to an appropriate topological number. With the increase of topological number, we can obtain the global transient evolution process and the ordered relation of all the transient cells including all the transient self-cycling sets.
- *Step 4.* Transient cells including the condensed vertices of transient self-cycling sets can be classified as transient sets $B(i)$, $B(i, j)$, $B(i, j, k)$, $B(i, j, k, l)$ and so on according to the number of domiciles they have.

Both common algorithms for depth-first search and broad-first search can be similarly adopted for determining all the strongly connected components of a digraph and a topological sorting of an acyclic digraph. For several corresponding algorithms, we refer to [13] or [11].

- Module 1. Constructing the generalized cell mapping of a dynamical system and forming the one-step transition probability matrix $\mathbf{P}$.
- Module 2. Obtaining the adjacency matrix $\mathbf{MR}_{a1}$ of the one-step mapping digraph.
- Module 3. Computing the transitive closure matrix $\mathbf{TR}_{a1}$ of the one-step mapping digraph.
- Module 4. Obtaining all the self-cycling sets.
- Module 5. Computing the period of a self-cycling set.
- Module 6. Determining all the persistent transient self-cycling sets.

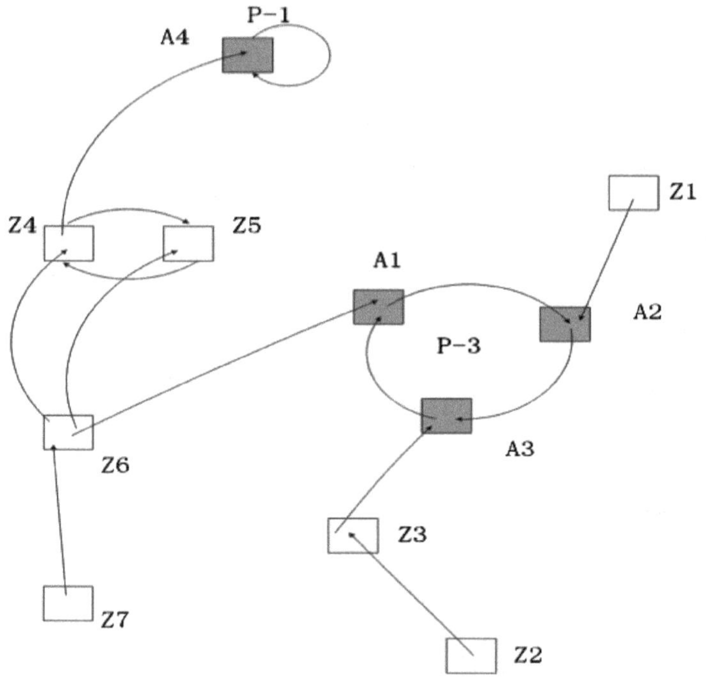

Fig. 10.1 The generalized cell mapping construction

- Module 7. Processing the topological sorting and determining the domains of attraction and the boundary regions.
- Module 8. Evaluating the quantitative properties of dynamical systems.

Modules 2–7 deal with the qualitative analysis of dynamical systems, and only adopt Boolean operations. After having explored the overall qualitative properties of dynamical systems, the whole evolution picture of the system becomes clear. Further, we can evaluate the quantitative properties such as the limiting probability distribution, the absorption probability or the expected absorption time.

10.5 Interior Crisis of the SD Oscillator

The generalized cell mapping method (GCMD), which is described in Fig. 10.1, is now used to investigate the interior crisis of the SD oscillator. Here we present the road to chaos, which is the collision of the periodic attractor and the chaotic saddle. Chaotic saddle is the transient self-cycling cell set in the cell state space. After the collision the chaotic saddle becomes a new portion of the chaotic attractor. The interior crisis does not change the shape of the basin of attraction. As used previously

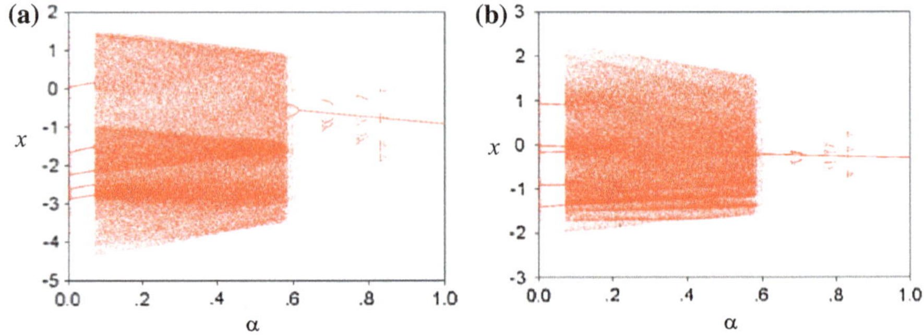

Fig. 10.2 Bifurcation diagram for $\omega_0 = 1.0606$, $f_0 = 0.83$, $\xi = 0.4$: **a** with initial data $(x_0, \dot{x}_0) = (1, 0)$ and **b** for $(x_0, \dot{x}_0) = (0, 1)$

the dimensionless equation of the SD oscillator with both viscous damping and external harmonic excitation is written one more time in the following form:

$$\ddot{x} + 2\xi\dot{x} + \omega_0^2\left(1 - \frac{1}{\sqrt{x^2 + \alpha^2}}\right) = f_0 \cos \omega t. \tag{10.11}$$

The computation for the bifurcation parameter $\alpha \in [0, 1]$ is performed using the Runge-Kutta scheme of fourth order taking $\omega_0 = 1.0606$, $f_0 = 0.83$ and $\xi = 0.4$ with initial data $(x_0, \dot{x}_0) = (1.0, 0.0)$ and $(x_0, \dot{x}_0) = (0.0, 1.0)$, respectively as shown in Fig. 10.2a, b. The bifurcation diagram can be constructed by collecting twenty periods of data of the Poincaré mapping points of $T/n = 2\pi/\omega_0$ after the 1980 periods of time, as shown in Fig. 10.2.

The path to chaos from periodic bifucation is found when parameter α passes the crisis point near $\alpha = 0.605$. The interior crisis phenomenon occurs when α decreases passing the crisis point near $\alpha = 0.065$ which enables the dimension of the attractor to be enlarged. The generalized cell mapping is constructed by choosing Poincaré section $\Sigma = (x, dx/dt, \theta)$ over the interested area $D : \{-5 < x < 2\} \times \{-2 < \dot{x} < 3\}$, which produces $N = 800 \times 800 + 1$ cells to cover the area including a sink cell. The cells can be ordered in one dimension from 1 to N, the imaginary points for each cell can be obtained by uniformly taking 10×10 samples in each cell to integrate over the period. The imaginary cell is marked $1 : j = map(i)$ if the transient probability is greater than zero and it is marked 0 otherwise.

For the fixed parameters $f_0 = 0.83$, $\zeta = 0.04$ and $\omega = 1.0606$, it can be seen by using the GCMD method that there exists a period one attractor (red), whose domain of attraction is white, there exist 4 non-periodic transient self-cycling cell sets, marked with dark cyan, green, pink and dark green, respectively, and 6 pseudo transient self-cycling cell sets, which are ignored in Fig. 10.3a for $\alpha = 0.61$. The cyan marks the domain of attraction of the infinite attractor. Figure 10.3b is the attractor obtained for $\alpha = 0.8$. The chaotic saddles plotted in Fig. 10.4a and b increase as the parameter α decreases over the interval [0.61, 1.0] and a self-cycling cell set is

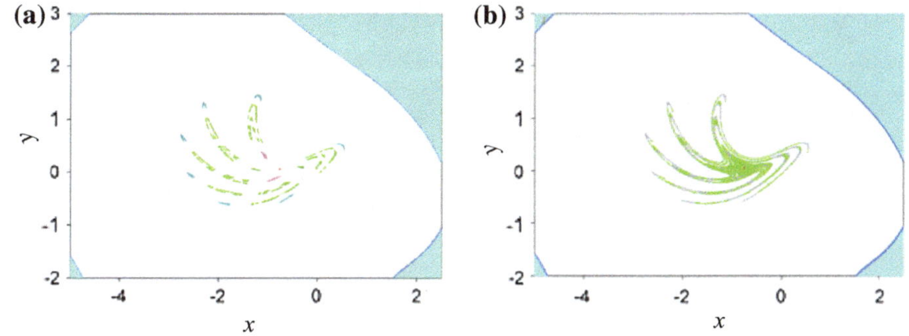

Fig. 10.3 Unstable self-cycling sets for the set of parameters $\omega_0 = 1.0606$, $f_0 = 0.83$, $\varepsilon = 0.04$ in the domain $D := \{-5 < x < 2.5\} \times \{-2 < \dot{x} < 3\}$: **a** chaotic saddle for $\alpha = 0.61$ and (**b**) an attractor for $\alpha = 0.80$

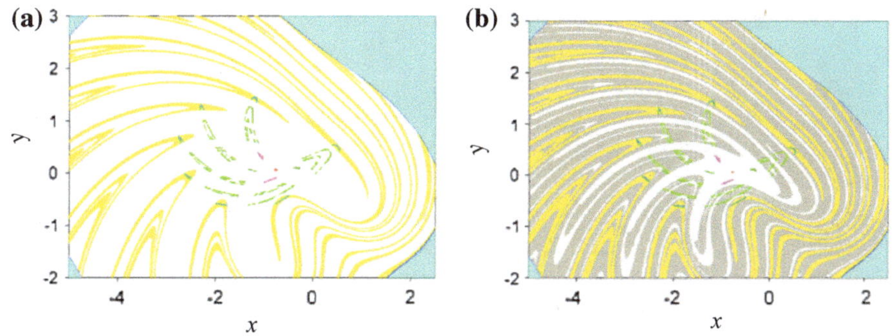

Fig. 10.4 Saddles obtained by the stable in **a** and unstable manifolds in **b**

formulated by several one, as shown in Fig. 10.5. Figure 10.5a is the transient state of collision of the attractor and the chaotic saddle. Figure 10.5b is the chaotic attractor induced by the crises of the periodic attractor expanding, as shown in Fig. 10.5a and b, respectively (Fig. 10.4).

The path to chaos from periodic motion during the interior crisis is found from the Homoclinic winding or Smale horseshoe structure. The periodic attractor collides the chaotic saddle within its domain attraction of which the volume is enlarging. This attractor is composed of the pre-crisis periodic attractor and the chaotic saddle, which is stable during the crisis without uncertainty happens when α changes from 0.6 to 0.61. This interior crisis is an invertible process and the domain of attraction remains unchanged, which illustrates the mechanism of the crisis of the oscillator. Another bifurcation value of α is located in the interval [0.065, 0.066]. Figure 10.6a for $\alpha = 0.065$ shows the attractor and the domain of attraction of itself and the infinite attractor with white and grey with a clear boundary for. Figure 10.6a for $\alpha = 0.066$ shows the phase portrait for the coexisted period 5 attractor and the transient self-cycling cell set. As α increases from 0.065 to 0.066 crossing the bifurcation point the

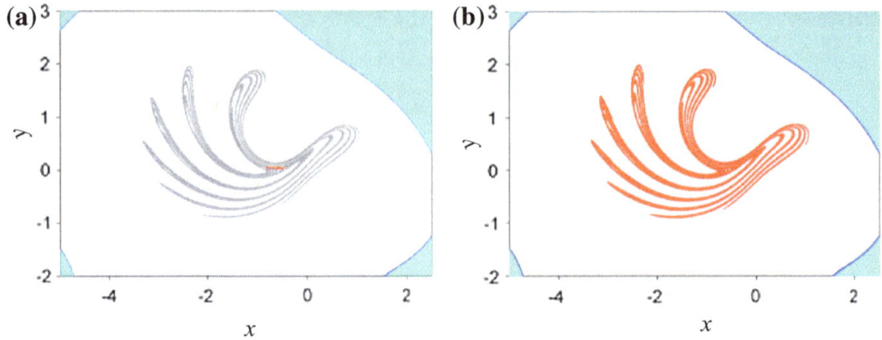

Fig. 10.5 Interior crisis of the SD oscillator: **a** for before crisis for $\alpha = 0.61$ and **b** for after crisis for $\alpha = 0.60$

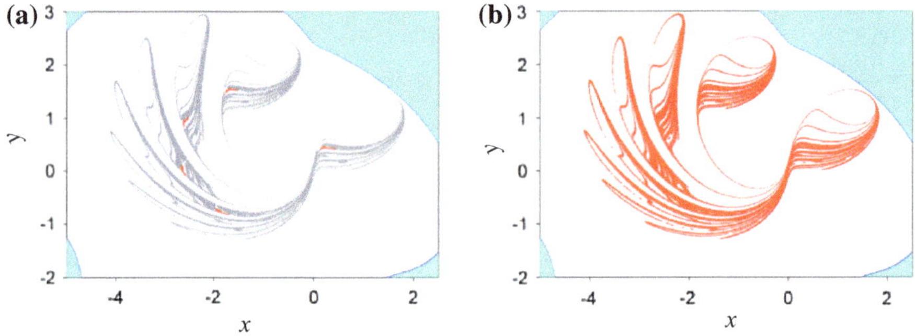

Fig. 10.6 Interior crisis of the SD oscillator: **a** before crisis for $\alpha = 0.065$, and **b** after crisis for $\alpha = 0.066$

periodic attractor will collide the transient self-cycling cell set, which results in the volume of the periodic attractor enlarged to a new attractor composed of the previous periodic attractor and the transient self-cycling cell set. The domain of attraction remains unchanged, which also shows the interior crisis invertible mechanism of the SD oscillator.

10.6 Summary

In this chapter, the generalized cell mapping method has been introduced and successfully used for the investigation of the qualitative behaviour of the proposed SD oscillator. The cell mapping and the generalized cell mapping has been briefly introduced and the theoretical proof of the correspondence between the generalized cell mapping of dynamical systems and digraphs has been given, which is based on a strictly mathematical proof. A digraph algorithm to determine the global properties of the systems has been given by Boolean operations (logical operations of the variables 0 and 1), which are absolutely accurate (without computational round-off errors), reliable and time-saving. This procedure has been used in the investigation of the properties of the SD oscillator to drive the interior crisis phenomena by means of the generalized cell mapping digraph method. Attractors, basins, basin boundaries and unstable solutions are obtained once through a global analysis at low computational cost. Moreover, the approach of a periodic attractor to a chaotic saddle in its basin attractor before an interior crisis, and the collision of the periodic attractor with the chaotic saddle when the crisis occurs, are explicitly demonstrated.

References

1. Hsu, C. S. (1987). *Cell-to-cell mapping: A method of global analysis for nonlinear systems*. New York: Springer.
2. Hsu, C. S. (1992). Global analysis by cell mapping. *International Journal of Bifurcation and Chaos*, *2*(4), 727–771.
3. Hsu, C. S. (1995). Global analysis of dynamical systems using posets and digraphs. *International Journal of Bifurcation and Chaos*, *5*(4), 1085–1118.
4. Jiang, J., & Xu, J. X. (1994). A method of point mapping under cell reference for global analysis of nonlinear dynamical systems. *Physics Letters A*, *188*(2), 137–145.
5. Jiang, J., & Xu, J. X. (1996). An iterative method of point mapping under cell reference for the global analysis of non!linear dynamical systems. *Journal of Sound Vibration*, *194*(4), 605–62.
6. Hong, L., & Xu, J. X. (1999). Crises and chaotic transients studied by the generalized cell mapping digraph method. *Physics Letters A*, *262*(4–5), 361–375.
7. Grebogi, C., Ott, E., & Yorke, J. A. (1982). Chaotic attractors in crisis. *Physical Review Letters*, *48*(22), 1507–1510.
8. Lai, Y. C., & Grebogi, C. (1994). Crisis and enhancement of chaotic scattering. *Physical Review E*, *49*(5), 3761–3770.
9. Issaacson, D. L., & Madsen, R. W. (1976). *Markow chains: Theory and applications*. New York: Wiley.
10. Helena, R. (1978). Introduction to modern mathematics. *The Journal of Symbolic Logic*, *43*(1), 153–154. Translated by O Wojtasiewicz.
11. Swamy, M. N. S., & Thulasiramann, K. (1981). *Graphs networks and algorithms*. New York: Wiley.
12. Hong, L., & Xu, J. X. (1999). Crisis and chaotic transients studied by the generalized cell mapping digraph method. *Physics Letters A*, *262*(4–5), 361–375.
13. Mott, J. L., Kandel, A., & Baker, T. P. (1986). *Discrete mathematics for computer scientists and mathematicians*. New Jersey: Prentice-Hall.
14. Chen, Y. S., & Leung, A. Y. T. (1998). *Bifurcation and chaos in engineering*. London: Springer.

Chapter 11
SD Oscillator with Friction and Impacts

This chapter extends the analysis of the SD oscillator to situations involving coupling with other nonlinearities or higher dimensional motions. The SD oscillator is a strictly one-dimensional dynamical system moving along a given line, now friction will be added as a constraint for the sliding along this line. If this constraint is removed, the system becomes two dimensional, which in turn implies that another constraint must be added to separate permanent sliding motions from jumping effects. This has important consequences on the solutions and on the mathematical tools that should be used, so that in the present chapter we shall only begin with the analysis. Both in the continuous and in the discontinuous cases, we present the set of equilibria, which on the one hand is very different from what was found in the previous chapters, and on the other hand includes the equilibria of the SD oscillator without jumps and friction as particular cases. Then notions and tools for the calculation of the dynamics and for the study of the stability are introduced as hints for further analysis. The main qualitative feature of this extension is the coupling of the nonlinearity which arises from the geometry, which can be either smooth or discontinuous, with those due to the modeling of contact and friction, which are always nonsmooth.[1]

11.1 Introduction

Let us now introduce the types of nonlinearities the extension of the SD oscillator is dealing with. Since the motion involves the sliding of a single mass, it was natural to add some friction. Moreover, specifically at large strains, it also seemed that

[1] This chapter relies on the common work with Professor Elaine Pratt, Department of Mathematics, Aix-Marseille University and CNRS, Laboratoire de Mécanique et d'Acoustique, 4, Impasse Nikola Tesla, CS 40006, 13453 Marseille Cedex 13, France, and Dr. Zhixin Li, Centre for Nonlinear Dynamics Research, School of Astronautics, Harbin Institute of Technology, Harbin 150001 China.

Q. Cao and A. Léger, *A Smooth and Discontinuous Oscillator*, Springer Tracts in Mechanical Engineering, DOI 10.1007/978-3-662-53094-8_11

the possibility of loosing contact should be introduced to have the most general framework for the motion of the mass. The corresponding coupling of the geometrical nonlinearity together with contact and friction will give a model for a very large range of mechanical situations. This coupling has however already been tackled in the case of buckling problems when the nonsmoothness was due to plasticity. In structural mechanics, a bifurcation diagram was explored for the equilibrium states of elastic plastic beams at large strains [1] and a more qualitative investigation had been given in the case of a simple 2 degrees-of-freedom model [2]. But no studies concerning this coupling exist from a theoretical point of view in the case of contact and friction, whether in the static case or in the dynamical one.

From a geometrical point of view, the mechanical system is exactly the same as the one described in the previous chapters. Of course adding the possibility of loosing contact implies that the kinematics is no longer strictly one dimensional in general but the position of the mass remains described in such a way that all the non linearity of the deformation of the spring is taken into account. Moreover the restoring force is either smooth or nonsmooth depending on the value of the geometrical parameter introduced since Chap. 1.

A model where a single mass is moving in a half-plane, undergoing shocks and friction whenever it attains the boundary of the half-plane, with linearly elastic springs and linearized deformations has been studied in [3] and references herein. This model, initially introduced for the study of quasi-static evolution, has been widely studied during the past ten years. The non linearity was due to the unilateral contact condition and to the Coulomb friction law, which are strongly non smooth. The theoretical foundations for the analysis of the corresponding differential systems have first been given in [4] or [5], and the main tools for their numerical analysis in [6]. The set of equilibrium states of this mass spring system, which generically contains infinitely many equilibria, has been studied in [7], the conditions for the well posedness of the dynamics arose from [8] and have been proved in detail in [9] for the case of linear strains. A first investigation and some new concepts of stability of the equilibrium states have been given in [10, 11].

11.2 The Contact and Friction Conditions

The first point consists in describing the contact and friction conditions that are added to the modeling of the SD oscillator, both from the mechanical and from the mathematical points of view. Using a model as simple as possible, these conditions are classically divided into two parts, one concerning the tangential components, and the other concerning the normal components, where tangential and normal refer in the space $\mathbb{R}^2$ to a local frame $(\overrightarrow{t}, \overrightarrow{n})$, in which $\overrightarrow{t}$ is attached to the line the particle is impacting or sliding on. Let U and R, with components (u_t, u_n) and (R_t, R_n), be respectively the displacement of the particle and the reaction of the line which is identified with an obstacle. These conditions read as follows:

- For the normal components:

$$u_n \leq 0, \quad R_n \leq 0, \quad u_n R_n = 0, \tag{11.1}$$

This set of relations is classically referred to as Signorini conditions or unilateral contact conditions. It means that the position of the mass must be on the right side of the obstacle, the reaction of the obstacle can only push the particle away, and a nonzero reaction is possible only when the particle is in contact with the obstacle.

- Concerning the tangential components, the modeling which is chosen here is the so-called Coulomb's friction law. It involves a positive real coefficient, the so-called friction coefficient μ, in the following way:

$$\begin{cases} \mu R_n \leq R_t \leq -\mu R_n, \\ \text{with} \quad \begin{cases} |R_t| < -\mu R_n \Longrightarrow \dot{u}_t = 0, \\ |R_t| = -\mu R_n \Longrightarrow \exists \lambda \geq 0 \text{ s.t. } \dot{u}_t = -\lambda R_t. \end{cases} \end{cases} \tag{11.2}$$

These conditions mean that no sliding velocity is possible as long as the tangential component of the reaction belongs strictly to an interval the ends of which are proportional to the normal component of the reaction, and that the particle can be set into sliding motion only if the tangential component of the reaction reaches one end of this interval.

Signorini conditions and Coulomb's friction conditions are represented on Fig. 11.1. In a mechanical model, the reaction of the obstacle is a force, which consequently will be introduced in the same way as the given external forces or the internal stresses in the right hand side of the equation of the dynamics. As it is quite clear on Fig. 11.1, we stress that, due to the vertical parts of the graphs, the corresponding sets of relations do not give a function connecting the reaction to the position or to the velocity. This particularity will make the dynamical system very different from usual ordinary differential equations, and different from the general case of the SD oscillator studied in the previous chapters, even in the discontinuous case. Many studies and corresponding softwares change the graphs of these conditions, respectively in the $\{u_n, R_n\}$ plane (for contact) and in the $\{\dot{u}_t, R_t\}$ plane (for friction)

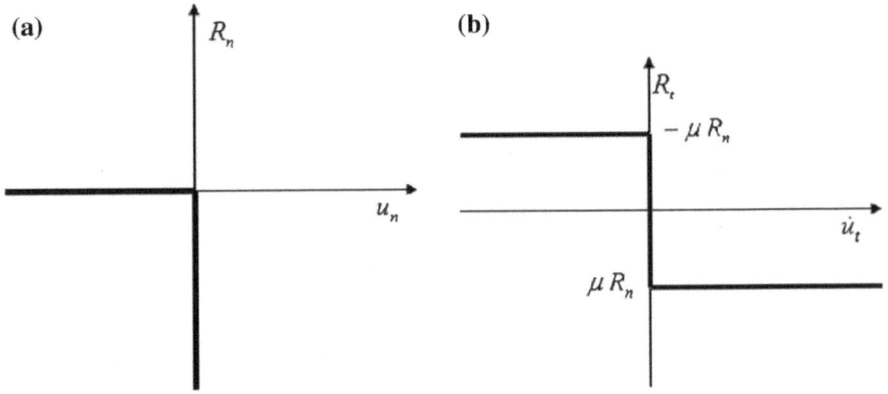

Fig. 11.1 **a** The unilateral contact condition, **b** The Coulomb's friction law

into graphs of ordinary functions by some smoothing procedure, but the outcome depends obviously on the smoothing one has adopted. Moreover, the nonregularized relations (11.1) and (11.2) have a true physical meaning (non penetrability, sliding only for large enough force, ...). The most common regularizations of the contact or friction laws are clearly not physical, as could be easily seen by looking at the examples used in the literature, but they remove all the mathematical difficulties, keeping in particular the framework of ordinary differential equations.

11.3 The Model and the Equations

We now come back to the basic equations of the dynamics of the SD oscillator but introducing the generalization mentioned above. A mass m is still connected with a rigid frame by linearly elastic springs but in addition to the effects of the spring, the mass is constrained to remain in a half-plane due to the presence of an obstacle represented by the horizontal line. On the one hand, the contact and friction with this obstacle is assumed to hold in a strictly unilateral way as given by equations (11.1) and (11.2). On the other hand, there are no approximations of the geometry, which means that the exact expression of large strains is kept, which in other words means from a mechanical point of view that the change of the length of the spring is associated with the exact change of its direction, as assumed in the foundation of the kinematics of the SD oscillator. The two vectors of the local frame are respectively tangent and normal to the obstacle. The system is represented on Fig. 11.2.

When the mass is subjected to an external force F, of components (F_t, F_n), its trajectory is the solution of a system of the form:

$$\begin{cases} m\ddot{U}(t) = \mathcal{N}(U) + F + R, \\ \text{Initial data}, \\ \text{Unilateral contact conditions}, \\ \text{Coulomb's friction conditions}, \end{cases} \tag{11.3}$$

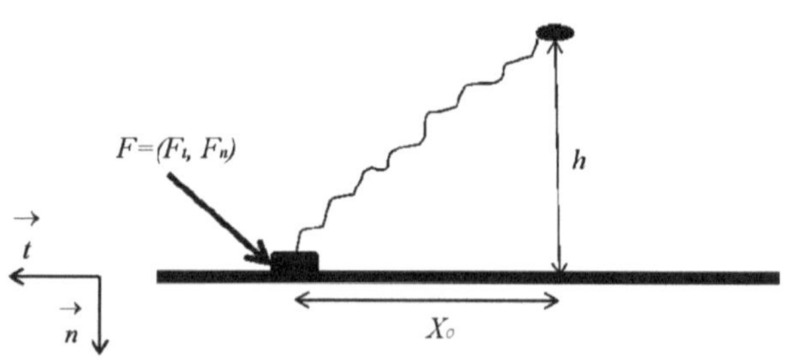

Fig. 11.2 The simple mass-spring model

where $\mathcal{N}(U)$, of components $\mathcal{N}_t(u_t, u_n)$, $\mathcal{N}_n(u_t, u_n)$, stands for the two-dimensional geometrical nonlinearity, while R includes the unilateral nonlinearities. Let us now write the dynamical problem in a more accurate way, which shows the very strong differences with the one-dimensional case without friction.

$$\begin{cases} i)\ m\ddot{u}_t = \mathcal{N}_t(u_t, u_n) + F_t + R_t, \\ \qquad\qquad\qquad\qquad\qquad\qquad\qquad t > 0 \\ ii)\ m\ddot{u}_n = \mathcal{N}_n(u_t, u_n) + F_n + R_n, \\ iii)\ u(0) = u_0,\ \dot{u}^+(0) = v_0, \\ iv)\ u_n \le 0,\ R_n \le 0,\ u_n R_n = 0, \\ v)\ \forall \varphi \in \mathbb{C}^0((0,T];\mathbb{R}),\ \displaystyle\int_{[0,T]} R_t(\varphi - \dot{u}_t^+) - \mu R_n(|\varphi| - |\dot{u}_t^+|) \ge 0, \\ vi)\ u_n(\tau) = 0 \implies \exists\, e\ \in [0,T] \text{ such that } \dot{u}_n^+(\tau) = -e\dot{u}_n^-(\tau). \end{cases} \tag{11.4}$$

Let us add some comments:

1. Due to the possibility of impacts all along the trajectory, it is not possible to expect the velocity $\dot{u}(t)$ to be a continuous function. So that the first and most fundamental comment is the following: the acceleration, which is the derivative of a function which is not continuous, should be taken in the sense of distributions, so that equations (11.4-*i*–*ii*) should be understood as equalities between distributions. Moreover, due to relations (11.1) the component R_n should satisfy a sign condition, which implies that (11.4-*ii*) is an equality between measures (see [4, 5]). Relations (11.2) imply that it is also the case for relations (11.4-*i*).
2. The next comment is a consequence of the previous one: since the equations of the dynamics should be understood in the sense of measures, they can be integrated over an interval, as a basic definition of a real valued measure. Although it would be out of the purpose of the present chapter, this property is at the foundation of a numerical method for the calculation of the trajectory (see [6]). Moreover, since the acceleration is a measure, its primitive, which is the velocity, is a function of bounded variation, of which a basic property is that it has a right and a left limit at any time, so that in particular $\dot{u}_n^+(0)$ is well defined and can be stated as equal to some given value v_0 (see [12]) (while $\dot{u}_n(0) = v_0$ would have no meaning).
3. Equation (11.4-*v*) is a reformulation of friction conditions, which is also required by the fact that the reaction is a measure and the velocity a function of bounded variation. As a matter of fact, conditions (11.2) are formal but (11.4-*v*) is perfectly defined for the reasons of the above comments:

   - integrations by parts show that (11.2) is actually recovered from (11.4-*v*) and conversely (11.2) leads to (11.4-*v*),
   - since the set of measures is the dual of the set of continuous functions, integrals over $[0, T]$ of terms as $\varphi \cdot R$ have the meaning of duality products,

- a function of bounded variation can be integrated with respect to any measure so that integrals over $[0, T]$ of terms as $R.\dot{u}^+$ are also well defined since R is a measure.

4. The impact condition (11.4-*vi*) is necessary as a closure statement for problem (11.4) to have a solution after an impact time in the case of any discrete system; it would not be introduced in the dynamics of continuous bodies due to the fact that an impact on the boundary leads to waves inside the body. But it is necessary for the idealization of a system by any collection of single points or rigid bodies because this idealization removes the possibility of waves inside the bodies. It involves a so-called restitution coefficient, denoted by e in system (11.4), the real value of which comes from physics, and is well defined for the same reason as comment 2 above. The case $e = 1$ is called perfectly elastic, while the case $e = 0$ defines the perfectly plastic impact.
5. The friction conditions induce a dissipative term behaviour. All the qualitative properties given here would not be qualitatively changed by some additional more classical damping term.

Attention is drawn to the fact that problem (11.4) has four unknowns which are u_t, u_n, R_t, R_n, while it involves two equations and a set of conditions mostly of the type of inequalities. This makes it obvious that the dynamical system has been changed qualitatively. Within the framework of the above comments, it is then natural to answer the question of the well-posedness of the dynamics since the classical conditions for a Cauchy problem to have a solution no longer apply. The existence and uniqueness results are relatively difficult and beyond the scope of the present book. The results are recalled without proof, referring to the corresponding articles or books for interested readers.

1. For any initial data compatible with the obstacle, there exists a trajectory which is a solution to problem (11.4) on any time interval $[0, T]$ as soon as the external force $F(t)$ is an integrable function over $[0, T]$. In the case where $\mathcal{N}(U)$ is linear, the proof has been first given in [13].
2. Even if the external force is extremely smooth, up to $\mathbb{C}^\infty([0, T])$, the impact problem, i.e. without friction, is ill-posed in the sense that more than one trajectory may start from the same initial data. This result has been proved by explicit counter-example to uniqueness [5].
3. For this restriction to a problem without friction, well-posedness is recovered if the external force is an analytic function of time ([14, 15]).
4. In the case with friction, well-posedness has been proved in [9] in the case of a linear term in place of a general $\mathcal{N}(U)$. In general, and for any number of particles, well-posedness has now been obtained (see [16]) under the same conditions.

11.3.1 The Geometrically Linear Case

In order to understand the effects of contact and friction, let us remove the large strains and discuss the set of equilibria for a short subsection. This means that we assume that the nonlinear term $\mathcal{N}(U)$ of equation (11.3) is changed into a simple stiffness term involving linear springs at small strains:

$$\mathcal{N}(U) = \mathbb{K}U = \begin{pmatrix} K_t & W \\ W & K_n \end{pmatrix} \begin{pmatrix} u_t \\ u_n \end{pmatrix},$$

and assume the external force F is constant of components F_t, F_n. Then the main qualitative results in such a case have been obtained in [7] and, after the suppression of the rigid translations of the spring, can be stated as:

- The structure of the set of equilibria depends on only two parameters which are $K_t F_n - W F_t$ and $\mu - \frac{K_t}{W}$.
- Equilibrium states out of contact always exist when $K_t F_n - W F_t < 0$. The equilibrium state without contact is then unique. Moreover, if in addition $\mu - \frac{K_t}{W} > 0$, then the equilibrium state without contact coexists with infinitely many equilibrium states in contact which therefore have a strictly negative normal component of the reaction.
- If $K_t F_n - W F_t \geq 0$, there are no equilibrium states out of contact. Moreover, if $K_t F_n - W F_t = 0$ and $\mu - \frac{K_t}{W} < 0$ the set of equilibria reduces to a single state in grazing contact.
- If $K_t F_n - W F_t > 0$ there always exist infinitely many equilibrium states in contact, which all have a strictly negative normal component of the reaction. If in addition $\mu - \frac{K_t}{W} \geq 0$ then this set fills completely a half-line in the $\{R_t, R_n\}$ plane, while it fills only a bounded interval if $\mu - \frac{K_t}{W} < 0$.

Although this case with a linear stiffness was simple, it involves infinitely many equilibria for several ranges of the parameters. Moreover, restricting our attention to equilibria in contact, the existence of infinitely many solutions hold for almost all the values of the parameters (in fact except for a single non generic case), and this is clearly a consequence of the friction law (11.2).

11.3.2 The Geometrical Nonlinearity in 2D

The calculation of functions $\mathcal{N}_t(u_t, u_n)$ and $\mathcal{N}_n(u_t, u_n)$ is essentially the same as that of the restoring force of the previous chapters, except that the two-dimensional dynamics requires two components of the force. A change of the position of the mass is represented on Fig. 11.3, which emphasizes the difference with the one dimensional sliding motion. Then we shall first describe the set of stationary solutions to problem (11.4) as this had been done in the case of linear strains recalled in Sect. 11.3.1. Assume the mass is initially in contact with the obstacle (this will in

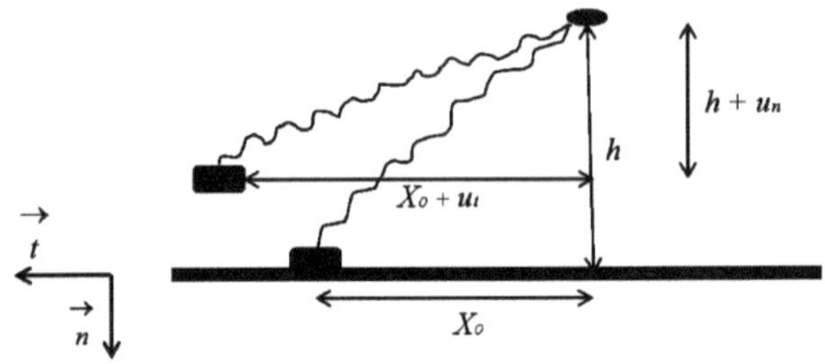

Fig. 11.3 Calculation of the non linear strain

no way be restrictive) and that $\sqrt{X_0^2 + h^2}$ is the natural length of the spring. A displacement of components u_t, u_n changes the length of the spring from $\sqrt{X_0^2 + h^2}$ to $\sqrt{(X_0 + u_t)^2 + (h + u_n)^2}$.

Assume in addition that the spring is linear with a stiffness k, then simple calculations give the components of the restoring force:

$$\begin{cases} \mathcal{N}_t(u_t, u_n) = -k(X_0 + u_t) + \dfrac{k(X_0 + u_t)\sqrt{X_0^2 + h^2}}{\sqrt{(X_0 + u_t)^2 + (h + u_n)^2}}, \\ \mathcal{N}_n(u_t, u_n) = -k(h + u_n) + \dfrac{k(h + u_n)\sqrt{X_0^2 + h^2}}{\sqrt{(X_0 + u_t)^2 + (h + u_n)^2}}, \end{cases} \tag{11.5}$$

so that the explicit form of the equilibrium equations reads:

$$\begin{cases} \mathcal{N}_t(u_t, u_n) + F_t + R_t = 0, \\ \mathcal{N}_n(u_t, u_n) + F_n + R_n = 0, \\ u_n \leq 0, \;\; R_n \leq 0, \;\; u_n R_n = 0, \\ \mu R_n \leq R_t \leq -\mu R_n, \end{cases} \tag{11.6}$$

where $\mathcal{N}_t(u_t, u_n)$ and $\mathcal{N}_n(u_t, u_n)$ are given by equations (11.5).

Remark 11.1 Whether in the case of linear strains or in the nonlinear case, contact and friction conditions are kept strictly in their non regularized form given by Eqs. (11.1) and (11.2).

The main difficulty for obtaining the solutions to problem (11.6) arises from the fact that this set of equilibrium equations involves several inequalities, so that we shall follow the idea which was successfully used in [7] and led to the results of Sect. 11.3.1. A first investigation of the existence of stationary positions out of contact is possible since the force is constant. This means, according to unilateral contact and friction conditions, that we will look for solutions of system (11.6) with $\{R_t, R_n\} = \{0, 0\}$ and a strictly negative normal component u_n of the solution. Then the investigation of the existence of stationary solutions in contact implies conversely that R_t and R_n should not be equal to zero in general, but u_n is equal to zero. The possible coexistence of solutions out of contact with solutions in contact will necessarily be discussed, as this coexistence of different types of solutions already appears in the case of a linear stiffness.

Although we are now studying a problem with a bi-dimensional kinematics, the model presented in Fig. 11.2 is the same as the SD oscillator. The SD oscillator has been introduced as continuous or discontinuous according to a parameter α which, using the current notations, reads $\alpha = h/\sqrt{X_0^2 + h^2}$. In order to remain in the same context as in the previous chapters, it will then be necessary to study separately the case $\alpha \neq 0$, which could be seen as the generic case for the effect of friction, and then the particular case $\alpha = 0$. The two following sections deal with the case $\alpha \neq 0$.

11.4 Stationary Solutions Out of Contact

If equilibrium solutions out of contact exist for some external data (k, h, X_0, F_t, F_n), R_n must be equal to zero so that these solutions must satisfy the following system:

$$\begin{cases} \mathcal{N}_t(u_t, u_n) + F_t = 0, \\ \mathcal{N}_n(u_t, u_n) + F_n = 0, \\ u_n < 0. \end{cases} \tag{11.7}$$

Remark 11.2 In fact the unilateral contact condition amounts to checking that eventual solutions of the two first algebraic equations of system (11.7) are compatible with the single inequality $u_n < 0$, in which case the Coulomb friction law would be automatically satisfied by $\{R_t, R_n\} = \{0, 0\}$.

System (11.7) can be rewritten into the explicit form:

$$\begin{cases} (X_0 + u_t)\left[1 - \dfrac{\sqrt{X_0^2 + h^2}}{\sqrt{(X_0 + u_t)^2 + (h + u_n)^2}}\right] = \dfrac{F_t}{k}, \\ \\ (h + u_n)\left[1 - \dfrac{\sqrt{X_0^2 + h^2}}{\sqrt{(X_0 + u_t)^2 + (h + u_n)^2}}\right] = \dfrac{F_n}{k}, \\ u_n < 0. \end{cases} \tag{11.8}$$

The method is identical to the one used in the linear case, that is solutions to the two equations of system (11.8) are computed, then the conditions under which these solutions are not in contact are given. However this time the algebraic system with unknowns u_t and u_n is relatively intricate so that the case $F_n \neq 0$ shall be looked into first, and the particular case $F_n = 0$ shall then be studied.

11.4.1 Non-zero Normal Component F_n

An elementary manipulation gives in this case

$$u_t + X_0 = \frac{F_t}{F_n}(u_n + h).$$

Introducing this into system (11.8):

$$u_n + h - \frac{F_n}{k} = \sqrt{\frac{X_0^2 + h^2}{(h+u_n)^2\left(1 + \left(\frac{F_t}{F_n}\right)^2\right)}}(h + u_n),$$

so that

$$(u_n + h - \frac{F_n}{k})^2 = h^2\alpha(F_t, F_n)^2, \tag{11.9}$$

where a real parameter $\alpha(F_t, F_n)$, depending on the geometry and on the loading, has been introduced $\alpha(F_t, F_n) = \sqrt{\left(1 + (\frac{X_0}{h})^2\right) \Big/ \left(1 + \left(\frac{F_t}{F_n}\right)^2\right)}$. The geometrical dimensionless parameter $\beta = \sqrt{1 + (\frac{X_0}{h})^2}$ is also introduced and $\alpha(F_t, F_n)$ depends on β.

Equation (11.9) has two solutions:

$$u_n = \frac{F_n}{k} - h + h\alpha(F_t, F_n) \text{ and } u_n = \frac{F_n}{k} - h - h\alpha(F_t, F_n).$$

These solutions shall be solutions to system (11.8) if and only if they are strictly negative.

The quantities h and $\alpha(F_t, F_n)$ being strictly positive, the second solution is strictly smaller than the first one. So that if the first solution is negative then the second one shall also be negative.

The conditions implying the existence of equilibrium solutions out of contact are given by:

$$\begin{cases} \frac{F_n}{kh} - 1 + \alpha(F_t, F_n) < 0 \implies \text{two solutions out of contact,} \\ \frac{F_n}{kh} - 1 + \alpha(F_t, F_n) > 0 \text{ and } \frac{F_n}{kh} - 1 - \alpha(F_t, F_n) < 0 \implies \text{one solution out of contact,} \\ \frac{F_n}{kh} - 1 - \alpha(F_t, F_n) > 0 \implies \text{no solution out of contact.} \end{cases} \tag{11.10}$$

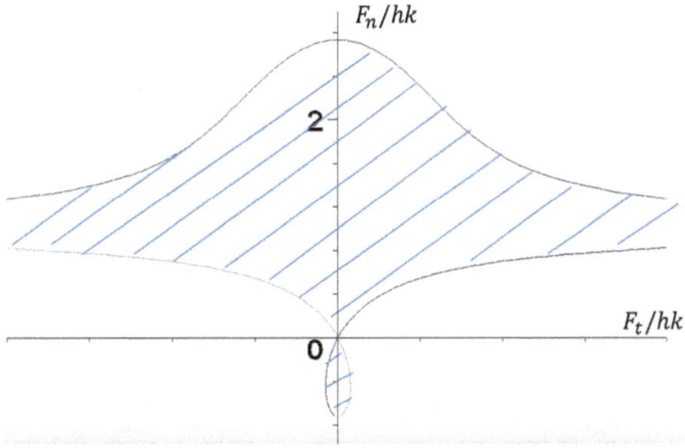

Fig. 11.4 Values of the force corresponding to stationary solutions out of contact

These conditions define three zones in the $\{\frac{F_t}{kh}, \frac{F_n}{kh}\}$ plane represented on Fig. 11.4. Zone 1 (above the hatched zone) where no solutions out of contact exist, zone 2 (the hatched zone) where a single solution out of contact exists (condition 1 or condition 2 is satisfied) and zone 3 (under the hatched zone) where two solutions out of contact co-exist (both conditions are satisfied).

Remark 11.3 Two solutions can be interpreted very simply since the spring is linearly elastic and the kinematics allows large displacements: one of the solutions corresponds to an extension of the spring and the other to a compression, both under the same force in the direction of the spring.

Remark 11.4 The direction of the force is given by F_t/F_n and the quantity X_0/h gives the direction of the reference configuration of the spring so that $F_t/F_n = X_0/h$ implies $\alpha = 1$ (in other words if the direction of the force is the direction of the reference configuration of the spring then $\alpha = 1$).

11.4.2 Zero Normal Component F_n

This section shall be completed by studying the case where the normal component of the force is equal to zero. This leads to the following:

- If $F_t \neq 0$, system (11.8) gives immediately the trivial solutions:

$$\begin{cases} u_n = -h \\ u_t = \dfrac{F_t}{k} - X_0 + h\beta \end{cases} \tag{11.11}$$

  and

$$\begin{cases} u_n = -h \\ u_t = \dfrac{F_t}{k} - X_0 - h\beta. \end{cases} \tag{11.12}$$

It is interesting to observe that this solution with $F_n = 0$ corresponds to a horizontal position of the spring at the level $-h$, which is quite trivial since only a horizontal force is applied.

- If $F_t = 0$, then system (11.8) gives

$$1 - \frac{\sqrt{X_0^2 + h^2}}{\sqrt{(X_0 + u_t)^2 + (h + u_n)^2}} = 0,$$

which is the equation of a circle of center $(-X_0, -h)$ and radius $\sqrt{X_0^2 + h^2}$ (the natural length of the spring), so that we get the trivial result that in the absence of external forces all the points belonging both to the circle of center $(-X_0, -h)$ and radius $\sqrt{X_0^2 + h^2}$ and to the open negative half plane are equilibrium solutions, which obviously results from the fact that our simple model has only one spring.

Remark 11.5 Changing from discrete systems to continuous media, it has already been observed that unloaded structures may have infinitely many equilibrium solutions within non linearized strains, while only the trivial shape of the structure corresponds to an equilibrium solution in the linear case (see e.g. [17] or [18]).

11.5 Stationary Solutions in Contact

The parameters are now μ, k, h, X_0, F_t, F_n. In order to find solutions in contact, the constraint $u_n = 0$ must be inserted into system (11.6) so that the remaining unknowns are u_t, R_t, and R_n. The equilibrium problem is now:

$$\begin{cases} (X_0 + u_t)\left[1 - \dfrac{h\beta}{\sqrt{(X_0 + u_t)^2 + h^2}}\right] = \dfrac{F_t + R_t}{k}, \\ \left[1 - \dfrac{h\beta}{\sqrt{(X_0 + u_t)^2 + h^2}}\right] = \dfrac{F_n + R_n}{kh}, \\ R_n \le 0, \ |R_t| \le -\mu R_n, \end{cases} \tag{11.13}$$

which can be rewritten as:

$$\begin{cases} i)\ (X_0 + u_t)(F_n + R_n) = h(F_t + R_t), \\ ii)\ \left[1 - \dfrac{h\beta}{\sqrt{(X_0 + u_t)^2 + h^2}}\right] = \dfrac{F_n + R_n}{kh}, \\ iii)\ R_n \le 0, \ |R_t| \le -\mu R_n. \end{cases} \tag{11.14}$$

Equation (11.14-*ii*) can be solved analytically, giving u_t as a function of R_n:

$$u_t(R_n) = -X_0 \pm h\sqrt{\frac{k^2h^2\beta^2}{(F_n + R_n - kh)^2} - 1}. \tag{11.15}$$

Inserting the expression of u_t obtained in (11.15) into equation (11.14-*i*) we obtain the following expression which defines a curve in the $\{R_t, R_n\}$ plane:

$$(R_t + F_t)^2 = (F_n + R_n)^2 \left(\frac{k^2h^2\beta^2}{(F_n + R_n - kh)^2} - 1\right). \tag{11.16}$$

Finding equilibrium solutions in contact then amounts to finding the intersection of this curve (11.16) with the Coulomb cone given by (11.14-*iii*).

11.5.1 Non-zero External Force

In general, the external force is not equal to zero, which can be taken as the generic case, and the sets of solutions are built in the following way, using an example of curve (11.16). This is represented on Fig. 11.5 in the $\{R_t, R_n\}$ plane, where five particular points, denoted by A, B, C, D and E have been added.

Remark 11.6 In the case of linearized strains, the same analysis with the same non-smooth contact and friction laws, led to a straight line instead of the curve given by formula (11.16).

Modifying the values of F_t and F_n has the effect of translating the curve in the $\{R_t, R_n\}$ plane, so that F_t and F_n will consequently be the parameters of the following

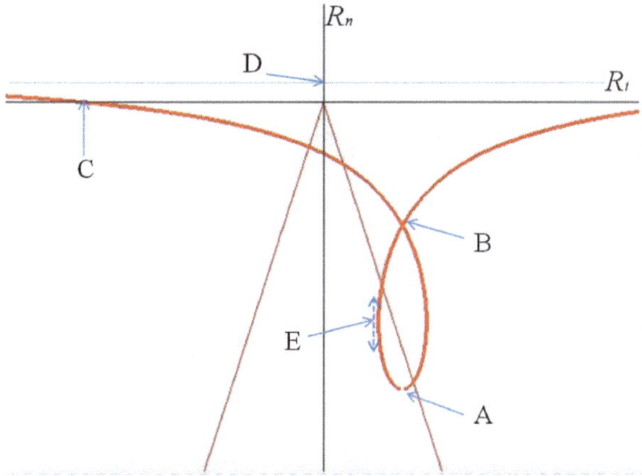

Fig. 11.5 An example of curve (11.16)

discussion for given values of X_0, h and μ. Moreover, modifying the values of X_0, h and μ does not lead to qualitative changes of Fig. 11.5. The coordinates of the points A, B, C, D and E are:

$$\begin{cases} A \left| \begin{array}{l} -F_t \\ kh(1-\beta-\frac{F_n}{kh}), \end{array} \right. \quad B \left| \begin{array}{l} -F_t \\ -F_n, \end{array} \right. \\ C \left| \begin{array}{l} -F_t - F_n\sqrt{\dfrac{\beta^2}{(\frac{F_n}{kh}-1)^2}-1} \\ 0, \end{array} \right. \\ D \left| \begin{array}{l} 0 \\ kh(1-\frac{F_n}{kh}), \end{array} \right. \quad E \left| \begin{array}{l} -kh(\frac{F_t}{kh}-(\beta^{2/3}-1)^{3/2}) \\ -kh(\frac{F_n}{kh}-1-\beta^{2/3}). \end{array} \right. \end{cases} \tag{11.17}$$

When the intersection of the curve with the Coulomb cone is empty there are no equilibrium solutions and conversely a non empty intersection corresponds to the existence of equilibrium solutions. So that we can determine the values of the parameters corresponding to the existence of equilibrium solutions. This phenomena are presented in the Figs. 11.5, 11.6a, b, 11.7a, b, and 11.8a, b, respectively.

1. $\dfrac{F_n}{kh} < 1-\beta$: Point A is in the positive half-plane so that there are no solutions in contact, whatever the value of μ or F_t. This situation is represented on Fig. 11.6a.
2. $\dfrac{F_n}{kh} \geq 1$: Point D (which corresponds to the intersection of the horizontal asymptote with the R_n axis) is in the negative half-plane so that the entire curve is in the negative half plane. There always exists a part of the curve of non zero length which is strictly inside the cone, so that there always exist infinitely many equilibrium solutions in contact. This situation is represented on Fig. 11.6b.

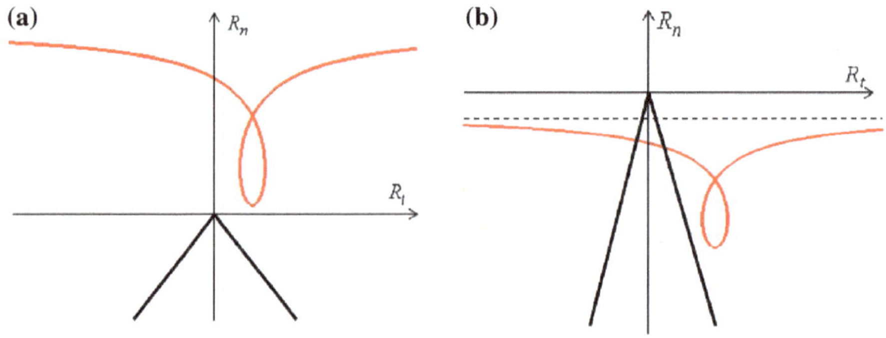

Fig. 11.6 **a** Non existence of equilibrium solutions in contact, **b** Existence of equilibrium solutions in contact for $\frac{F_n}{kh} \geq 1$

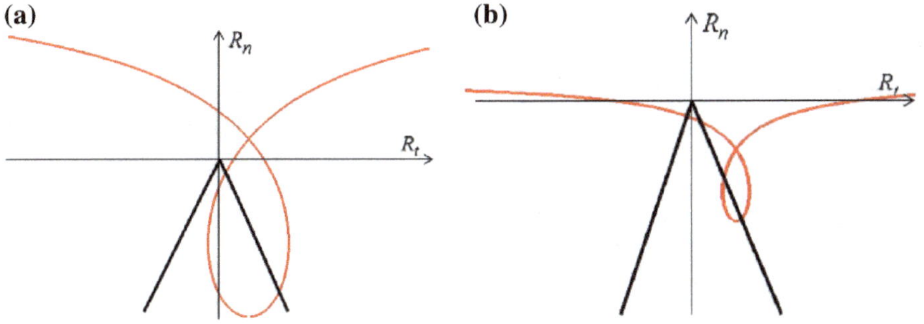

Fig. 11.7 Unconditional existence of equilibrium states in the case $1-\beta \leq \dfrac{F_n}{kh} < 1$: **a** for $\dfrac{|F_t|}{kh} < (\beta^{2/3}-1)^{3/2}$ and **b** for $|F_t| > F_n\sqrt{\dfrac{k^2h^2\beta^2}{(F_n-kh)^2}-1}$

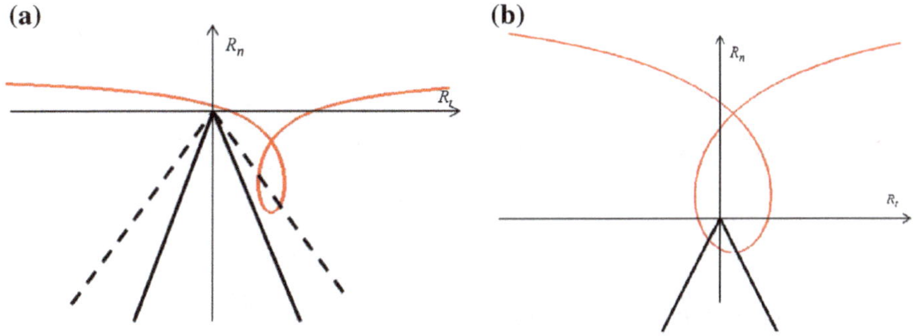

Fig. 11.8 **a** Conditional existence of equilibrium solutions in contact for $\dfrac{F_n}{kh} \geq 1$: thick continuous border of the cone: $\mu < \mu_c$, dashed border of the cone $\mu > \mu_c$. **b** Unconditional existence of equilibrium solutions in contact for $1-\beta \leq \dfrac{F_n}{kh} \leq 1-\beta^{2/3}$

3. $1-\beta \leq \dfrac{F_n}{kh} < 1$. In this range, the discussion is more intricate. In the $\left\{\frac{F_t}{kh}, \frac{F_n}{kh}\right\}$ plane, we shall determine a boundary above which there always exist equilibrium solutions in contact and under which there always exists a critical value of the coefficient of friction μ_c such that solutions in contact exist only for $\mu \geq \mu_c$.
   - Let $1-\beta^{2/3} < \dfrac{F_n}{kh} < 1$. Point E is in the negative half-plane, then equilibrium solutions exist unconditionally as soon as either one of the following conditions is satisfied (see Fig. 11.7): (a) $\dfrac{|F_t|}{kh} < (\beta^{2/3}-1)^{3/2}$, or, (b) $|F_t| > F_n\sqrt{\dfrac{k^2h^2\beta^2}{(F_n-kh)^2}-1}$, and only conditionally otherwise (see Fig. 11.8a).

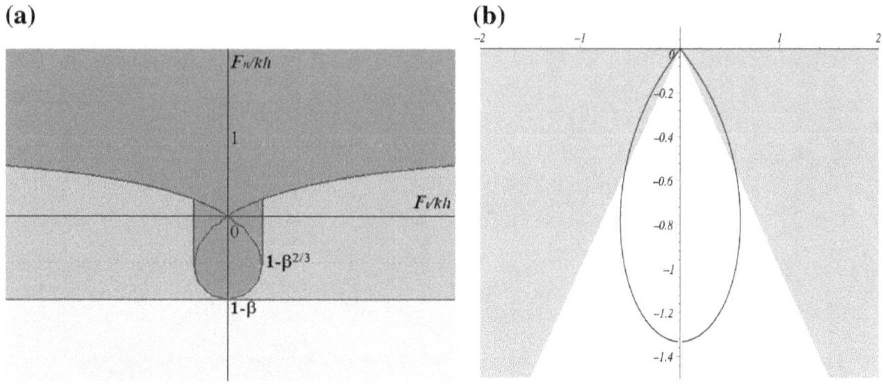

Fig. 11.9 **a** Values of the force corresponding to equilibrium solutions in contact, **b** The equilibria strictly in contact in the $\{R_t, R_n\}$ plane without external forces

- Let $1-\beta \leq \dfrac{F_n}{kh} \leq 1-\beta^{2/3}$. Point E is in the positive half-plane. Equilibrium solutions exist unconditionally if $|F_t| > F_n\sqrt{\dfrac{k^2h^2\beta^2}{(F_n-kh)^2}-1}$ (i.e. if the curve intersects the R_t axis either side of the R_n axis) and only for μ large enough otherwise, see Fig. 11.8b.

There may exist a unique solution in contact, for example when $\dfrac{F_n}{kh} = 1-\beta,\ F_t = 0$, in which case it is the vertex of the cone, or whenever the intersection of the curve and the cone is reduced to a single point where the curve is tangent to a border of the cone. However, there are no solutions in contact for $\dfrac{F_n}{kh} = 1-\beta,\ F_t \neq 0$.

Three different zones of the $\left\{\frac{F_t}{kh}, \frac{F_n}{kh}\right\}$ plane are represented in Fig. 11.9a. Zone 1 (light grey) where no equilibrium solution in contact exist, zone 2 (grey) where equilibrium solutions in contact exist when the coefficient of friction μ is sufficiently large, and zone 3 (dark grey) where equilibrium solutions in contact exist whatever the value of μ.

11.5.2 Zero External Force

When no external force is applied to the system. As stated at the end of Sect. 3, the equilibrium solutions out of contact are found on the part of the circle of center $(-X_0, -h)$ and radius $\sqrt{X_0^2+h^2}$ which is in the negative half plane. The two extremities of this arc ($(0, 0)$ and $(0, -2X_0)$) are therefore in grazing contact. We obtain from Eq. (11.16) that there still exist infinitely many equilibrium states represented on Fig. 11.9b in the $\{R_t, R_n\}$ plane.

Let us comment on this figure. The part of the curve in the $\{R_t, R_n\}$ plane which is in the Coulomb cone with a nonzero reaction corresponds to tangential positions which fill a closed interval included in the segment $[0, -2X_0]$ whose length depends on the value of μ. The set of equilibria in contact without any external force is the following.

- For $\mu < h/X_0$ it consists of the closed interval of positions $[-X_0 + h/\mu, -X_0 - h/\mu]$ strictly included in the interval $(0, -2X_0)$ on the u_t axis, and the points $(u_t = 0, u_n = 0)$ and $(u_t = -2X_0, u_n = 0)$.
- When μ increases the extremities of the interval $[-X_0 + h/\mu, -X_0 - h/\mu]$ end up reaching the points $(u_t = 0, u_n = 0)$ and $(u_t = -2X_0, u_n = 0)$ so that the set of equilibria consists of the closed interval $[0, -2X_0]$ on the u_t axis; and it is immediate that this interval cannot be extended by continuing to increase μ since any point outside this interval corresponds to an extended length of the spring which is not possible without external force.
- Observe that in the limit case $\mu = 0$, the whole curve represented on Fig. 11.9b shrinks to the single point $\{R_t = 0, R_n = kh(1-\beta)\}$, so that in this limit case there are only three equilibrium positions, $(u_t = 0, u_n = 0)$, $(u_t = -X_0, u_n = 0)$ and $(u_t = -2X_0, u_n = 0)$. We can moreover observe that the spring is compressed at the point $(u_n = 0, u_t = -X_0)$, which obviously means instability in the case without friction, while the spring is at rest in the two other positions. So that we recover exactly the set of equilibria of the SD oscillator without friction.

Remark 11.7 Whenever there are no equilibrium solutions in contact, there exist equilibrium solutions out of contact, and conversely whenever there are no equilibrium solutions out of contact, there exist equilibrium solutions in contact. Moreover, equilibrium solutions in contact and out of contact may exist simultaneously, as can be seen by superimposing Figs. 11.4 and 11.9a.

Only as a qualitative comment, we observe the difference between this analysis of the set of equilibrium states, which lasts for the about 10 pages, and the corresponding analysis of equilibria for the SD oscillator of the previous chapters. We shall comment later on this point and obtain the correlation between the results.

11.6 Discontinuous Oscillator with Contact and Friction

In order to avoid difficulties due to $h = 0$, we start from the initial two-dimensional equations, without introducing parameter β:

$$
\left\{
\begin{aligned}
\mathcal{N}_t(u_t, u_n) &= -k(X_0 + u_t) + \frac{k(X_0 + u_t)\sqrt{X_0^2 + h^2}}{\sqrt{(X_0 + u_t)^2 + (h + u_n)^2}}, \\
\mathcal{N}_n(u_t, u_n) &= -k(h + u_n) + \frac{k(h + u_n)\sqrt{X_0^2 + h^2}}{\sqrt{(X_0 + u_t)^2 + (h + u_n)^2}},
\end{aligned}
\right.
\tag{11.18}
$$

from which the equilibrium system is written as

$$
\left\{
\begin{aligned}
& k(X_0 + u_t)\left(1 - \frac{\sqrt{X_0^2 + h^2}}{\sqrt{(X_0 + u_t)^2 + (u_n + h)^2}}\right) = F_t + R_t, \\
& k(u_n + h)\left(1 - \frac{\sqrt{X_0^2 + h^2}}{\sqrt{(X_0 + u_t)^2 + (u_n + h)^2}}\right) = F_n + R_n, \\
& u_n \leq 0, \;\; R_n \leq 0, \;\; u_n R_n = 0, \;\; |R_t| \leq -\mu R_n.
\end{aligned}
\right.
\tag{11.19}
$$

We aim at studying the solutions to system (11.19) in the case $h = 0$. Recalling that X_0 is a positive number, this gives:

$$
\left\{
\begin{aligned}
& k(X_0 + u_t)\left(1 - \frac{X_0}{\sqrt{(X_0 + u_t)^2 + u_n^2}}\right) = F_t + R_t, \\
& k u_n\left(1 - \frac{X_0}{\sqrt{(X_0 + u_t)^2 + u_n^2}}\right) = F_n + R_n, \\
& u_n \leq 0, \;\; R_n \leq 0, \;\; u_n R_n = 0, \;\; |R_t| \leq -\mu R_n,
\end{aligned}
\right.
\tag{11.20}
$$

11.6.1 Existence of Solutions Out of Contact

In the same way as what has been done in the previous sections, we use the condition that solutions out of contact should satisfy $R_t = R_n = 0$, $u_n < 0$. This gives the following system

$$\begin{cases} k(X_0 + u_t)\left(1 - \dfrac{X_0}{\sqrt{(X_0 + u_t)^2 + u_n^2}}\right) = F_t, \\ ku_n\left(1 - \dfrac{X_0}{\sqrt{(X_0 + u_t)^2 + u_n^2}}\right) = F_n, \\ u_n < 0. \end{cases} \tag{11.21}$$

11.6.1.1 The Case $F_n \neq 0$

Elementary calculations change system (11.21) into:

$$\begin{cases} k(X_0 + u_t)\left(1 - \dfrac{X_0}{\sqrt{(X_0 + u_t)^2 + u_n^2}}\right) = F_t, \\ X_0 + u_t = \dfrac{F_t}{F_n} u_n, \\ u_n < 0, \end{cases} \tag{11.22}$$

which in turn gives, taking $u_n < 0$ into account

$$\begin{cases} u_t = \dfrac{F_t}{k} - X_0\left(1 + \dfrac{1}{\sqrt{1 + \left(\dfrac{F_t}{F_n}\right)^2}}\right) \\ u_n = \dfrac{F_n}{k} - \dfrac{X_0}{\sqrt{1 + \left(\dfrac{F_t}{F_n}\right)^2}}. \end{cases} \tag{11.23}$$

11.6.1.2 The Case $F_n = 0$

Again we start from system (11.21). Since u_n cannot be equal to zero for we are looking for solutions out of contact, we have necessarily:

$$1 - \frac{X_0}{\sqrt{(X_0 + u_t)^2 + u_n^2}} = 0$$

which implies that a solution is possible only if F_t is also equal to zero. In such a case there is a set of equilibria given by:

$$\begin{cases} u_t = -X_0 \left(1 \pm \sqrt{1 - \left(\frac{u_n}{X_0}\right)^2}\right), \\ -X_0 \leq u_n < 0. \end{cases} \tag{11.24}$$

It is interesting to observe that even in the case $h = 0$, there exist one solution out of contact in general for any negative F_n and for F_n positive sufficiently small, and there also exist equilibrium solutions out of contact when both components of the external force vanish.

11.6.2 Solutions in Contact

In this case equilibrium solutions involve a nonzero reaction, but the displacement is such that $u_n = 0$. The equilibrium system is consequently:

$$\begin{cases} k(X_0 + u_t)\left(1 - \dfrac{X_0}{\sqrt{(X_0 + u_t)^2}}\right) = F_t + R_t, \\ 0 = F_n + R_n, \\ |R_t| \leq -\mu R_n, \end{cases} \tag{11.25}$$

which can be changed into:

$$\begin{cases} k(X_0 + u_t) \; - kX_0 \, sgn(X_0 + u_t) = F_t + R_t, \\ 0 = F_n + R_n, \\ |R_t| \leq -\mu R_n, \end{cases} \tag{11.26}$$

from which we easily get that there exist two intervals of equilibrium solutions in contact for any positive F_n, which read:

$$\begin{cases} R_n = C^{st} = -F_n \\ \text{when } (X_0 + u_t) > 0 : \; \dfrac{F_t - \mu F_n}{k} \leq u_t \leq \dfrac{F_t + \mu F_n}{k}, \\ \text{when } (X_0 + u_t) < 0 : \; \dfrac{F_t - \mu F_n}{k} - 2X_0 \leq u_t \leq \dfrac{F_t + \mu F_n}{k} - 2X_0. \end{cases} \tag{11.27}$$

In particular when $F_t = 0$, these intervals reduce to:

$$\begin{cases} R_n = C^{st} = -F_n \\ \text{when } (X_0 + u_t) > 0: \ -\dfrac{\mu F_n}{k} \leq u_t \leq \dfrac{\mu F_n}{k}, \\ \text{when } (X_0 + u_t) < 0: \ -\dfrac{\mu F_n}{k} - 2X_0 \leq u_t \leq \dfrac{\mu F_n}{k} - 2X_0, \end{cases} \tag{11.28}$$

which means that for given F_n positive, all the equilibria in contact have the same normal component of the reaction and fill two intervals of length $2\mu F_n/k$, respectively around the trivial equilibrium positions $u_t = 0$ and $u_t = -2X_0$.

There are no equilibrium states in contact when F_n is negative, but in the same way as when h was different from zero, we shall have to keep in mind that the sets of equilibria in contact and without contact, overlap when associated with some values of the force.

11.7 Introduction to the Dynamics

The state of the analysis after the previous sections can be summarized as follows.

- The dynamics of the simple model which simply extends the kinematics and the mechanical conditions of the SD oscillator to the case where impacts and friction may occur, is well posed, due to the well-posedness of Cauchy problem (11.4) with geometrical nonlinearity (11.5).
- The investigation of the set of equilibrium states of problem (11.4) has been completed for any admissible values of the parameters.
- The qualitative analysis of the dynamics is widely open due to the fact that the Cauchy problem does not deal with a system of ordinary differential equations.

We shall first give some preliminary tools or results for the calculation of a trajectory and then apply to some hints for stability analysis. The dynamics under external oscillating loads, existence of equilibrium states or of periodic solutions, will be only suggested together with new concepts of stability.

11.7.1 Preliminaries

Although tools for going through some impact times in general motions have been given and introduced in a software (see [6]), we now restrict our attention to the case of sliding motion. This choice leads to some simplifications and the corresponding motions will remain closer to the SD oscillator. So let us specify Eq. (11.4) to the case of trajectories that remain continuously in contact. We obtain

$$\begin{cases} i)\ m\ddot{u}_t = \mathcal{N}_t(u_t, 0) + F_t + R_t, & \\ & t > 0 \\ ii)\ 0 = \mathcal{N}_n(u_t, 0) + F_n + R_n, & \\ iii)\ u_t(0) = u_0,\ \dot{u}_t(0) = v_0, & \\ iv)\ u_n \equiv 0, & \\ v)\ \forall \varphi \in \mathbb{C}^0((0,T];\mathbb{R}),\ \displaystyle\int_{[0,T]} R_t(\varphi - \dot{u}_t^+) - \mu R_n(|\varphi| - |\dot{u}_t^+|) \geq 0. & \end{cases} \tag{11.29}$$

No matter a possible motion is now strictly one-dimensional, the dynamics still involves two equations, namely (11.29-*i*) and (11.29-*ii*). The motion is submitted to the constraint $u_n \equiv 0$, but as a difference with the effect of the same constraint in the calculation of the equilibrium states, sliding occurs by assumption so that the reaction remains on the border of the friction cone all the motion long except when the trajectory reaches an equilibrium state. In other words, as long as we are trying to solve system (11.29), the reaction satisfies

$$R_t = \pm \mu R_n.$$

Straightforward calculations imply consequently that equations (11.29-*i*) and (11.29-*ii*) can be rewritten as

$$m\ddot{u}_t + k(X_0 + u_t - \mu h)\left[1 - \frac{h\beta}{\sqrt{(X_0 + u_t)^2 + h^2}}\right] = F_t - \mu F_n, \tag{11.30}$$

for sliding to the right, and

$$m\ddot{u}_t + k(X_0 + u_t + \mu h)\left[1 - \frac{h\beta}{\sqrt{(X_0 + u_t)^2 + h^2}}\right] = F_t + \mu F_n, \tag{11.31}$$

for sliding to the left. After a change of variable these equations will be more easily tackled in numerical calculations. By introducing a new unknown $y := \dfrac{u_t + X_0}{h}$, they become:

$$\ddot{y} + \frac{k}{m}(y - \mu)\left[1 - \frac{\beta}{\sqrt{y^2 + 1}}\right] = \frac{F_t - \mu F_n}{mh}, \tag{11.32}$$

$$\ddot{y} + \frac{k}{m}(y + \mu)\left[1 - \frac{\beta}{\sqrt{y^2 + 1}}\right] = \frac{F_t + \mu F_n}{mh}. \tag{11.33}$$

Remark 11.8 Assume that one of these equations is chosen and fixed, (11.32) for example, and let us study the corresponding dynamical system. We would immediately obtain that in the unloaded case this system has three equilibrium states, namely:

$$\begin{cases} y = \mu, \\ y = \pm\sqrt{\beta^2 - 1}, \end{cases} \tag{11.34}$$

which is the result of the SD oscillator without friction! System (11.29) means that even in cases involving no loss of contact, the set of three equilibria given by (11.34) is completely different from the actual set of equilibria of system (11.29). The reason for this huge difference is that the choice of one condition among $R_t = \mu R_n$, $R_t = -\mu R_n$, or $|R_t| < \mu |R_n|$ to be inserted into system (11.29) is not a data but an intrinsic part of the solution, the only data is the set of equations (11.1) rewritten as condition (11.29-*iv*).

Equations (11.32) and (11.33) are classical ordinary differential equations, each of which can consequently be integrated using classical tools, e.g. a Runge–Kutta method. So that the only remaining question is: which of these two equations must be used at each time to calculate the trajectory? As summarized in the comments in Sect. 11.3, the dynamics is certainly well-posed as long as the external force is smooth enough, so that only one choice is possible. Due to the Coulomb's friction law, the equation to be used at the initial time to start the calculation is given by the sign of the initial data. What happens after at each time is deduced from the following result which we only present qualitatively here. The detailed proof can be found in [19].

Proposition 11.1 *When the tangential velocity under constant loading becomes equal to zero after a sliding phase, either the velocity stays equal to zero for all time or the velocity changes sign (i.e. the sliding changes direction).*

The idea of the proof of Proposition 11.1 consists in checking the potential compatibility of each possibility for the trajectory to continue, either staying at rest, or sliding in the same direction as before the stop, or sliding in the other direction i.e. jumping to the other side of the cone, etc., and referring to the well-posedness result when a compatible possibility is found.

11.7.2 Some Steps Towards a Stability Analysis

11.7.2.1 Solutions Strictly in Contact

We are now in position to have a first hint about stability properties. These properties will be obtained from a direct calculation of the trajectories. In the simplest case, the external force is assumed to have only a nonzero normal component which ensures the existence of a set of equilibria strictly in contact, and no tangential component so that the set of equilibrium states is symmetrical with respect to the origin in the phase space. Keeping the normal component of the force as a constant, the motion only results from the choice of initial data out of equilibrium. As a consequence of the dissipative character of friction, all the trajectories will converge to equilibrium, as represented in the phase space on Fig. 11.10.

Adding any nonzero tangential component to the external force, this figure looses its symmetry and is for example changed into Fig. 11.11.

Remaining in the symmetric case, when the subsets of equilibrium states collapse, as it has been investigated due to the changes of the parameters, Fig. 11.10 is qualitatively changed into Fig. 11.12. Of course the same collapse of equilibrium subsets has been obtained for some ranges of the parameters in the case $F_t \neq 0$, with the consequence that all the trajectories go to the single set of equilibrium states.

Let us come back to Fig. 11.10 and to the observation that the set of equilibrium states attained by the trajectory depends on the initial data. As a matter of fact, Fig. 11.10 displays, by a grey zone, the parts of the phase space from where trajectories lead either to one set of equilibria or to another one. This suggests that there exists some ranges in the phase space where the trajectory is extremely sensitive to the initial data and that there is not continuous dependance on data in general at least in the cases where the subsets of equilibrium states are strictly disjointed. This can be observed more clearly on the following Fig. 11.13.

Moreover, let us choose any admissible initial data and follow the trajectory. We can calculate that the trajectory attains one equilibrium in finite time. This is not only due to the dissipative character of Coulomb's friction, but also to the fact that the set of equilibrium states is made of intervals and not of isolated points. These qualitative features suggest on the one hand a stronger notion of stability than the classical ones, and on the other hand a strong sensitivity to initial data.

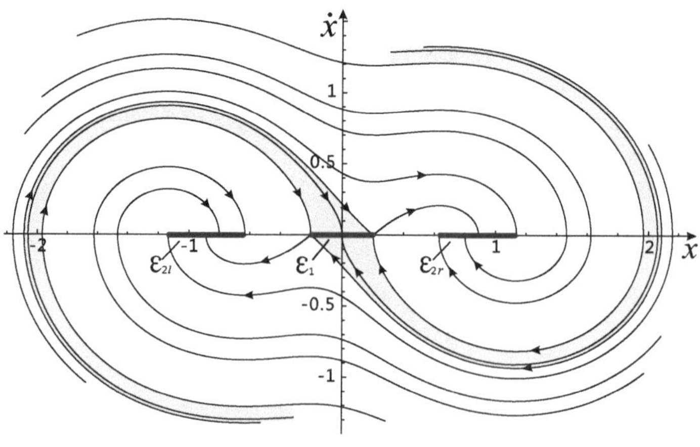

Fig. 11.10 Trajectories plotted in phase space when F_n is positive and F_t is equal to zero for $0 < \alpha < 1$

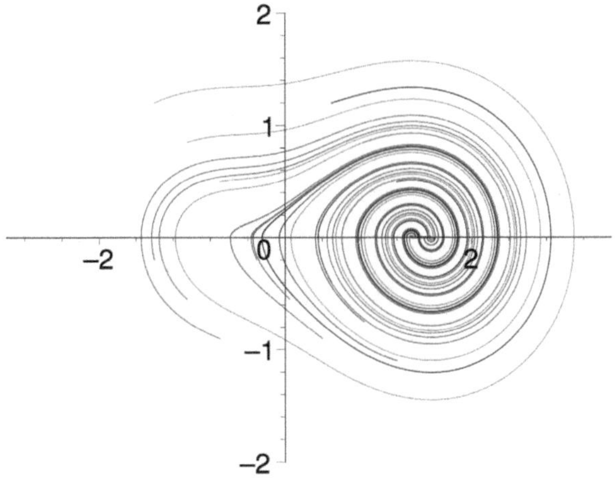

Fig. 11.11 Trajectories plotted in the phase space in the case where both F_n and F_t are strictly positive

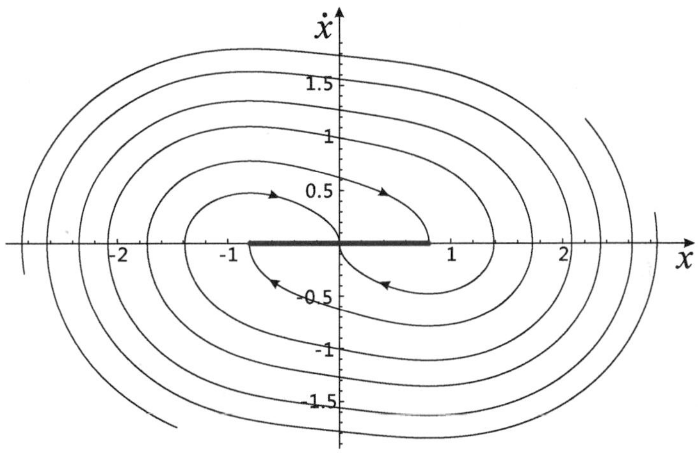

Fig. 11.12 Phase plane of the SD oscillator with dry friction for $\alpha = 1$

It is interesting for a complete understanding to observe such trajectories in the plane of the reactions. Not only two very different final states are reached in the phase space, with initial data very close to each others, but Fig. 11.14 shows that these final states are also strongly different in the plane of the reactions.

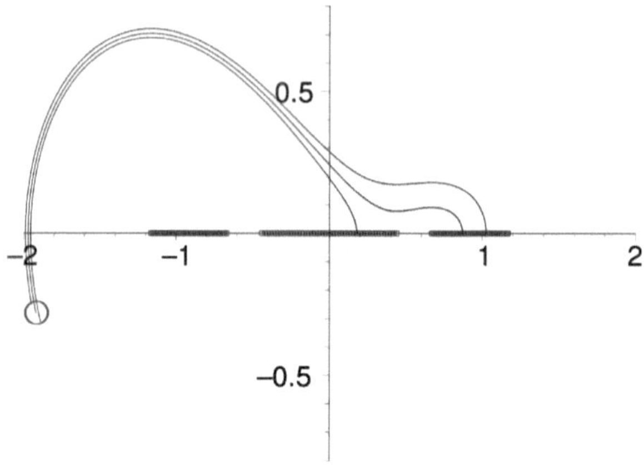

Fig. 11.13 Sensitivity to initial data: trajectories starting from initial data very close to each others in the small circle in the down left quarter of the phase space diverge and reach very different equilibrium states

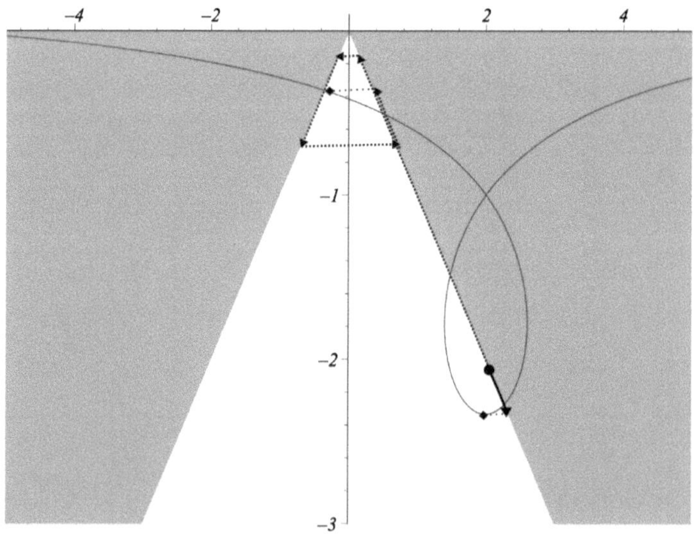

Fig. 11.14 Two trajectories starting from initial data very close to each others in the $\{R_t, R_n\}$ plane

11.7.2.2 Short Note About the Stability of Equilibrium States Out of Contact

Equilibrium states out of contact result only of the large strains and in particular nothing involve the presence of the obstacle. This means that the nonlinearity is only included in functions $\mathcal{N}_t(u_t, u_n)$ and $\mathcal{N}_n(u_t, u_n)$ which derive from a potential energy associated with the deformation of the spring. It is then known from Lejeune-Dirichlet theorem that an equilibrium configuration is stable if it achieves a strict local minimum of this potential. A simple and very classical calculation shows that a strict local minimum is obtained for $\bar{u}_t$, $\bar{u}_n$ if

$$\frac{\sqrt{X_0^2 + h^2}}{\sqrt{(X_0 + \bar{u}_t)^2 + (h + \bar{u}_n)^2}} < 1,$$

and it is easily obtained:

- *The following equilibrium solutions are stable:*
  - $(\bar{u}_t = \frac{F_t}{k} - X_0 + h\beta, \bar{u}_n = -h)$ *for* $F_n = 0$ *if* $F_t > 0$,
  - $(\bar{u}_t = \frac{F_t}{k} - X_0 - h\beta, \bar{u}_n = -h)$ *for* $F_n = 0$ *if* $F_t < 0$,
  - *all the equilibrium solutions satisfying condition 1 if* $F_n < 0$,
  - *all the equilibrium solutions satisfying condition 2 if* $F_n > 0$.
- *The following equilibrium solutions are unstable:*
  - $(\bar{u}_t = \frac{F_t}{k} - X_0 + h\beta, \bar{u}_n = -h)$ *for* $F_n = 0$ if $F_t < 0$,
  - $(\bar{u}_t = \frac{F_t}{k} - X_0 - h\beta, \bar{u}_n = -h)$ *for* $F_n = 0$ if $F_t > 0$,
  - *all the equilibrium solutions satisfying condition 1 if* $F_n > 0$,
  - *all the equilibrium solutions satisfying condition 2 if* $F_n < 0$.

11.7.2.3 The Discontinuous Case of the SD Oscillator

In the same way, we now focus our attention on the equilibrium sets in the discontinuous case when $\alpha = 0$ with friction. In this case, the equilibrium sets ε_1 in Fig. 11.10 shrinks to the single point (0, 0), while the two other stable equilibrium sets ε_{2l} and ε_{2r} remain centered at $(\pm 1, 0)$ with unchanged stability. The structure of the phase portrait of the discontinuous system with friction is shown in Fig. 11.15.

11.7.3 Qualitative Dynamics Under a Harmonic Excitation

Assume the external excitation is made of two parts: a constant one having a normal component which would ensure the existence of equilibrium states if it were alone, and harmonic part added on the tangential component, of period T and amplitude ε. We then have the following results, dealing with the continuous case of the SD oscillator with friction.

- *Assume the harmonic part of the load have a sufficiently small amplitude. Then there exists a whole range in the* $\{T, \varepsilon\}$ *plane of the excitation where there remain infinitely many equilibrium states, and where all the trajectories lead to equilibrium states in finite time.*

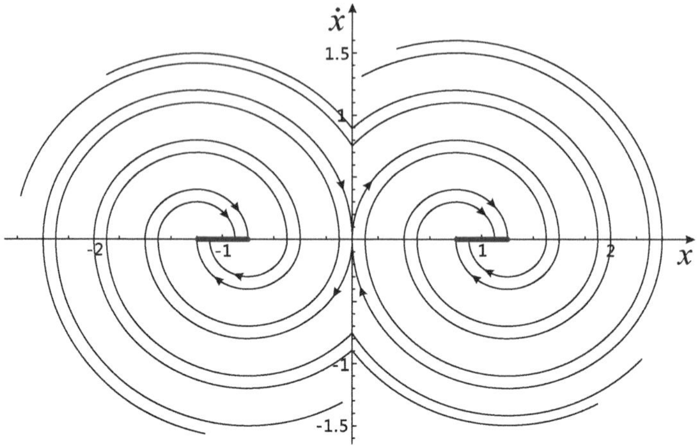

Fig. 11.15 Phase plane of the SD oscillator with dry friction in the discontinuous case where the sets of equilibrium states are symmetrical with respect to the origin of the phase space

Remark 11.9 This result looks very much like the one observed when dealing with the stability of equilibrium states, but now the dynamical system have a harmonic right-hand side!

- *Assume the harmonic part of the load have a sufficiently large amplitude. Then there no longer exist equilibrium states, but there exist periodic solutions for any period of the excitation. Moreover*
  (i) for small periods there exists everywhere a single periodic solution which is of small amplitude, let T^* be the upper bound of this range;
  (ii) for large periods there exists everywhere a single periodic solution which is still of small amplitude but different from the one of the case of small period, let T^{**} be the lower bound of this range;

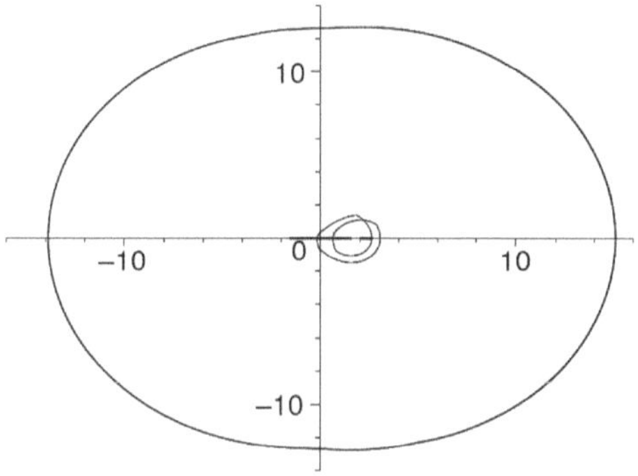

Fig. 11.16 Two periodic trajectories for $T \in [T^*, T^{**}]$: the periodic one solution with large amplitude and the periodic two solution with a small amplitude

(iii) the interval between these two ranges does not reduce to the empty set, and everywhere in this interval there exist at least two periodic solutions, one of large amplitude and of the period of the excitation, and another one of small amplitude and of a period at least twice the period of the excitation.
The later situation is represented on Fig. 11.16.

11.7.4 Suggestions for Further Stability Analyzes

Previous stability analyzes of the equilibrium states of systems involving non regularized contact and friction conditions consisted in perturbing an equilibrium state in a classical phase space and then calculating the evolution in time of the distance between the initial equilibrium state and the trajectory having the perturbed state as initial data (see e.g. [10]). These analyzes consisted in perturbing an equilibrium state in a classical phase space and then calculating the evolution in time of the distance between the initial equilibrium state and the trajectory having the perturbed state as initial data. If any neighborhood of the equilibrium contains a point which, taken as initial data of the dynamical problem, leads to a trajectory which diverges from the equilibrium, then the equilibrium is said to be unstable. On the contrary, if taking any point of a neighborhood of the equilibrium as initial data one gets a trajectory which tends to the equilibrium or remains in a tubular neighborhood of the equilibrium, then the equilibrium is said to be asymptotically stable, or stable in the sense of Lyapunov. But it is clear that such an analysis can be undergone only if the system is sufficiently simple, and this has been investigated only for systems made of one or two particles. Moreover the elementary and classical notion of stability which justifies the analysis may seem not really satisfactory in view of the graph of the Coulomb law. Indeed, perturbing a given strictly stuck equilibrium by a tangential velocity may require a large perturbation of the external forces no matter how small the velocity is. This means in turn that it is quite possible that an equilibrium defined by ($U = U^{eq}$, $\dot{U} = 0$) is not modified by adding any relatively small external forces, which corresponds exactly to what have just been stressed in the present section.

We consequently give a conjecture which can be formulated in the following way:

Conjecture 1 *Let a discrete system with any finite number of degrees of freedom be submitted to unilateral contact and Coulomb friction. Assume the data are such that there exists an equilibrium state in which some reactions are strictly inside the Coulomb cone while the other reactions are in imminent sliding. Then the trajectory produced by any sufficiently small perturbation of the data leads to a new equilibrium where the number of reactions strictly inside the cone is larger than before the perturbation.*

Comments

- *A state is said to be in imminent sliding to mean that the particle is motionless but the reaction is at the border of the cone, so that an infinitely small perturbation may set the particle into motion.*
- *It is equivalent to saying that an equilibrium state* ($U = U^{eq}, \dot{U} = 0$) *is not perturbed by a small enough external force, or to saying that its reaction is strictly inside the Cone. This is the foundation of the conjecture.*
- *This conjecture has been partially justified, in fact completely proved in the case of simple systems and backed up by numerical computations for larger size systems (see e.g.* [11]*).*

This conjecture might give hints for stability analysis of systems involving non regularized unilateral contact and Coulomb friction. In the case of any simple mass-spring system in which an equilibrium solution is perturbed by a sufficiently small force, it corresponds exactly to what has been presented here, namely:

Conjecture 1bis *Let a discrete mass-spring system with unilateral contact and Coulomb friction at equilibrium be perturbed by a sufficiently small constant force. Then the trajectory leads in finite time to an equilibrium where all the reactions are strictly inside the Coulomb cone.*

11.8 Summary

This chapter has introduced contact and friction in the SD oscillator, which will enable to understand the behaviour of systems which become very close to physics due to sliding and jumping effects. As it is seen from the graph given at the beginning of this chapter, the contact and friction laws that have been chosen seem rather simple, but they are really as close to physics as we can. Of course they lead to mathematical difficulties but they already gave very interesting results, specially about the sets of equilibria.

To address the future study for this system, some short comments are necessary:

The nonlinearity of contact and friction alone lead to a very unusual feature of the set of equilibrium states. For any classical dynamical system, the set of equilibrium states is made of a discrete set of points, but here, due to the fact that the right hand sides involve mathematical objects which are not functions, the set of equilibrium states is generically made of an interval, possibly unbounded; only the case without contact involves a single point. This situation can nevertheless be calculated and is completely determined by the knowledge of the stiffness parameters, the friction coefficient, the components of the external force.

Coupling contact and friction with the nonlinearity of the SD oscillator, that is with a geometrical nonlinearity, changes drastically the set of equilibria. By comparison with the nonlinearity of the SD oscillator without friction alone, the set of equilibria is changed from three separated points to infinitely many points which fill completely one or several intervals. By comparison with the case of contact and friction alone with linear geometry, the intervals of equilibria are strongly deformed, becoming parts of curves, possibly not connected. Although very intricate, these features have now been completely studied.

Concerning the equation of the dynamics, things are not as clear. Only the case of the linear stiffness have been, more or less, completely explored in the case of persistent contact, but it is not presented here since out of the topics of the present book which is concentrated on the SD oscillator (see [3, 19] for details). The dynamics of the SD oscillator with impacts and friction is far from being completely understood. It is now only partially investigated in the case where contact is maintained for any time (see [20]). The investigation is completely open and probably difficult when the trajectories involve impacts, which means that generically the stability questions are also open [21].

References

1. Elkoulani, A., & Léger, A. (1996). Bifurcated solutions of the initial velocity problem for an elasto-plastic beam. *Comptes Rendus de l Académie des Sciences - Series I - Mathematics*, *322*(10), 1007–1013.
2. Cimetiére, A., & Léger, A. (1996). Some problems about elastic-plastic post-buckling. *International Journal of Solids Structures*, *32*(10), 1519–1533.
3. Pratt, E., & Léger, A. (2011). Qualitative analysis of a forced nonsmooth oscillator with contact and friction. *Annals of Solid and Structural Mechanics*, *2*(1), 1–17.
4. Moreau, J. J. (1988). *Unilateral contact and dry friction in finite freedom dynamics*. Vienna: Springer.
5. Schatzman, M. (1978). A class of differential equations of second order in time. *Nonlinear Analysis*, *2*(3), 355–373.
6. Radjai, Farhang, & Richefeu, Vincent. (2009). Contact dynamics as a nonsmooth discrete element method. *Mechanics of Materials*, *41*(6), 715–728.
7. Basseville, S., Léger, A., & Pratt, E. (2003). Investigation of the equilibrium states and their stability for a simple model with unilateral contact and coulomb friction. *Archive Applied Mechanics*, *73*(5–6), 409–420.
8. Brezis, H. (1973). *Opérateurs maximaux monotones et semi-groupes de contraction dans les espaces de Hilbert, North-Holland Mathematics Studies, No. 5. Notas de Matematica (50)*. Amsterdam and Lodon: North-Holland Publishing Company Ltd.
9. Ballard, P., & Basseville, S. (2005). Existence and uniqueness for dynamical unilateral contact with coulomb friction: a model problem. *Mathematical Modelling and Numerical Analysis*, *39*(1), 57–77.
10. Basseville, S., & Léger, A. (2006). Stability of equilibrium states in a simple system with unilateral contact and coulomb friction. *Archive Applied Mechanics*, *76*(7/8), 403–428.
11. Pratt, E., Léger, A., & Jean, M. (2010). About a stability conjecture concerning unilateral contact with friction. *Nonlinear Dynamics*, *59*(1–2), 73–94.
12. Moreau, J. J. (1988). Bounded variation in time. In J. J. Moreau, P. D. Panagiotopoulos, & G. Strang (Eds.), *Topics in nonsmooth mechanics* (pp. 1–14). Basel: Birkhäuser.
13. Monteiro Marques, M. D. P. (1993). *Differential inclusions in nonsmooth mechanical problems*. Basel: Birkhäuser.
14. Percivale, D. (1985). Uniqueness in the elastic bounce problem. *Journal of Differential Equations*, *56*(2), 206–215.
15. Klepp, H. J. (2010). Existence and uniqueness of solutions for accelerations for multi-body systems with friction. *Zamm Journal of Applied Mathematics and Mechanics Zeitschrift Fur Angewandte Mathematik Und Mechanik*, *75*(9), 679–689.

16. Charles, A., & Ballard, P. (2014). Existence and uniqueness of solutions to dynamical unilateral contact problems with coulomb friction: the case of a collection of points. *Mathematical Modelling and Numerical Analysis*, *48*(1), 1–25.
17. Geymonat, G., & Léger, A. (1999). Nonlinear spherical caps and associated plate and membrane problems. *Journal of Elasticity*, *57*(3), 171–200.
18. Léger, A., & Miara, B. (2005). On the multiplicity of solutions to marguerre von kármán membrane equations. *Journal de Mathématiques Pures et Appliquées*, *84*(3), 357–374.
19. Léger, A., Pratt, E., & Zhang, X. (2013). Study of a transition in the qualitative behavior of a simple oscillator with coulomb friction. *Nonlinear Dynamics*, *74*(3), 517–531.
20. Pratt, E., Léger, A., & Jean, M. (2009). Stability in unilateral contact problems with dry friction. *Lecture Notes in Applied and Computational Mechanics*, *56*, 13–30.
21. Léger, A., & Pratt, E. (2016). *Qualitative analysis of nonsmooth dynamics*. London: ISTE Editions.

Chapter 12
Experimental Investigations

Attention is now turned on experimental investigations in order to show the engineering applications of the SD oscillator, which might actually work well and usefully in engineering. In the first part of this chapter we setup a mechanical experimental rig for $\alpha > 0$ involving rigid foundation, guiding bar and springs which is close to the theoretical model itself. This device is designed to be adjustable to detect the complex nonlinear dynamical phenomena when parameters changes. The results obtained experimentally are in a good agreement with the results presented in the previous chapters. The second part deals with building the corresponding electronic devices of nonlinear circuits for both smooth when $\alpha > 0$ and discontinuous when $\alpha = 0$. Although this investigation is at the very beginning by the analogy of the model with electric or electronic systems, the results obtained are useful for a better understanding of the nonlinear phenomena of the SD oscillator.[1]

12.1 Introduction

Little attention has been paid to the experimental investigation of this kind of nonlinear systems due to the barrier of irrational nonlinearity [1–3] to the conventional nonlinear methodologies. Single-well nonlinear oscillator connected with a vibrating mass was constructed to study nonlinear vibration responses under sinusoidal frequency excitation [4]. Dynamical behaviours of vibration absorber with cubic nonlinear rigidity and two degrees of freedom were studied by experiments and the vibration absorbing characteristics of this system were discussed in [5, 6]. Resonance responses of a two degree of freedom system consisting of linear and nonlinear rigid

[1] This chapter relies on a common work with Professors Enli Chen, Cunzhi Pan, Mr. Ming Feng, Mr. Kai Guo and Dr. Xinwei Yang, School of Mechanical Engineering, Shijiazhuang Tiedao University, Shijiazhuang 050043 China.

Q. Cao and A. Léger, *A Smooth and Discontinuous Oscillator*, Springer Tracts in Mechanical Engineering, DOI 10.1007/978-3-662-53094-8_12

bars had been given, and nonlinear vibrations of visco-elastic conveyor belt was analyzed experimentally in [7].

We first set up a way to investigate the complex phenomena that occur in engineering applications with both large displacement and micro-scale vibrations for nonlinear system with irrational type [3, 8]. The experimental rig for the case of $\alpha > 0$ and the data collecting system is built for the parameters of the system to be adjustable and measurable. Harmonic excitation is introduced to explore the occurrence of periodic solutions, bifurcations, chaos and transitions between these regimes, which is in a good agreements with that obtained theoretically. Using the nonlinear circuits elements, the square root, the derivatives and the power function and the Sign function as well to build up the devices for both $\alpha > 0$ and $\alpha = 0$, respectively. The investigation via both the mechanical experimental rig and the nonlinear circuit devices are a good start to show the nonlinear phenomena of the SD oscillator.

Recall the dimensionless equation of the SD oscillator written in the following form.

- In the smooth case of $\alpha > 0$:

$$\ddot{x} + 2\xi\dot{x} + \omega_0^2\left(1 - \frac{1}{\sqrt{x^2 + \alpha^2}}\right) = f_0 \cos\omega t, \tag{12.1}$$

- In the discontinuous case of $\alpha = 0$:

$$\ddot{x} + 2\xi\dot{x} + \omega_0^2\,(x - Sign(x)) \;=\; f_0 \cos\omega t. \tag{12.2}$$

The meaning of the parameters has been given in the previous chapters. The following sections show the experimental investigations via mechanical and electronic devices.

12.2 Mechanical Experiment for $\alpha > 0$

12.2.1 The Experimental Rig

The experimental rig is built according to the mechanical model of the SD oscillator shown in Fig. 2.1. As shown in Fig. 12.1, the oscillator is mounted on an electrodynamic vibration shaker with a data collecting system, shown in Fig. 12.2.

More precisely, the mechanical oscillator comprises a mass with a pillar which connects two pairs of symmetrical springs. The other ends of the springs are connected to the moving mass as shown in Fig. 12.1. The system involves a horizontal shaker, and a level guide for the mass to slide with a reasonably small friction along a direction parallel to the shaker. The data collecting system involves a high speed camera connected to a computer as shown in Fig. 12.3. The system parameter in this

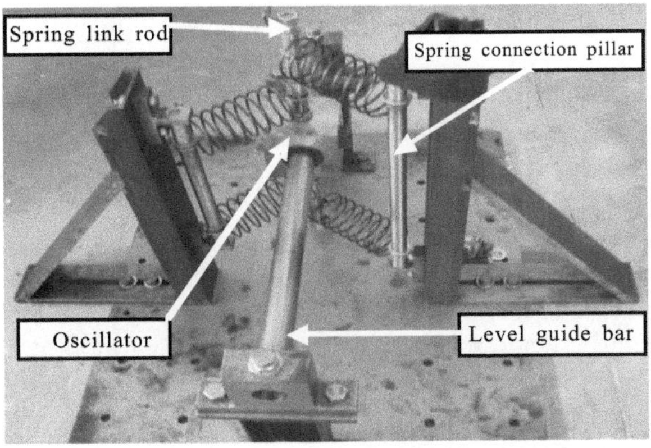

Fig. 12.1 Mechanical SD oscillator for experimental investigations with a mass vibrating along a guide bar with line-bearing

Fig. 12.2 Experiment system with the SD oscillator, shake and the data collecting system

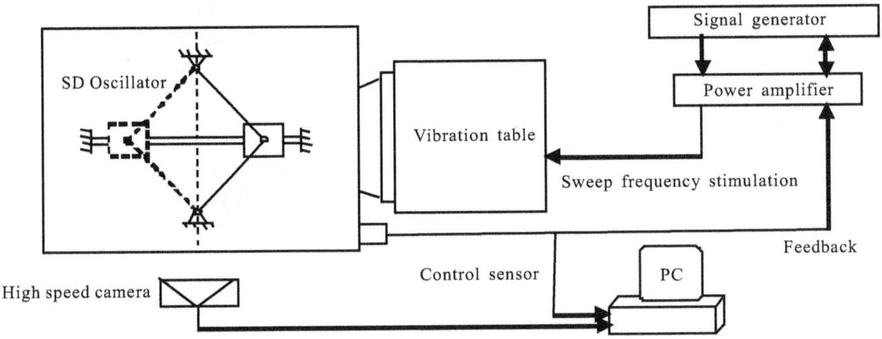

Fig. 12.3 Experiment system ($\alpha > 0$)

experimental rig is adjustable by changing the length l of the half distance between the rigid rods.

The experimental device is fixed on the level plane of an electrodynamic vibration shaker, which can slide along the level guide bar. The main devices are ES-10 electrodynamic vibration shaker, SD-PUMA vibration controller and CA-YD-186 acceleration transducer. The displacements of the oscillator are measured by a laser displacement meter.

The whole experimental process is divided into three procedures: the first one is the dynamical parameter identification for the experimental rig, such as the damping effects, which can be determined from the free damping motion; the second one deals with the dynamical responses of the oscillator under the different excitation frequencies in a broad range of frequency sweepings. The last procedure concerns the collected data during the experimental investigations motion status under different working conditions.

The responses of the oscillator at different excitation frequencies are collected, analyzed by time histories, frequency spectra, phase portraits and also Poincaré sections to identify the behaviours of the motion. During the experiment, the data collecting is done under the stable state by ignoring the initial status on the experimental results to obtain the steady state motion.

12.2.2 Experimental Rig Formulation

The part of the SD oscillator, shown in Fig. 12.4, consists of the clamped mass m and the pair of linear springs of stiffness k mounted on the shaker, which can be regarded as the SD oscillator with a vibrating base.

The absolute displacement of the mass is described as follows

$$Z = X + Y, \tag{12.3}$$

where X is the relative displacement of the oscillator to the shaker, Y is the displacement of the shaker. It is assumed that the viscous damping of the system is of

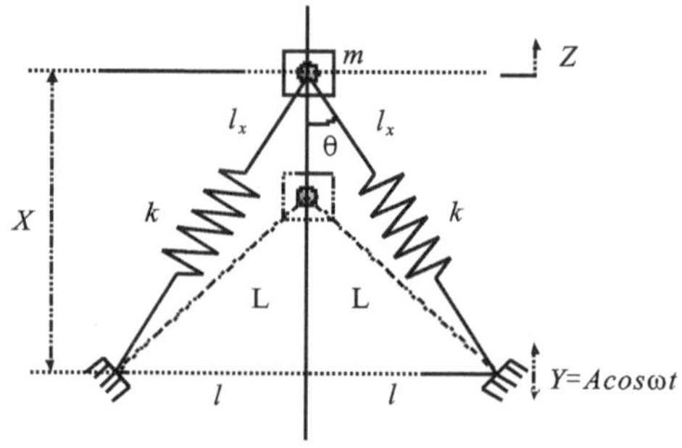

Fig. 12.4 SD oscillator with a vibrating base

coefficient c so that the equation of motion reads

$$m(\ddot{X} + \ddot{Y}) + c\dot{X} + 2k(\sqrt{X^2 + l^2} - L)\frac{X}{\sqrt{X^2 + l^2}} = 0. \tag{12.4}$$

Suppose that the vibration of the shaker is of the form of $Y = Acos(\omega t)$, it follows that

$$m\ddot{X} + c\dot{X} + 2kX(\sqrt{X^2 + l^2} - L)\frac{1}{\sqrt{X^2 + l^2}} = mA\omega^2 cos(\omega t), \tag{12.5}$$

which can be set in a dimensionless form by the same changes of variables and parameters as previously, leading to

$$x'' + \frac{c}{m\omega_0}x' + 2x\left(1 - \frac{1}{\sqrt{x^2 + \alpha^2}}\right) = \frac{A\omega^2 cos(\frac{\omega\tau}{\omega_0})}{L\omega_0^2}, \tag{12.6}$$

where ω_0 is the natural frequency. Rewriting Eq. (12.6) by letting $\eta = \dfrac{c}{m\omega_0}$, $F_0(\omega) = \dfrac{A\omega^2}{L\omega_0^2}$ it follows that

$$x'' + \eta x' + 2x\left(1 - \frac{1}{\sqrt{x^2 + \alpha^2}}\right) = F_0 cos(\frac{\omega\tau}{\omega_0}), \tag{12.7}$$

which is changed into a two dimensional system of the first order by using $x' = y$.

$$\begin{cases} x' = y, \\ y' = -\eta y - 2x\left(1 - \dfrac{1}{\sqrt{x^2 + \alpha^2}}\right) + F_0 cos(\frac{\omega\tau}{\omega_0}). \end{cases} \tag{12.8}$$

The unperturbed equilibrium stability can be obtained as the previous chapters. Recalling the property of the SD oscillator, we have that the unique stable equilibrium branch exists for the corresponding system when parameter $\alpha > 1$; two stable equilibrium branches exist for system when parameter $0 \leq \alpha < 1$; equilibrium bifurcation occurs when parameter $\alpha = 1.0$.

12.2.3 Data Analysis

Although the analysis has been carried out successfully with the dimensionless equations, the difficulties arise from the experimental investigation with the set up rig. The equilibrium lengths of the springs have been taken as 120 mm and the level distance $2l$ between the spring ends is 80.4 mm which implies that the dimensionless parameter $\alpha = 0.67$. A laser device is used to measure the displacements of the oscillator and the data can be collected and analyzed by using Matlab. The amplitude of the excitation and the frequency of the shaker are adjustable to get a deep insight of the complex dynamics including the time history, the spectrum, the trajectory and also the Poincaré sections.

The shaker on which the SD oscillator is mounted has an acceleration of 1 g, and an excitation amplitude Ap-p of 30 mm, the frequency band is adjustable from 1.5 to 50 Hz. The experimental data are collected when the shaker vibrates with sinusoidal status at a given frequency and the SD oscillator is stabilized. The first experimental investigation is detected for the set up rig when the excitation frequency is 2.75 Hz. A period one motion can be obtained, as shown in Fig. 12.5, with the time history, the spectrum, the phase portrait and the Poincaré section. These pictures show the vibration status with period one vibrating around the initial position with small amplitude. The next data collection is for the excitation frequency at 2.93 Hz, the oscillation behaves obviously as a regular vibration as shown in Fig. 12.6 where (a) is the time history, (b) the Spectrum, (c) is the Phase portrait and (d) is the Poincaré section. It can be seen that the oscillator vibrates with a large amplitude and meanwhile revolves around the pair of centers with a period five motion.

Figure 12.7 shows the period two motion when $f = 3.22$ Hz where again a is the time history, b the Spectrum, c the Phase portrait and d the Poincaré section.

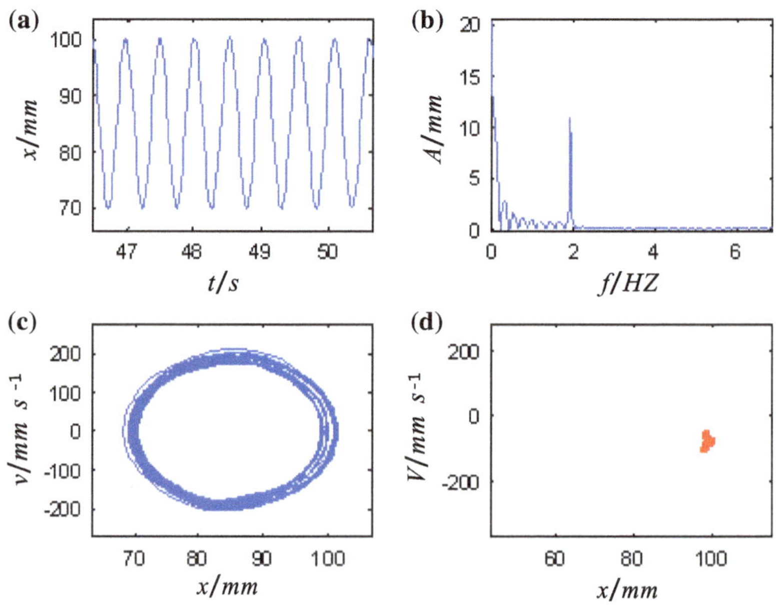

Fig. 12.5 Measured period one motion for $f = 2.75$ Hz: **a** Time history, **b** Spectrum, **c** Phase portrait and **d** Poincaré section

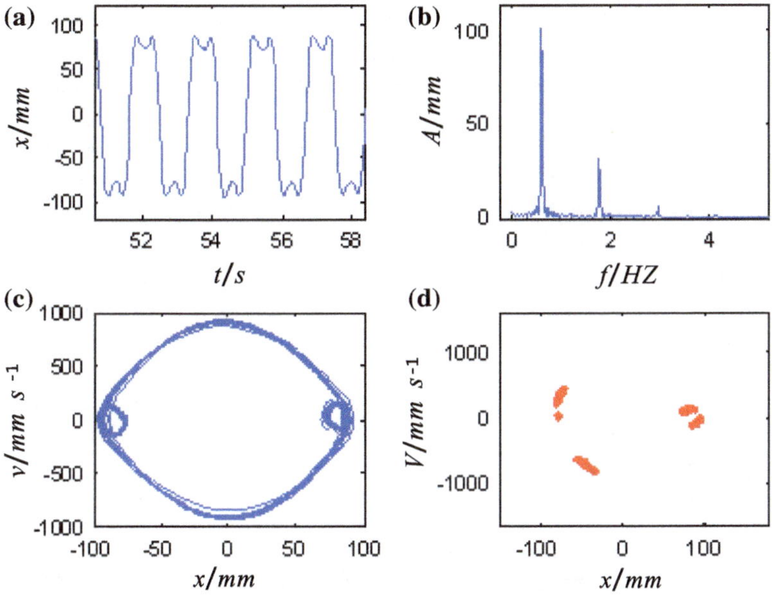

Fig. 12.6 Measured period five motion for $f = 2.93\,\text{Hz}$: **a** Time history, **b** Spectrum, **c** Phase portrait and **d** Poincaré section

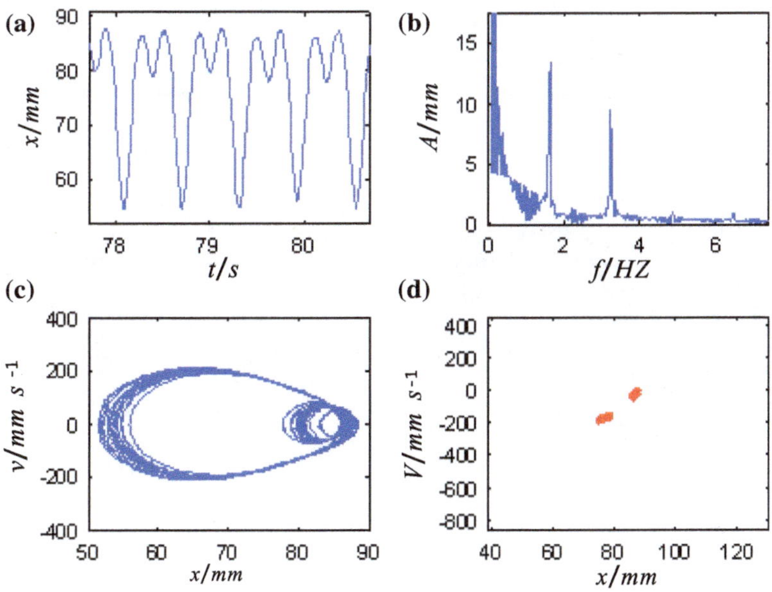

Fig. 12.7 Measured period two motion for $f = 3.22\,\text{Hz}$: **a** Time history, **b** Spectrum, **c** Phase portrait and **d** Poincaré section

A period two solution is demonstrated by the time history, the spectrum and the Poincaré section. It is clearly seen that the oscillator vibrates with two different amplitudes at this excitation and for the $\frac{1}{2}$ frequency, respectively, around one of the center point.

Chaotic behaviour can also be found when the excitation frequency is $f = 3.62\,\text{Hz}$, as shown in Fig. 12.8 with the same (a)–(d) as previously. The trajectory executes irregular motions around the pair of centers with large amplitude.

More complicated motions can be detected using the experimental rig under an excitation frequency of $f = 4.37\,\text{Hz}$. This is shown in Fig. 12.9 with (a) Time history, (b) Spectrum, (c) Phase portrait and (d) Poincaré section. The measured motion of the mass includes 2/7 and 4/7 of the excitation frequency. There are seven points in the Poincaré section, which means that the period seven motion is achieved. In the same way, a period three solution is observed via this device at the frequency $f = 5.35\,\text{Hz}$, as shown in Fig. 12.10 with (a) Time history, (b) Spectrum, (c) Phase portrait and (d) Poincaré section. It is clear that there are three points plotted in the Poincaré section.

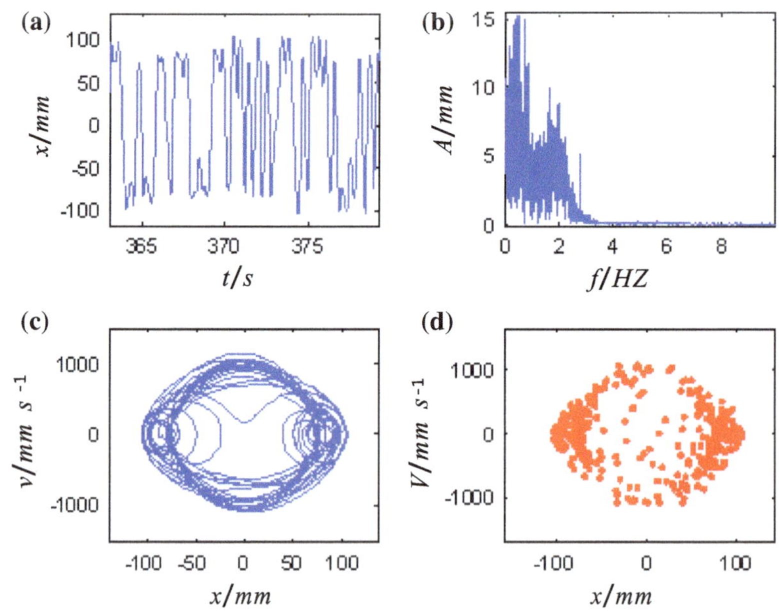

Fig. 12.8 Measured chaotic motion for $f = 3.62\,\text{Hz}$: **a** Time history, **b** Spectrum, **c** Phase portrait and **d** Poincaré section

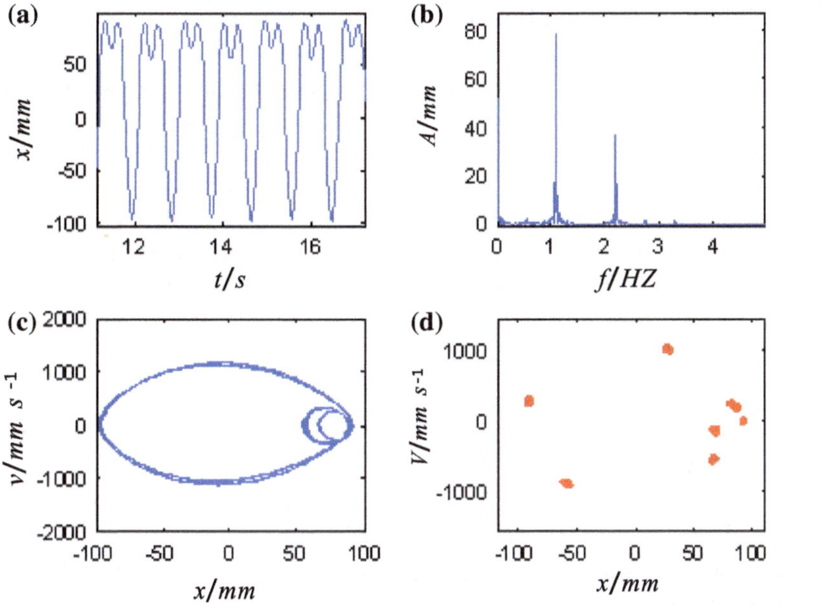

Fig. 12.9 Measured period seven motion for $f = 4.37\,\text{Hz}$: **a** Time history, **b** Spectrum, **c** Phase portrait and **d** Poincaré section

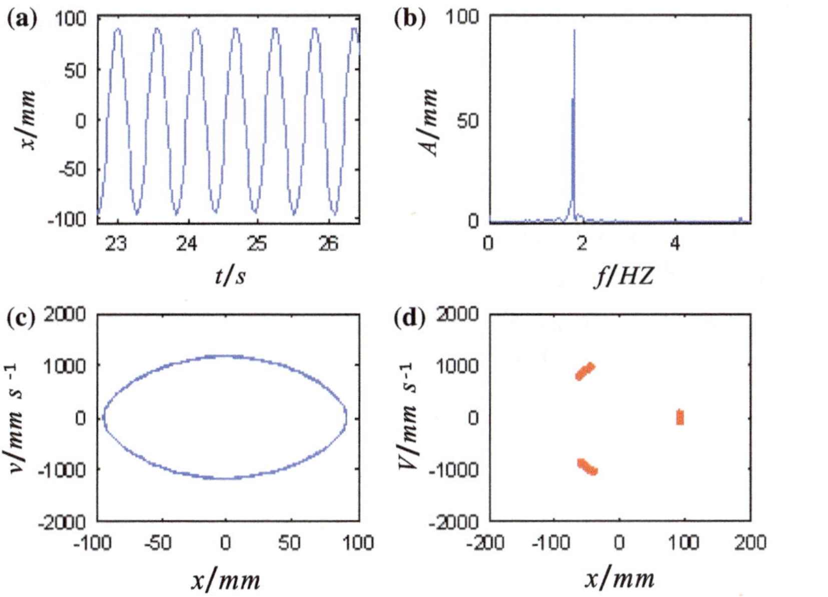

Fig. 12.10 Measured period three motion for $f = 5.35\,\text{Hz}$: **a** Time history, **b** Spectrum, **c** Phase portrait and **d** Poincaré section

12.3 An Electronic Device

Here we present an electronic device to investigate the behaviour of the SD oscillator. Using a modular design method the system is made of a specific combination of nonlinear circuits. The result of the modularity is, from the experimental point of view, a single printed circuit board with acquisition and processing devices completed by a symbolic calculation software. From the theoretical point of view, it allows to ensure that the system is in line with the SD oscillator. Lots of experiments have been performed showing, either in the unperturbed case or in the presence of damping and forcing, that the properties of the SD oscillator actually arise in electronic devices. Periodic solutions, bifurcations, transitions to chaos, have been observed as well.

12.3.1 Building the System

The SD oscillator has been built with elementary components which are springs, lumped mass and dampers with external forcing. It is suggested that the classical analogue method can be used to simulate the mechanical systems using electronic circuits to show the same behaviour as the mechanical system. We first recall the equations of SD oscillator. There exist constants, functions of square root, cosine, Sign, and motion x with its derivatives $\dot{x}$ and $\ddot{x}$, which can be modified by the corresponding circuits to built the system for both smooth and discontinuous oscillators.

It begins with the basic idea that the mechanical mass-spring oscillator with damping, as seen in Fig. 12.11a, can be transposed to a circuit oscillator of a RLC system, shown in Fig. 12.11b, in which the mass m corresponds to the inductance L, the external force $f(t)$ to the voltage $u(t)$, the damping coefficient c to the resistance R and the stiffness k refers to the reciprocal of the capacitance $\frac{1}{C}$. The motion x, the derivative $\dot{x}$ and the acceleration $\ddot{x}$ are converted respectively into the charge q, the intensity $i = \frac{dq}{dt}$ and the voltage $u = \frac{d^2q}{dt^2} = \frac{di}{dt}$.

The corresponding equations of both the mechanical oscillator and the circuit oscillator are written as follows

$$m\ddot{x} + c\dot{x} + kx = f(t), \tag{12.9}$$

and

$$L\frac{d^2q}{dt^2} + R\frac{dq}{dt} + \frac{1}{C}q = u(t). \tag{12.10}$$

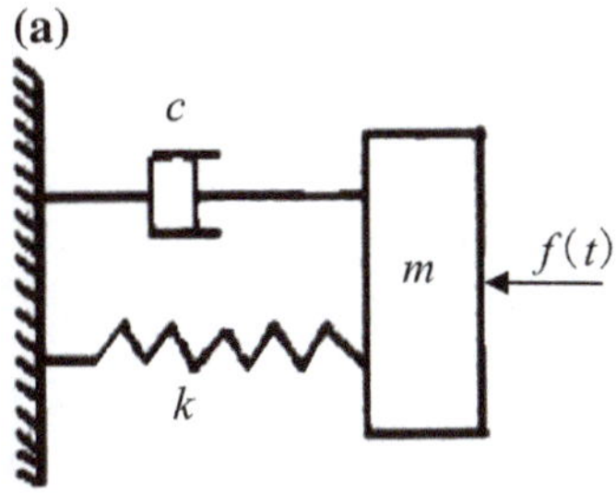

Fig. 12.11 Transition from **a** the mechanical mass spring oscillator with damping to **b** the RLC system

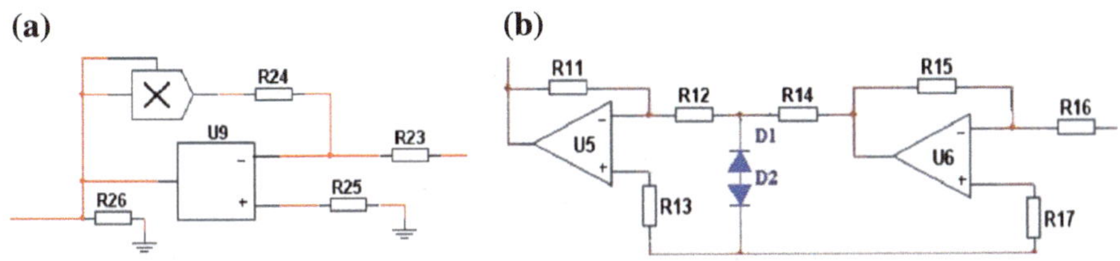

Fig. 12.12 **a** The circuit for nonlinearity with **square root** function and **b** The circuit for the nonlinearity with **Sign** function

Let's come back to the SD oscillator expessed by system (12.1) and (12.2) for the smooth case $\alpha > 0$ and discontinuous case $\alpha = 0$, respectively. Comparing with the mass spring system (12.9), there exists nonlinear terms involving **square root** function and **Sign** function of x.

In these way, we introduce the corresponding circuits for the components dealing with these two nonlinear functions, as represented in Fig. 12.12a for the **square root** function and Fig. 12.12b for the **Sign** function. This is given here without detailed explanation, the interested readers can refer to the references, see [9–11] for instance. Then we present the circuit units for both integration and differentiation which aim at capturing the frame structure information about the circuit SD oscillator. The corresponding circuits for the integration and differentiation units are shown in Fig. 12.13a and b, respectively.

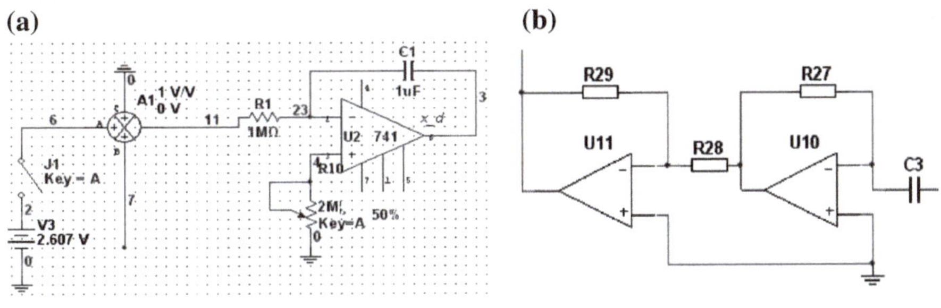

Fig. 12.13 **a** The integration subcircuit, **b** The differentiation subcircuit

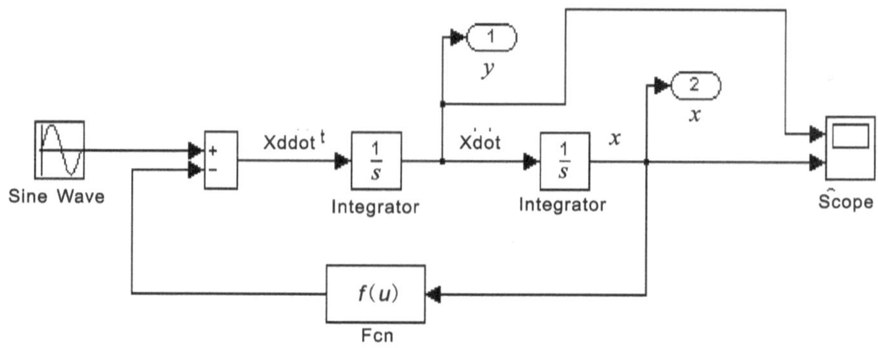

Fig. 12.14 The principle of the electronic modeling of the SD oscillator

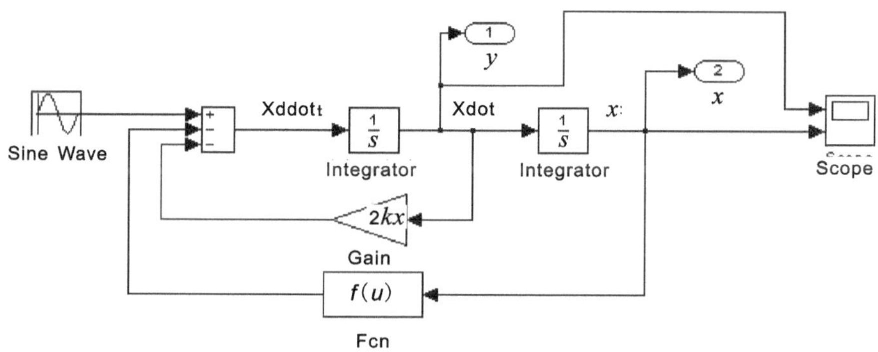

Fig. 12.15 The principle of the electronic modelling of the SD oscillator with a feedback of $2kx$

To get the best understanding of the nonlinear circuit for this SD oscillator, we employ the Matlab Simulink function to build the principal circuit structures by putting the units of arithmetics operations of addition, subtraction, multiplication and division together with the reciprocal to compile the units of differentiation, integration and the Sign, cosine and square root functions to both the smooth system, seen in Fig. 12.15, and the discontinuous system, as shown in Fig. 12.14, respectively.

Putting all these components together gives Figs. 12.18 and 12.19. Furthermore, the powerful software Multisim10, which is developed by the National Instruments of America, to build the circuit simulation system, shown in Figs. 12.16 and 12.17, respectively. With the help of simulation results the analog circuits for both smooth and discontinuous system can be designed, modified and improved successfully to reach the final version of the devices. The analog circuit are shown in Figs. 12.18 and 12.19, and the corresponding physical devices are shown in Figs. 12.20 and 12.21, respectively.

Although we have not got into more details about the design and the working process of the circuit, this gives a clear idea of how to build the system to change the mechanical oscillator into the circuit oscillator. The important advantage of the

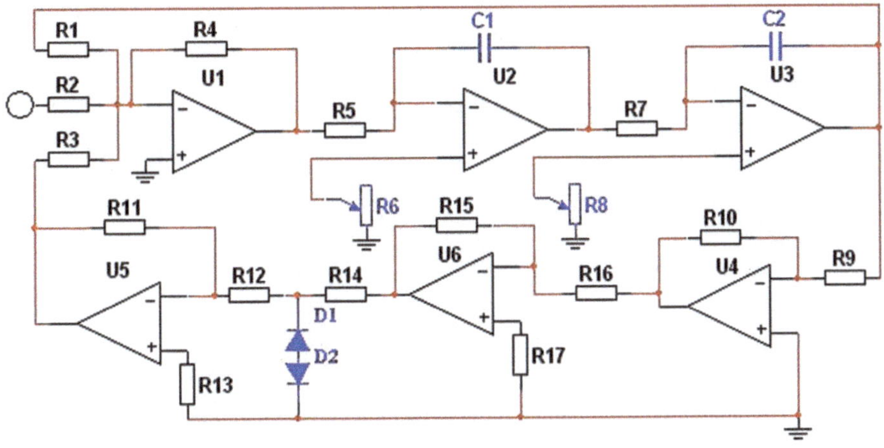

Fig. 12.16 Multisim 10 circuit of the SD oscillator

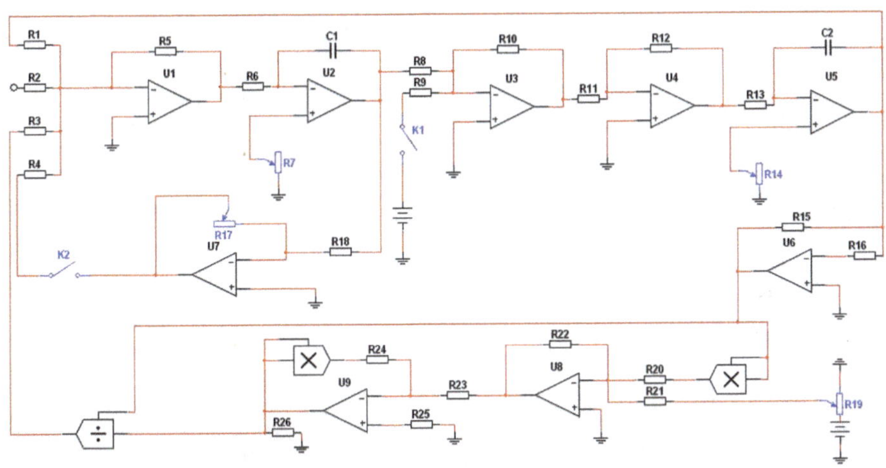

Fig. 12.17 Multisim 10 circuit of the SD oscillator with a feedback

circuit oscillator which is addressed here is that the discontinuous case can successfully avoid the barrier of the mechanical experimental investigations. These circuit oscillator provides the experimental possibility to go further into the understanding of nonlinear dynamics for both the smooth and the discontinuous systems.

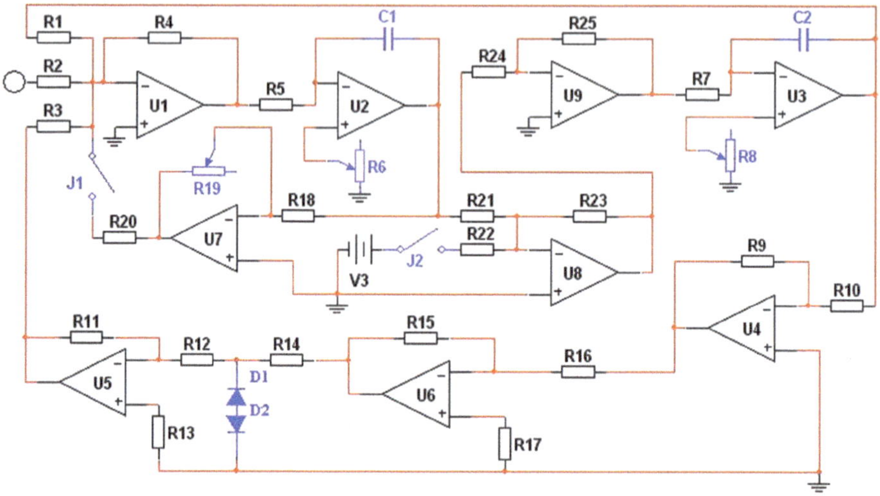

Fig. 12.18 Analog simulation circuit of the SD oscillator for discontinuous case

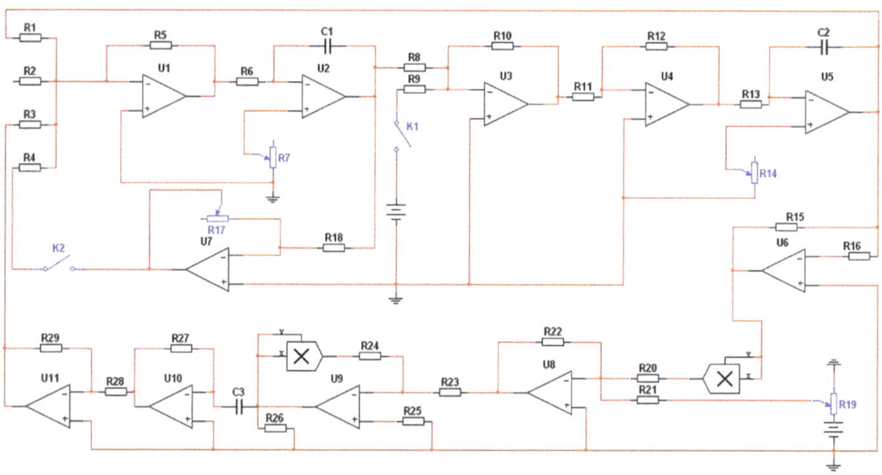

Fig. 12.19 Analog simulation circuit of the SD oscillator for smooth case

Fig. 12.20 The printed circuit for the modeling of the SD oscillator in the discontinuous case

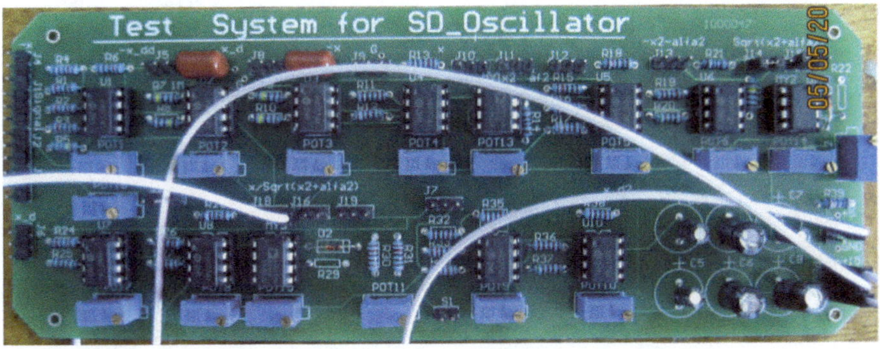

Fig. 12.21 The printed circuit for the modeling of the SD oscillator in the smooth case

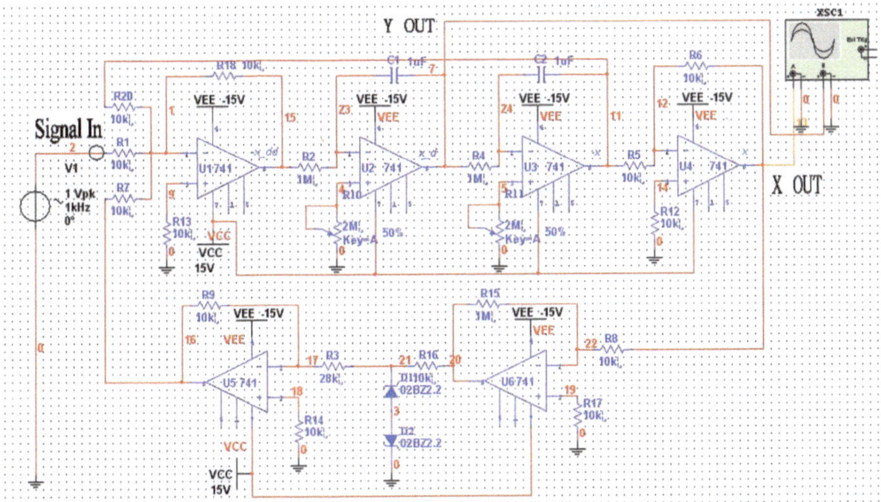

Fig. 12.22 Experimental setup for the SD discontinuous circuit with PC, data collecting system and the signal creator

12.3.2 The Experimental Results

In this subsection, we carry out the experimental investigations for the designed nonlinear circuit, based on system (12.2) to provide a primary experimental result with a very basic methodology.

We consider the circuit oscillator of system (12.2) with the corresponding circuits of Simlink circuit, seen in Fig. 12.14, Multisim10 circuit, seen in Fig. 12.16 and analog circuit, seen in Fig. 12.18 with the SD circuit oscillator, as seen in Fig. 12.20, using the modular design ideas to create SD analog circuits and finish its PCB manufacturing. The experimental device setup consists in the discontinuous analog circuit

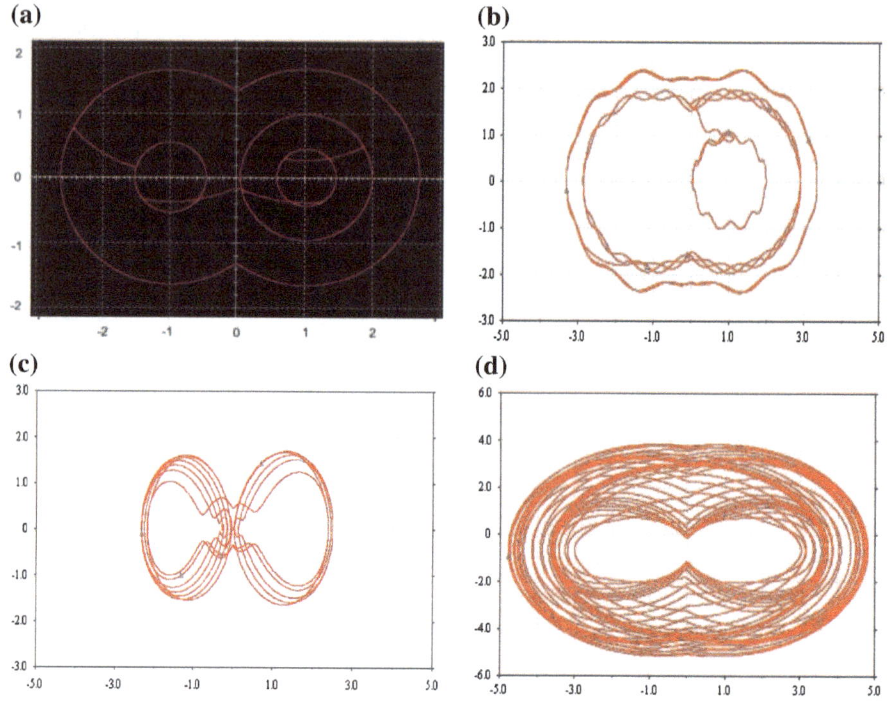

Fig. 12.23 The analog results for discontinuous circuit oscillator at natural frequency $\omega_0 = 0.1667\,\text{Hz}$: **a** for free resistance and external signal, **b** for the extensive of the external voltage $f = 0.7\,\text{V}$ with frequency $\omega_0 = 1.63\,\text{Hz}$, **c** for $f = 0.7\,\text{V}$ $\omega_0 = 0.489\,\text{Hz}$ and **d** for $f = 0.32\,\text{V}$ and $\omega_0 = 0.667\,\text{Hz}$

of the test system for SD oscillator, shown in Fig. 12.20, powered by an adaptor and a PC to analyze the data collected by a single chips.

We set the PC and peripheral as master-slave structure form, according to the experimental system visualization and operation requirement. The peripherals are designed as a signal acquisition and processing system based on C8051F020 to acquire multi-circuits signal and communicate with the PC through serial port RS232.C. LabVIEW8.6 is used to design the PC program to measure binary channel signal and to observe the portrait directly. It can be clearly seen, in Fig. 12.22, that the device can get the input signal created by a signal creator from the pot of **Signal In** and the response of the system can be collected from both **X OUT** and **Y OUT** pots respectively via the single chip of C8051F020. These voltage signals are analyzed via a developed software to get the electric quantity and the current through the system. The resulting voltage is presented at the XSC1 screen, and is similar with the phase portrait of the nonlinear dynamics.

In the following, some experimental results derived from the designed circuit oscillator are presented to show the potential to investigate the discontinuous dynamics using nonlinear circuit, which provides the possibility of getting into the discontinuous dynamics experimentally.

The analogue analysis results are plotted in Fig. 12.23 for the discontinuous circuit oscillator at natural frequency $\omega_0 = 0.1667\,\text{Hz}$, in Fig. 12.23a for free resistance and external signal, in Fig. 12.23b for the extensive of the external voltage $f = 0.7\,\text{V}$ with frequency $\omega_0 = 1.63\,\text{Hz}$, in Fig. 12.23c for $f = 0.7\,\text{V}$ $\omega_0 = 0.489\,\text{Hz}$ and in Fig. 12.23d for $f = 0.32\,\text{V}$ and $\omega_0 = 0.667\,\text{Hz}$.

12.4 Summary

Experimental systems had been built to observe the properties of the SD oscillator. The first kind is a mechanism very close to the abstract model described in this book, using mass, springs and bars, installed on a rigid or vibrating support to represent the forcing. The second kind is an electronic device, using specific sub-circuits to represent each abstract component of the differential equation, such as derivative, square root, sign function, and its integration, together with a external input of forcing. Both systems were coupled with specific data acquisition devices, to simulate both smooth and non-smooth regimes for the oscillator. These experimental results suggested that engineering applications can be built whatever in the scale of the mechanism or in the electronic circuit level.

References

1. Thompson, J. M. T., & Hunt, G. W. (1973). *A general theory of elastic stability*. London: Wiley.
2. Cao, Q. J., Wiercigroch, M., Pavlovskaia, E. E., Grebogi, C., & Thompson, J. M. T. (2006). Archetypal oscillator for smooth and discontinuous dynamics. *Physics Review E*, *74*(046218), 1–5.
3. Cao, Q., Wang, D., Chen, Y., & Wiercigroch, M. (2012). Irrational elliptic functions and the analytical solutions of the sd oscillator. *Journal of Theoretical and Applied Mechanics*, *50*(3), 701–715.
4. Gatti, G., Brennan, M. J., & Kovacic, I. (2010). On the interaction of the responses at the resonance frequencies of a nonlinear two degree-of-freedom sysytem. *Physica D-Nonlinear Phenomena*, *239*(10), 591–599.
5. Gourdon, E., Alexander, N. A., & Taylor, C. A. (2007). Nonlinear energy pumping under transient forcing with strongly nonlinear coupling: Theoretical and experimental results. *Journal of Sound of Vibration*, *300*(3), 522–551.
6. Gatti, G., Kovacic, I., & Brennan, M. J. (2010). On the response of a harmonically excited two degree-of-freedom system consisting of a linear and a nonlinear quasi-zero stiffness oscillator. *Journal of Sound of Vibration*, *329*(3), 1823–1835.
7. Liu, C., Zhang, Y., Han, Q. K., Wen, B., Liu, C., Zhang, Y., et al. (2005). Han, and B Wen. Experimental research on nonlinear vibration characteristics of rotor bearing system with coupling fault of rub-impact and oil whirl. *American Society of Mechanical Engineers*, 1155–1160.
8. Chen, E. L., Cao, Q. J., Feng, M., & Tian, R. L. (2012). The preliminary investigation on design and experimental research of nonlinear characteristics of sd oscillator. *Chinese Journal of Theoretical and Mechanics*, *44*(3), 584–590.

9. Ashwin, P. (1995). *Symmetric Dynamics and Electronic Circuits, chapter 2* (pp. 35–53). Singapore: World Scientific.
10. Kia, B., Lindner, J. F., & Ditto, W. L. (2015). Nonlinear dynamics based digital logic and circuits. *Frontiers in Computational Neuroscience*, *9*, 49.
11. Enns, R. H., & Mcguire, G. (1997). *Nonlinear LRC Circuit*. Boston: Birkhäuser.

Chapter 13
Rotating SD Oscillator: An Ancient Application

In the previous chapters the SD oscillator as a geometrical model has been investigated in different ways including the fundamental theory and the methodologies. Generally, this model consists of a pair of linear springs and a lumped mass vibrating along a straight line, but its structure can be simplified into a single mass spring system considering the symmetry. Another geometrical model is the pendulum which is regarded as the first geometrical nonlinear system in the history of science, since it concerns, as recalled in the introduction of this book, a mass suspended from some point by an inextensional rod which swings or rotates along the circle centered at the point of suspension. These equations of motion are formulated by Pythagorean theorem presenting abundant nonlinear phenomena. This chapter presents an ancient application of the SD oscillator in the case where it is coupled with a pendulum, which is named the rotating SD oscillator. Incidentally this system will also present an example of cylindrical dynamical system with resonances and chaotic attractors which are typical of the pendulum or the SD type and also the coupled attractors.[1]

13.1 Introduction

As far as the authors are aware, a seismograph [1] coming from ancient China is the first mechanical system using a behaviour close to that of the SD oscillator. It is made of an oblique spring pinned to a rigid support, which is the SD oscillator providing irrational nonlinearity, linked by a pendulum which is the earliest well known nonlinear system [2–4]. This is a cylindrical dynamical mass-spring system [5] having either a smooth or a discontinuous restoring force according to the geometrical positions of the spring and of the pendulum. The nonlinear dynamics is presented in both the unperturbed case and the free vibration behaviours. Complicated resonant structures, involving periodic, quasi periodic and stochastic phenomena are presented for

[1]This chapter relies on a common work with Dr. Ning Han, Centre for Nonlinear Dynamics Research, School of Astronautics, Harbin Institute of Technology, Harbin 150001 China.

Q. Cao and A. Léger, *A Smooth and Discontinuous Oscillator*, Springer Tracts in Mechanical Engineering, DOI 10.1007/978-3-662-53094-8_13

the system when submitted only to harmonic perturbations, while a chaotic behaviour is observed when, in addition to the harmonic excitation, the system has a viscous damping.

13.2 Equation of Motion and the Unperturbed Dynamics

The device involves an oblique spring, which resists both in tension and compression, pinned to a rigid support, linked by a rotating pendulum. The stiffness of the spring is linear and the restoring force supplied to the system is nonlinear due to the geometrical configuration. Everything that have been described about the SD oscillator applies to this system, in particular the fact that the restoring force can be either smooth or discontinuous when a geometrical parameter varies. This section describes the complicated phenomena of equilibrium bifurcations and dynamics on a cylinder with separatrices of periodic solutions when the system is unperturbed, and periodic resonant behaviours for the time-dependent Hamiltonian system and chaotic attractors when the system is perturbed. Let us consider the system shown in Fig. 13.1c, d where L is the length of the pendulum, m a mass submitted to its weight, k and l are respectively the elastic stiffness and the natural length of the inclined elastic spring which is fixed to the rigid frame.

The governing equation of this pendulum is the following:

$$mL\ddot{x} - mg\sin x + kh\sin x\left(1 - \frac{l}{\sqrt{L^2 + h^2 - 2Lh\cos x}}\right) = 0, \tag{13.1}$$

where x is the angular displacement. Using the notations of Fig. 13.1, h is the height from A to B, which can be positive or negative.

Now suppose that the system (13.1) is perturbed by a viscous damping and an external harmonic excitation of amplitude F_0 and frequency ω. This leads to the equation:

$$mL\ddot{x} + \delta L\dot{x} - mg\sin x + kh\sin x\left(1 - \frac{l}{\sqrt{L^2 + h^2 - 2Lh\cos x}}\right) = F_0\cos\omega t. \tag{13.2}$$

As in the previous chapters we use a dimensionless form of Eq. (13.1) obtained by letting $L = 1, l = 1$, and $\omega_0^2 = \frac{k}{m}$.

$$\ddot{x} - g\sin x + \omega_0^2 h\sin x\left(1 - \frac{1}{\sqrt{1 + h^2 - 2h\cos x}}\right) = 0, \tag{13.3}$$

which is smooth for $|h| \neq 1$ and discontinuous, at $x = 0$ for $h = 1$ and at $x = \pi$ for $h = -1$. The dimensionless form for the perturbed system (13.2) can be obtained by letting $f_0 = \frac{F_0}{mL}$ and $\xi = \frac{\delta}{2m}$.

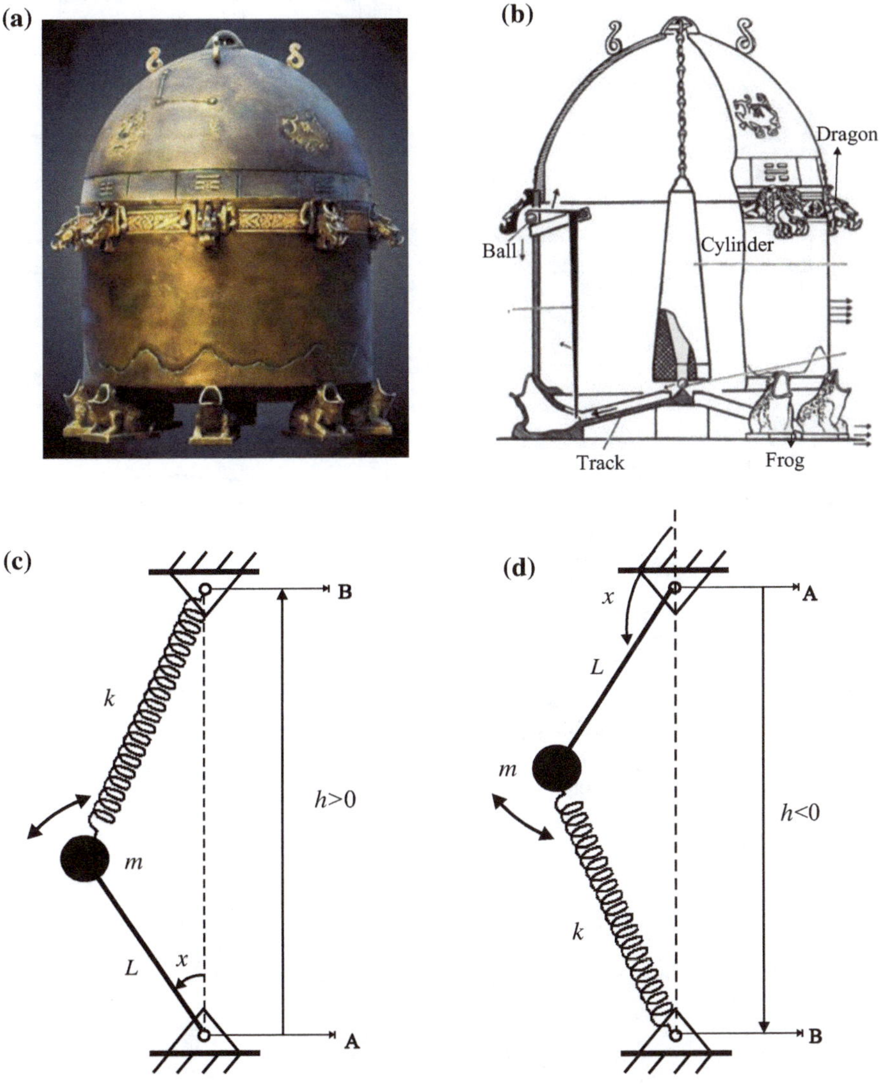

Fig. 13.1 **a** Restoration chart of the ancient seismograph [6], **b** internal structure of the seismograph, **c** and **d** physical models for a rotating SD oscillator: pendulum linked by a spring pinned its fixed support

$$\ddot{x} + 2\xi\dot{x} - g\sin x + \omega_0^2 h \sin x\left(1 - \frac{1}{\sqrt{1 + h^2 - 2h\cos x}}\right) = f_0\cos\omega t. \quad (13.4)$$

The graphs of the restoring force in system (13.3), say $F(x)$, are plotted in Fig. 13.2a–c, for $\omega_0 = \sqrt{20}$ and for the different values of parameter h.

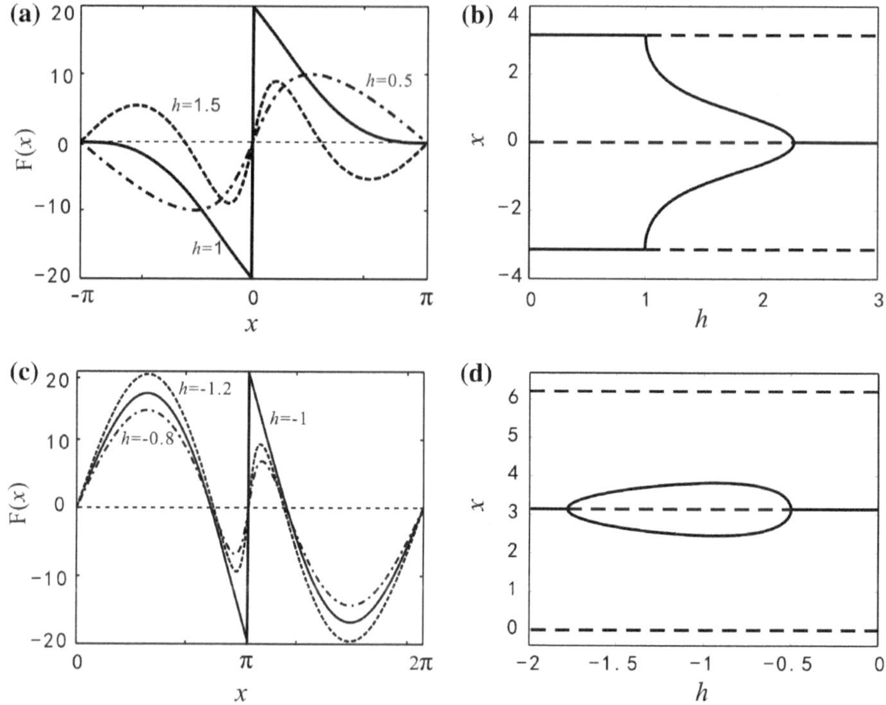

Fig. 13.2 Restoring forces and the equilibrium bifurcations for $\omega_0 = \sqrt{20}$: **a** and **c** nonlinear restoring force $F(x)$ for $h > 0$ and $h < 0$ respectively, **b** and **d** bifurcation diagrams for x versus h, for $h \geq 0$ and $h \leq 0$, respectively

$$F(x) = g \sin x - \omega_0^2 h \sin x \left(1 - \frac{1}{\sqrt{1 + h^2 - 2h \cos x}}\right), \tag{13.5}$$

The equilibria of system (13.3) are the solutions of equation $F(x) = 0$ i.e.

$$x_1 = 0,\ x_{2,3} = \pm\pi,\ x_{4,5} = \pm \arccos\left(\frac{1 + h^2}{2h} - \frac{\omega_0^4 h}{2(g - \omega_0^2 h)^2}\right),$$

where $x_{4,5}$ exists only for $\left|\frac{1+h^2}{2h} - \frac{\omega_0^4 h}{2(g-\omega_0^2 h)^2}\right| < 1$.

Figure 13.2b–d plot the branches of equilibria for $\omega_0 = \sqrt{20}$ and show the pitchfork equilibrium bifurcations, where the solid lines denote the stable branches while the dashed lines indicate the unstable ones.

Letting $\dot{x} = y,\ \dot{y} = F(x)$ in system (13.3) yields the Hamiltonian

$$H(x, y) = \frac{y^2}{2} + (g - \omega_0^2 h) \cos x - \omega_0^2 \sqrt{1 + h^2 - 2h \cos x} - (g - \omega_0^2 h) + \omega_0^2 |1 - h|, \tag{13.6}$$

with the help of which the trajectories of system (13.3) can be classified and analyzed. Let us calculate the transient sets on the surface of $F(x)$ in the parameter space (h, ω_0) (see e.g. [7]).

$$\sum = \mathcal{B} \cup \mathcal{H} \cup \mathcal{D} = \mathcal{B}_1 \cup \mathcal{B}_2 \cup \mathcal{H}, \tag{13.7}$$

where $\mathcal{B}$ and $\mathcal{H}$ correspond to a bifurcation set and an hysteresis set respectively, while $\mathcal{D}$ refers to the double limit set. We have

$$\begin{cases} \mathcal{B}_1 = \left\{ \omega_0^2 h \left(1 - \dfrac{1}{|1+h|}\right) - g = 0 \right\}, \\ \mathcal{B}_2 = \left\{ g - \omega_0^2 h \left(1 - \dfrac{1}{|h-1|}\right) = 0 \right\}, \\ \mathcal{H} = \{\omega_0^2 (h-1) - g = 0\}, \\ \mathcal{D} = O. \end{cases}$$

so that the parameter space (h, ω_0) is divided into four persistent regions, marked I, II, III and IV, as shown in the part Σ of Fig. 13.3, $\mathcal{B}_1, \mathcal{B}_2$ and $\mathcal{H}$ are plotted with solid blue, yellow and red respectively, while the solid lines, marked grey, correspond to the discontinuous regimes for $|h| = 1$. All possible phase portraits of system (13.3) in the parameter space, as shown in Fig. 13.3 Σ, are plotted in Fig. 13.3a–l, where they are marked with the same symbols as in the corresponding parameter space in Fig. 13.3 Σ.

The bold dots denote the center and saddle points of the smooth equilibria connecting the corresponding heteroclinic, see Fig. 13.3g for example, homoclinic orbits, marked red, of the first, and the second kind in Fig. 13.3, respectively. While small cycles indicate nonstandard or saddle-like equilibria connecting the corresponding homoclinic-like orbits, marked blue, as shown in Fig. 13.3a–c, k, l, respectively. Those dynamics present the global bifurcations for both the smooth and discontinuous cases [8, 9] of pitchforks for equilibria and transitions for homoclinic and heteroclinic orbits. The dynamics in the smooth case, i.e. $|h| \neq 1$, is displayed in Fig. 13.3d–j, showing two types of pitchfork bifurcations, seen in Fig. 13.3 from Fig. 13.3d, e to Fig. 13.3f and from Fig. 13.3i, j to Fig. 13.3h, and the transition from a pair of homoclinic orbits, Fig. 13.3f, to a double homoclinic orbit, Fig. 13.3g, bifurcating to a pair of homoclinic orbits, a pair of homoclinic orbits, seen in Fig. 13.3h. Bifurcations in the discontinuous case, i.e. for $|h| = 1$ are also given in Fig. 13.3a–c, k, l, which reveal a pitchfork bifurcation of the saddle-like equilibrium $(0, 0)$ in Fig. 13.3a, as shown in Fig. 13.3, from Fig. 13.3a–c, and a saddle-like equilibrium transition from $(0, 0)$ to $(\pi, 0)$, as described in Fig. 13.3k, l as well.

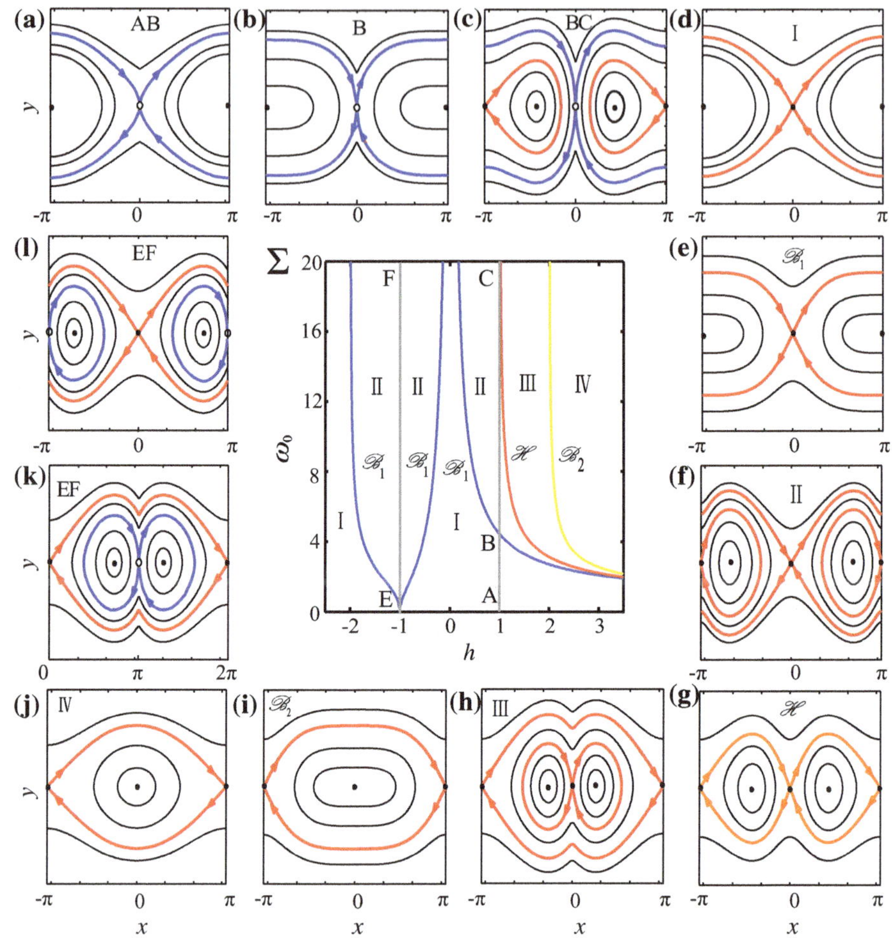

Fig. 13.3 Bifurcation diagram, and a–l the corresponding phase portraits, the details seen in the context

13.3 Resonances of the Time Dependent Hamiltonian Dynamics

The time-dependent Hamiltonian can also be derived for the undamped system (13.4) (i.e. $\xi = 0 =$). It is obtained as:

$$H(x, y, t) = \frac{y^2}{2} + (g - \omega_0^2 h)\cos x - (g - \omega_0^2 h) \\ - \omega_0^2\sqrt{1 + h^2 - 2h\cos x} + \omega_0^2|1 - h| - f_0 x \cos\omega t, \qquad (13.8)$$

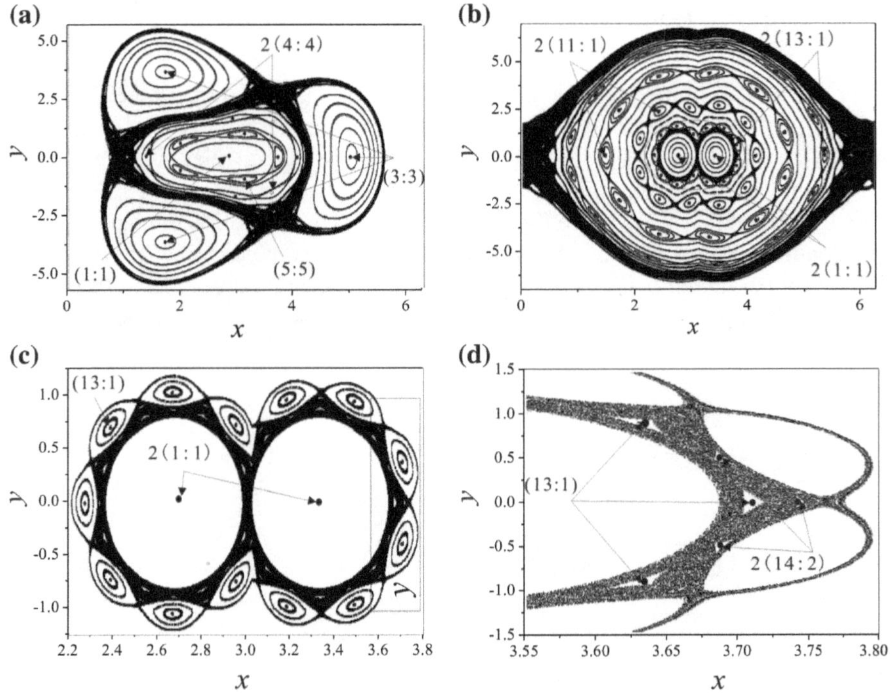

Fig. 13.4 Chaotic sea and the quasi-periodic trajectories (islands) for $\xi = 0$: **a** Poincaré section for $f_0 = 10, \omega = 5.5, h = 1, \omega_0 = \sqrt{19.5}$, **b** Poincaré section for $f_0 = 20, \omega = 30, h = -1, \omega_0 = \sqrt{5}$, **c** the amplified area for the center part in **b**, **d** locally amplified for the boxed area in **c**

in which the time t is regarded as a parameter, while system (13.4) acts as an one and half degree of freedom system with natural frequency ω_0 and excitation frequency ω. The motions of this system are restricted to lie on a torus with both action and angular variables [10]. The most interesting situation is the resonant behaviour [11] which occurs when $m\omega_0 - n\omega = 0$, m and n being related prime integers, which implies that the resonant motion can return to the starting point after m cycles of longitude and n cycles of latitude along the torus. The winding number of the resonance is defined as $\alpha = \frac{n}{m}$; the period of the motion might be integer times of the winding number. As examples, we show the Poincaré section, or a resonance web, for $f_0 = 10$, $\omega = 5.5$, $h = 1$, $\omega_0 = \sqrt{19.5}$ in Fig. 13.4a with resonant solutions of $(1:1)$, $(3:3)$, $(5:5)$, and a pair of $(4:4)$. Figure 13.4b shows the Poincaré section for $f_0 = 20$, $\omega = 30$, $h = -1$, $\omega_0 = \sqrt{5}$ with resonant solutions of a pair of $(1:1)$, a pair of $(11:1)$, a pair of $(13:1)$, a pair of $(14:2)$ and $(15:3)$ respectively.

13.4 Chaotic Behaviours of the Perturbed System

In the following, numerical calculations are carried out to detect the behaviour of system (13.4) in the presence of viscous-damping and external excitation of amplitude f_0 and frequency ω. Figure 13.5a shows a bifurcation diagram constructed for x sampled stroboscopically versus control parameter h for $f_0 = 10$, $\xi = 0.24$, $\omega = 4.75$ and $\omega_0 = \sqrt{50}$, while Fig. 13.5b–d present the transition of chaotic attractors. Again a bifurcation diagram is constructed for x sampled stroboscopically versus control parameter h for $f_0 = 10$, $\xi = 0.1$, $\omega = 5.5$ and $\omega_0 = \sqrt{28}$ in Fig. 13.5e. Figure 13.5f–h present the chaotic transition between the smooth and the discontinuous cases.

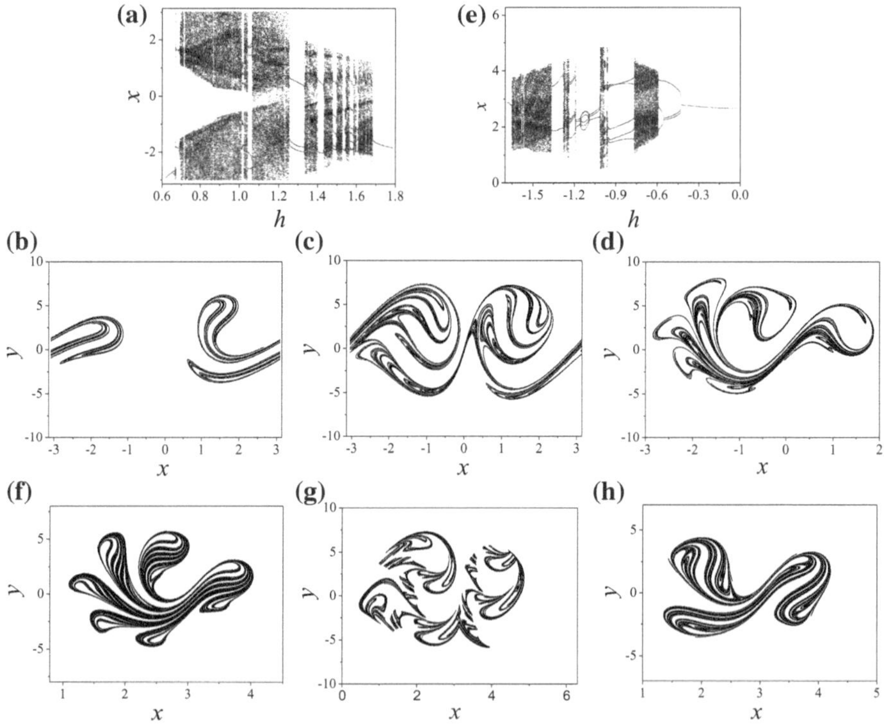

Fig. 13.5 **a** Bifurcation diagrams for x versus h constructed for: $f_0 = 10$, $\xi = 0.24$, $\omega = 4.75$ and $\omega_0 = \sqrt{50}$, **b**, **c**, **d** chaos for $h = 0.8, 1.22, 1.5$ with the largest Lyapunov exponents 0.536155, 0.464603, 0.314516 and the corresponding Lyapunov dimension 1.52763, 1.49214, 1.39591, respectively, **e** bifurcation diagrams for x versus h constructed for $f_0 = 10$, $\xi = 0.1$, $\omega = 5.5$ and $\omega_0 = \sqrt{28}$, **f**, **g**, **h** chaos for $h = -1.47, -1, -0.68$ with the largest Lyapunov exponent 0.545585, 0.400991, 0.412826 and the Lyapunov dimension 1.73175, 1.67245, 1.67364, respectively

13.5 Summary

As an earlier application of the SD oscillator, we have modeled the mechanism used in the ancient seismograph by the SD oscillator coupled with a pendulum which constructs the rotating SD oscillator consisting of a pendulum linked to a rigid frame by an elastic spring. This short study has presented the set of equilibria and the qualitative properties of complicated bifurcation and chaos. The behaviour of the system appears to be close to that of SD oscillator, the pendulum or the coupled behaviours depending on the coupling of the SD oscillator and the pendulum. This rotating SD oscillator leads to a new cylindrical example which is that the equilibrium branches and the dynamics hold on a cylinder. It is interesting to observe that such a device, which had been used in the ancient times can be understood as an the application of the SD oscillator together with the pendulum.

References

1. Fan, Y. *History records of the eastern Han dynast: Zhang Heng biography*, Chap. 59 (p. 445).
2. Galileo, G. (1952). De motu (on motion). *Collection of Galileo Galilei's manuscripts and related translations*, Retrieved.
3. Matthews, M. R. (2001). How pendulum studies can promote knowledge of the nature of science. *Journal of Science Education and Technology*, *10*(4), 359–368.
4. Liang, Y., & Feeny, B. F. (2008). Parametric identification of a chaotic base-excited double pendulum experiment. *Nonlinear Dynamics*, *52*(1–2), 181–197.
5. Barbashin, E. A. (1967). The classification of trajectories of a dynamical system with cylindrical phase space. *Differencial'nye Uravnenija*, 2015–2020.
6. Feng, R., & Yu, Y. X. (2006). Zhang heng's seismometer and longxi earthquake in ad 134. *Acta Seismogica Sinica*, *19*(6), 104.
7. Golubitsky, M., Schaeffer, D. G., & Stewart, I. (1985). *Singularities and groups in bifurcation theory* (Vol. I). New York: Springer.
8. Lamba, H., & Budd, C. J. (1994). Scaling of lyapunov exponents at nonsmooth bifurcations. *Physical Review E*, *50*(1), 84–90.
9. Bressloff, P. C., & Stark, J. (1990). Neuronal dynamics based on discontinuous circle maps. *Physics Letters A*, *150*(3), 187–195.
10. Chernikov, A. A., Sagdeev, R. Z., Usikov, D. A., Yu Zakharov, M., & Zaslavsky, G. M. (1987). Minimal chaos and stochastic webs. *Nature*, *326*, 559–563.
11. Lichtenberg, A. J., & Lieberman, M. A. (1992). *Regular and chaotic dynamics*. New York: Springer.

Chapter 14
An Isolator: Dynamics and Vibration Isolation

As an application of the SD oscillator, in this chapter, a nonlinear isolator based on a spring-mass system with a single degree of freedom is proposed, which can be designed as a nonlinear supporting system for ground vibration tests (GVT) for large scale machines like aircrafts and for vibration isolation due to a stable quasi-zero-stiffness (SQZS). The SQZS structure is constructed with a mass supported by a vertical spring component to support the weight of the mass and a pair of horizontal linear springs providing an equal vertical negative stiffness. This is typical of an irrational restoring force due to the geometrical configuration. The unperturbed dynamics is studied with the equilibrium bifurcations. A parameter optimization is used to obtain SQZS over the maximum interval of low frequency isolation. The dynamical analysis of the perturbed system near the optimized parameters and the isolation performance are presented in this chapter.[1]

14.1 Introduction

Much attention has been paid to the low frequency isolation design for GVT (ground vibration test) of large scale aircraft [1], isolation and other applications [2–6]. For GVT, a low frequency supporting system is required to provide a free-free mode, where the ratio of the support's natural frequency and the fundamental frequency of the aircraft to be tested should be lower than 1:3 [7]. As we know that the fundamental frequency of a large thin-wing aircraft can be lower than 1 Hz, even than 0.7 Hz, this requires the supporting frequency to be lower than 0.3 Hz [8]. Even earlier efforts have been done to achieve the lowest supporting frequency, the conventional isolation design cannot achieve such a required low frequency, such as an undercarriage support (with 2 Hz), a linear spring support (with 1 Hz), or a pneumatic support (with 0.5 Hz) [1, 8].

[1]This chapter relies on a common work with Dr. Zhifeng Hao, Centre for Nonlinear Dynamics Research, School of Astronautics, Harbin Institute of Technology, Harbin 150001 China.

Q. Cao and A. Léger, *A Smooth and Discontinuous Oscillator*, Springer Tracts in Mechanical Engineering, DOI 10.1007/978-3-662-53094-8_14

The first conceptual design of low frequency was proposed by W.G. Molyneux [1] with a geometrical arrangement of springs to provide a zero or very small stiffness. D.G. Blair et al. [9] considered ultra low frequency isolation applied to laser interferometers for gravitational wave detection. Platus adopted the negative stiffness mechanism in [10] for isolation against sub-Hertz vibrations. Zhang et al. [11] used an isolator comprising elastic components and Euler columns providing the frequency down to 0.5 Hz. Alabuzhev et al. [12] introduced a class of vibration protection designs with quasi-zero stiffness consisting of load-bearing elastic elements with constant positive stiffness as well as devices with negative stiffness. A. Carrella et al. [13, 14] studied the force transmissibility of a quasi-zero-stiffness isolator by using a Duffing system. Most of the investigations focus on the static design and experiments of low frequency system.

14.2 Unperturbed Dynamics

Consider a new model, as shown in Fig. 14.1, which comprises a weighty mass m linked by three linear springs: a vertical spring of stiffness K_1 and natural length H, and a pair of inclined springs of stiffness K_2 and natural length L. The springs are capable of tension and compression and are pinned to each rigid support.

The governing equation of motion is the following

$$m\ddot{Z} + \left(K_1 + 2K_2\left(1 - \frac{L}{\sqrt{a^2+Z^2}}\right)\right)Z - K_1(h+d) + mg = 0, \tag{14.1}$$

where Z is the displacement of the mass with respect to the point O, $h + d$ is the end of the vertical spring and a the half distance between the inclined springs and the rigid supports.

As it has been done in the previous chapter, system (14.1) is put in a dimensionless form by the following changes of variables and parameters. $X = Z/L$, $\tau = \omega_0 t$, $\omega_0^2 = (K_1 + 2K_2)/m$, $\alpha = a/L$ ($\alpha \in [0, \infty)$), $\gamma = 2K_2/(K_1 + 2K_2)$ ($\gamma \in [0, 1]$) and $\eta = [K_1(h+d) - mg]/ml_1\omega_0^2$, written as

$$\ddot{X} + X - \frac{\gamma X}{\sqrt{\alpha^2+X^2}} - \eta = 0, \tag{14.2}$$

which behaves either smooth ($\alpha > 0$) or discontinuous ($\alpha = 0$) in which case system (14.2) becomes

$$\ddot{X} + X - \gamma sign(X) - \eta = 0. \tag{14.3}$$

Again for $\alpha > 0$, system (14.1) can be rewritten in the following form by letting $x = X/\alpha$, $\mu = \gamma/\alpha$, $\lambda = \eta/\alpha$,

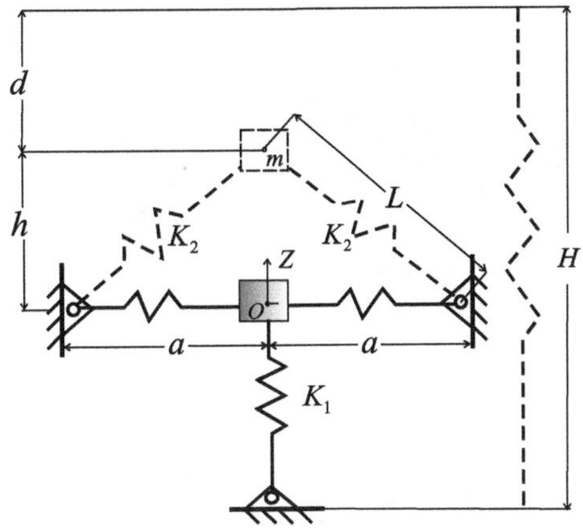

Fig. 14.1 Dynamical model for the mechanism of supporting system with low frequency comprising a vertical and a pair of oblique springs

$$\ddot{x} + x - \frac{\mu x}{\sqrt{1 + x^2}} - \lambda = 0. \tag{14.4}$$

This equation is rewritten as a first-order differential equations by $\dot{x} = y$,

$$\begin{cases} \dot{x} = y \\ \dot{y} = -x + \dfrac{\mu x}{\sqrt{1 + x^2}} + \lambda, \end{cases} \tag{14.5}$$

with the potential

$$V(x) = \frac{1}{2}x^2 - \mu\sqrt{1 + x^2} - \lambda x \tag{14.6}$$

and the Hamiltonian

$$H(x, y) = \frac{1}{2}x^2 + \frac{1}{2}y^2 - \mu\sqrt{1 + x^2} - \lambda x. \tag{14.7}$$

Bifurcations and stability are now studied in the case of the unperturbed system (14.4). The equilibrium surface of system (14.4) is given by

$$\{(x, \mu, \lambda) \mid F(x, \mu, \lambda) = f(x, \mu) - \lambda = 0\}. \tag{14.8}$$

To understand the structure of the equilibrium surface, $F(x, \mu, \lambda) = 0$, marked pink in Fig. 14.2a, we introduce two groups of planes $\mu = constant$ and $\lambda = constant$ to cut the equilibrium surface. The sections of $F(x, \mu, \lambda) = 0$ with $\mu < 1$, $\mu = 1$ and $\mu > 1$ can been seen in Fig. 14.2a, marked blue, and also on Fig. 14.2b1–b3 in the (λ, x) plane. The sections of $F(x, \mu, \lambda) = 0$ with $\lambda < 0$, $\lambda = 0$ and $\lambda > 0$ are marked in Fig. 14.2a with white, black and green, respectively, and are also presented by Fig. 14.2c1–c3 in (μ, x) plane.

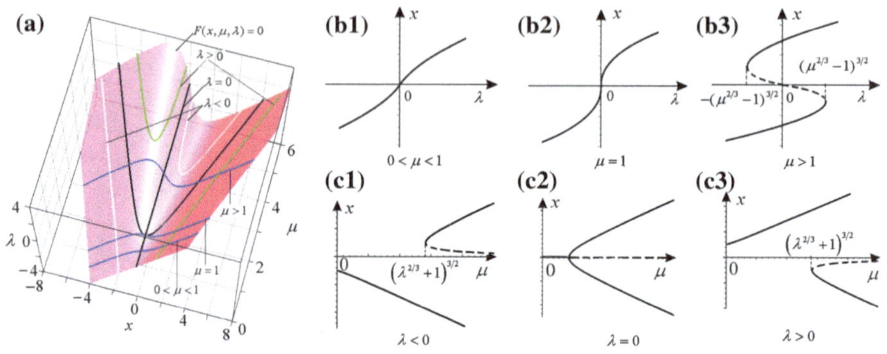

Fig. 14.2 **a** The equilibrium surface (*pink*) in space (x, μ, λ), **b1**–**b3** the sections cut by $\mu = 0.5$, 1, 3, **c1**–**c3** the sections cut by $\lambda = -1.5$, 0, 1.5

Furthermore, to classify the nonlinear behaviour of the system, we define the generalized transition sets of double parameters (μ, λ) as follows,

$$\Sigma_g = \mathscr{B} \cup \mathscr{H} \cup \mathscr{S}, \tag{14.9}$$

where

$$\begin{aligned} \mathscr{B} &= \left\{ \begin{array}{c} (\mu, \lambda) \mid \exists x, \mu \in \mathbb{R}, s.t., F(x, \mu, \lambda) = \\ F_x(x, \mu, \lambda) = 0, \mu \geq 0, \lambda \in \mathbb{R} \end{array} \right\}, \\ \mathscr{H} &= \left\{ \begin{array}{c} (\mu, \lambda) \mid \exists x, \lambda \in \mathbb{R}, s.t., F(x, \mu, \lambda) = \\ F_x(x, \mu, \lambda) = F_{xx}(x, \mu, \lambda) = 0, \\ \mu \geq 0, \lambda \in \mathbb{R} \end{array} \right\}, \\ \mathscr{S} &= \left\{ \begin{array}{c} (\mu, \lambda) \mid \exists x, \mu \in \mathbb{R}, s.t., F(x, \mu, \lambda) = 0, \\ \forall \delta > 0, x_a = x - \delta, x_b = x + \delta, \\ F_x(x_a) = F_x(x_b), \mu \geq 0, \lambda \in \mathbb{R} \end{array} \right\}. \end{aligned} \tag{14.10}$$

Here $\mathscr{B}$, $\mathscr{H}$ and $\mathscr{S}$ are the bifurcation set, the hysteresis set and the symmetric set, respectively, which read as

$$\begin{aligned} \mathscr{B} &= \{(\mu, \lambda) \mid \mu = (\lambda^{2/3} + 1)^{3/2}, \lambda \in \mathbb{R}\}, \\ \mathscr{H} &= \{(\mu, \lambda) \mid \mu = 1, \lambda \in \mathbb{R}\}, \\ \mathscr{S} &= \{(\mu, \lambda) \mid \mu \in \mathbb{R}, \lambda = 0\}. \end{aligned} \tag{14.11}$$

Meanwhile, the bifurcation sets can be classified as

$$\mathscr{B} = \mathscr{B}_0 \cup \mathscr{B}_1 \cup \mathscr{B}_2, \tag{14.12}$$

where

$$\begin{aligned} \mathscr{B}_0 &= \{(\mu, \lambda) \mid \mu = 1, \lambda = 0\}, \\ \mathscr{B}_1 &= \{(\mu, \lambda) \mid \mu = (\lambda^{2/3} + 1)^{3/2}, \lambda > 0\}, \\ \mathscr{B}_2 &= \{(\mu, \lambda) \mid \mu = (\lambda^{2/3} + 1)^{3/2}, \lambda < 0\}. \end{aligned} \tag{14.13}$$

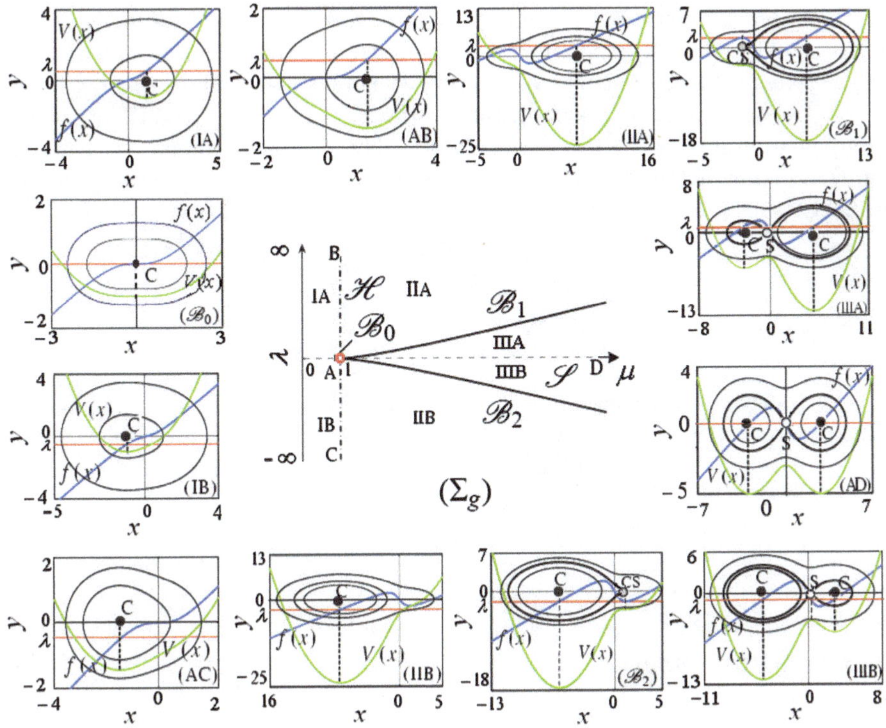

Fig. 14.3 Diagrams for the generalized transition set, (Σ_g), and the related phase portraits, the centers marked with *black bullets*, the saddles marked with *black hollow circles* and the saddle-centers with *grey bullets*, with the corresponding restoring forces $f(x)$ (*blue curves*), λ (*red lines*) and the potentials $V(x)$ (*green curves*)

Here $\mathscr{B}_1$ and $\mathscr{B}_2$ are a pair of saddle-centers [15] bifurcation sets. Hence, the generalized transition can be described as

$$\Sigma_g = \mathscr{B}_0 \cup \mathscr{B}_1 \cup \mathscr{B}_2 \cup \mathscr{H} \cup \mathscr{S}, \tag{14.14}$$

which bifurcates at the catastrophe point $A(\mu, \lambda) = (1, 0)$ and divides the parameter plane (μ, λ) into six persistent regions marked by IA, IIA, IIIA, IB, IIB, IIIB, as shown in Fig. 14.3(Σ_g). In each of these persistent regions, the system is structurally stable, while on Σ_g the system is structurally unstable. The corresponding phase portraits (black closed curves), the restoring forces $F(x, \mu, \lambda) = f(x, \mu) - \lambda\ (f(x, \mu)$ and λ (marked in blue and red respectively) and the potentials $V(x)$ (marked green) are plotted in Fig. 14.3(IA)–($\mathscr{B}_0$). The stability of the equilibria of system (14.4) is given on Fig. 14.3(IA)–($\mathscr{B}_0$), in which the stable equilibrium points (centers) correspond to the bottoms of the potential "well" while the unstable ones (saddles or saddle-centers) correspond to the summits of the potential "obstacle". The complex equilibrium bifurcations and the transitions are shown in Fig. 14.3 when the parameters vary over the (μ, λ) plane, seen in Fig. 14.3(Σ_g). It is found that system (14.4) behaves as a pair of saddle-center bifurcations from a single stable system to an asymmetrical bi-stable system via a saddle-center bifurcation when the parameter (μ, λ) moves from region IIA to region IIIA, or symmetrically from region IIB to region IIIB, crossing the bifurcation sets $\mathscr{B}_1$, $\mathscr{B}_2$, respectively. A pitchfork bifurcation occurs when the pair (μ, λ) moves along the symmetric set $\mathscr{S}$ from $\mu < 1$ to $\mu > 1$ passing through the catastrophe point A indicated on Fig. 14.3($\mathscr{B}_0$) and (AD).

14.3 Perturbed Dynamics

The dynamical phenomena are now investigated for system (14.4) near the optimized parameter $\mu = 1, \lambda = 0$ when the system is subjected to a viscous damping and an external harmonic excitation of amplitude f and frequency ω, that is for system:

$$\ddot{x} + 2\zeta\dot{x} + x - \frac{\mu x}{\sqrt{1+x^2}} - \lambda = f\cos(\omega\tau). \tag{14.15}$$

14.3.1 Resonances

The first step is the resonance behaviour of system (14.15) without damping ($\zeta = 0$), the Hamiltonian of which is given by

$$H_f(x, y, \tau) = \frac{1}{2}x^2 + \frac{1}{2}y^2 - \mu\sqrt{1+x^2} - fx\cos(\omega\tau), \tag{14.16}$$

which is an one and half degree of freedom system and where the time τ is regarded as a parameter.

To construct an extended phase space (x, y, E, τ), $E = -H_f$ is treated as another additional coordinate, acting as the conjugate momentum of τ, while the flow in the extended space is described by an auxiliary parameter ξ playing the role of the time [16].

A new Hamiltonian H_E is then obtained via the generating function $F_{2(1)}(x, y, E, \tau) = xy + E\tau$ and written as follows

$$H_E(x, y, E, \tau) = \frac{1}{2}x^2 + \frac{1}{2}y^2 + E - \mu\sqrt{1+x^2} - fx\cos(\omega\tau) = \mathscr{E}, \tag{14.17}$$

which is similar to a two-degree-of-freedom autonomous system still non-integrable owing to lack of other first integral besides (14.17) $\mathscr{E} = constant$ [17].

The first two terms in the above extended Hamiltonian denote a harmonic oscillator, for which a pair of action and angle variables J, θ defined from $x = \sqrt{2J}\sin\theta$ and $y = \sqrt{2J}\cos\theta$ is introduced and the Hamiltonian (14.17) is written as

$$H_a(J, E; \theta, \tau) = J + E - \mu\sqrt{1 + 2J\sin^2(\theta)} - f\sqrt{2J}\sin(\theta)\cos(\omega\tau). \quad (14.18)$$

A dynamical torus is then obtained by the introduction of another canonical transformation by means of the generating function $F_{2(2)}(J_\theta, J_\varphi; \theta, \varphi) = \omega_\varphi \tau J_\varphi + \theta J_\theta$, such that the phase space (J, θ, E, τ) is transformed into $J_\theta, J_\varphi, \theta, \varphi$ defined by $J_\theta = J, \theta = \theta, J_\varphi = E/\omega$ and $\varphi = \omega\tau$.

The new Hamiltonian in the action-angular space $(J_\theta, J_\varphi, \theta, \varphi)$ reads as follows

$$\begin{aligned} H(J_\theta, J_\varphi; \theta, \varphi) &= H_0(J_\theta, J_\varphi) + H_1(J_\theta, J_\varphi; \theta, \varphi) \\ &= \omega_\theta J_\theta + \omega_\varphi J_\varphi - \mu\sqrt{1 + 2J_\theta \sin^2(\theta)} - f_0\sqrt{2J}\sin(\theta)\cos(\varphi), \end{aligned}$$

which describes the motions of the system on the extended torus.

For convenience it is assumed that $\varphi = \omega_\varphi \tau \bmod 2\pi$ and $\varphi = 0$ is equivalent to $T = 2\pi/\omega$. There exist two frequencies

$$\omega_\varphi = \frac{\partial H_{a0}}{\partial J_\varphi} = \omega, \omega_\theta = \frac{\partial H_{a0}}{\partial J_\theta}, \quad (14.19)$$

which are the external frequency and the natural frequency of the system, respectively. The behaviour of the forced system is restricted to a torus with both action and angular variables ω_φ and ω_θ which can be described by the definition of winding number $\alpha = \omega_\theta/\omega_\varphi$. Here ω_θ and ω_φ are the rotating frequencies along the cycles of latitude and longitude along the torus while J_θ and J_φ retain two constants corresponding to ω_θ and ω_φ for a given initial value in (14.19).

The most interesting case is the resonance, which happens when $m\omega_\theta - n\omega_\varphi = 0$, where m and n denotes a pair of related prime integers. The behaviour is a harmonic resonance if $m = n = 1$, a sub-harmonic resonance of m order if $m > 1, n = 1$, a super-harmonic resonance of n order if $m = 1, n > 1$, and a super-sub-harmonic resonance if $m > 1, n > 1$ [16]. The winding number $\alpha = m/n$ is such that m and n are the number of fixed points and the number of twists of the trajectories dividing by their greatest common factor in the Poincaré section. In addition, if the winding number α is irrational, the trajectory never closes on itself, filling a torus densely, for which the map in the Poincaré section exhibits a generic KAM curve, namely a quasi-periodic motion.

The following numerical simulations are presented in the form of Poincaré sections taken at zero phase angle by a stroboscopic time-T map of $T = 2\pi/\omega$ for different initial values and trajectories on phase plane for the optimized system depicted by Eq. (14.16). Figure 14.4a presents the Poincaré section calculated for $\zeta = 0, \mu = 1, \lambda = 0, f = 0.5, \omega = 1.086$, and the complex KAM structure [16]. A peculiar orbit identified as a chaotic sea or stochastic web can be found, which fills a finite area and connects the isolated islands encircling an associated series of fixed points (one prime resonant of period 1 solution, marked purple, secondary resonances of period 3, marked grey; a pair of period 2, marked black and dark red, with the corresponding secondary resonances of period 6, marked light blue and green,

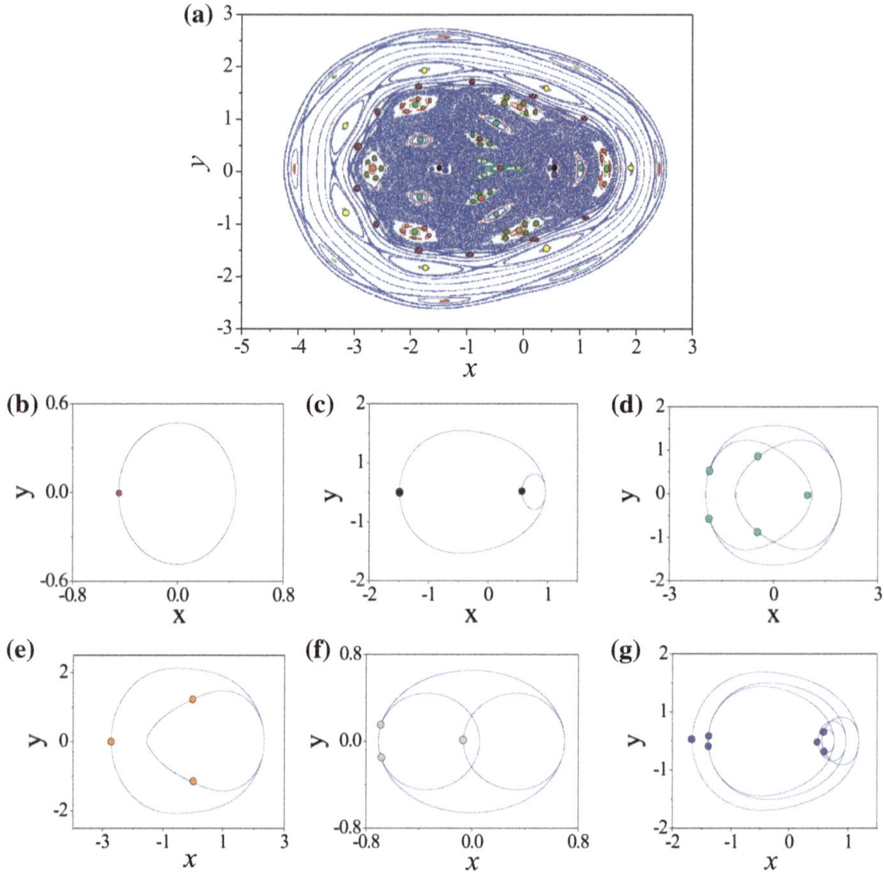

Fig. 14.4 **a** Global KAM structure ($\zeta = 0$, $\mu = 1$, $\lambda = 0$, $f = 0.50$, $\omega = 1.086$). **b–e** the resonant fixed points and the corresponding trajectories of period 1, 2, 5 and period 3 respectively, **f** and **g** the secondary resonances of period 3 and period 6 associated with the period 1 and period 2 plotted in (**b**) and (**c**) respectively

respectively; one period 5, marked cyan and a pair of period 3, marked light green and orange, with the corresponding secondary resonances of period 6, marked light blue and dark green, respectively).

As can be seen from this figure, there exist KAM curves and series of fixed points outside the finite region (a period 13 closed to the chaotic sea, a pair of period 4 of the utmost resonances, marked green and red respectively, and a series period 7 fixed points) with the associated quasi-periodic islands connected by chaotic orbits. As examples, Fig. 14.4b–g plot some of the resonant trajectories and the corresponding Poincaré sections given in Fig. 14.4a: Fig. 14.4b–e for prime resonances of period 1, period 2, period 5 and period 3 solutions respectively, and Fig. 14.4f, g for the secondary resonances of period three and period six associated period one and period

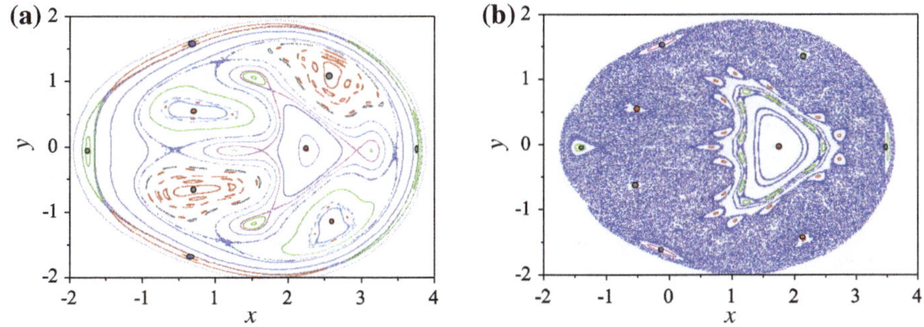

Fig. 14.5 Global KAM structure on Poincaré sections: **a** for $\zeta = 0, \mu = 1, \lambda = 0, f_0 = 0.40, \omega = 0.50$, **b** for $\zeta = 0, \mu = 1, \lambda = 0, f_0 = 0.45, \omega = 0.50$

two resonances respectively. Figure 14.5 shows the existence of invariant curves with hyperbolic fixed points surrounding elliptic fixed points [18], which results in the sudden change to chaotic sea when the parameter varies. Figure 14.5a presents the KAM structure with KAM curves, hyperbolic fixed points connecting invariant curves of homoclinic and heteroclinic types surrounding the elliptic fixed points for $\mu = 1, \zeta = 0, \lambda = 0, f = 0.40, \omega = 0.50$. Figure 14.5b plots the KAM structure, for $\mu = 1, \zeta = 0, \lambda = 0, f = 0.45, \omega = 0.50$, with a chaotic sea burst from the broken invariant curves together with one period one and two pairs of period two resonances preserved from the former KAM structure. This implies a very complicated bifurcation process for further investigation.

14.3.2 Responses with Viscous Damping and External Excitation

We now present some results concerning the behaviour of period doubling, multiple stability, chaos crisis [19, 20], saddle-node bifurcation [21] and transient chaos [22, 23], by examining the bifurcation diagram of system (14.15). The diagram is plotted in Fig. 14.6a for the stroboscopically sampled displacement x versus the control parameter μ near the optimized parameters $\mu = 1, \lambda = 0$ with $\zeta = 0.027, f = 0.40$, and $\omega = 0.26$, for which the corresponding Lyapunov exponent [24] diagram is given on Fig. 14.6b, where rectangular areas are enlarged on Fig. 14.6c, d, respectively.

From Fig. 14.6a, or more clearly on Fig. 14.6c, a pair of coexisting period doubling (marked red and blue) bifurcates from a period 1 solution via a symmetry-breaking bifurcation at $\mu = 0.9302$, leading to the corresponding coexisting chaotic attractors. The phase portraits for the period doubling (marked in blue) shown on Fig. 14.6c are plotted in Fig. 14.7a–d for periods 1, 2 and 4, and a small chaotic attractors for $\mu = 0.960, 0.977, 0.980$ and 0.984, respectively, while the corresponding coexisting chaotic attractor for $\mu = 0.984$ is shown on Fig. 14.7e.

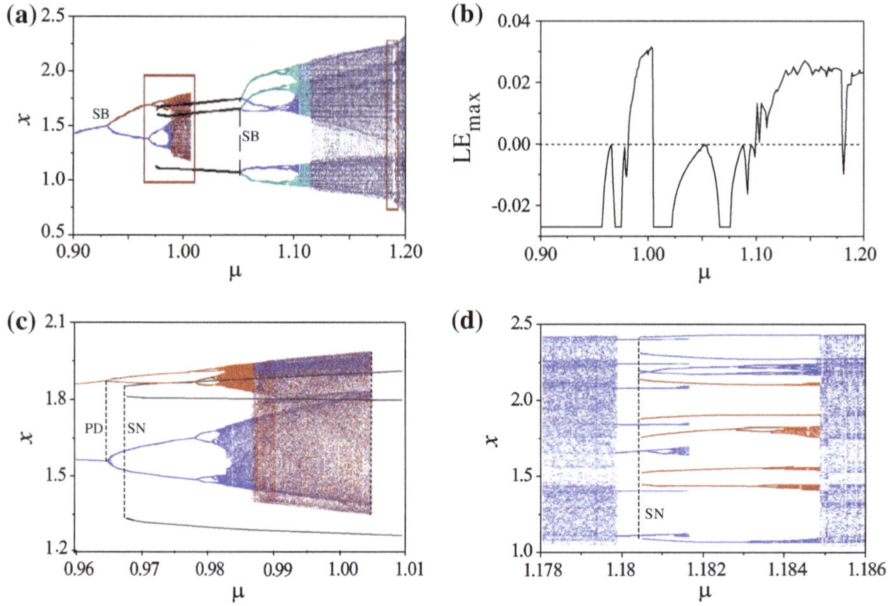

Fig. 14.6 **a** Bifurcation diagram of x versus μ with symmetry-breakings, marked with SB, for $\zeta = 0.027$, $\lambda = 0$, $f = 0.40$, $\omega = 0.26$, **b** the corresponding maximum Lyapunov exponent for (**a**) with initial condition (2, 0); **c**, **d** the enlarged diagram of the left and right rectangle area in Fig. 14.6a, respectively, with period doubling (PD) and saddle-node bifurcations (SN)

As parameter μ increases up to 0.9884, the interior crises burst out, which arise from the collisions between the coexisting chaotic attractors and the saddles within the corresponding basins of attraction. The corresponding phase portraits and Poincaré sections are plotted in Fig. 14.7f. In addition, Fig. 14.6c also shows the boundary crisis of chaos as μ increases again to $\mu = 1.0042$, which originates from the collision between the chaotic attractor and the unstable periodic attractor on the boundary of its basin.

It can also be seen from Fig. 14.6a that a period 3 solution (marked black) occurs at $\mu = 0.9689$, which coexists with the above pair of period doubling, bifurcates into a pair of period 3 solutions at $\mu = 1.0521$ and leads to chaos (marked blue and cyan) via the period doubling scheme. The interior crises appears at a larger chaotic attractor for $\mu = 1.1122$.

The basin analysis [25, 26] demonstrates the coexistence of multiple stable solutions using Poincaré sections (coloured bullets), as shown in Fig. 14.8. The coexistence of a pair of period 3 solutions, one stable and the other unstable, which denotes the burst of a saddle-node bifurcation at $\mu = 0.9689$, is shown in Fig. 14.6c. The basins (marked cyan and pink) of a pair of period 1 solutions (red and yellow) are shown for $\mu = 0.96$ in Fig. 14.8a. Figure 14.8b plots the basins (marked cyan, pink and white respectively) of a pair of period 2 (red and yellow) and a period

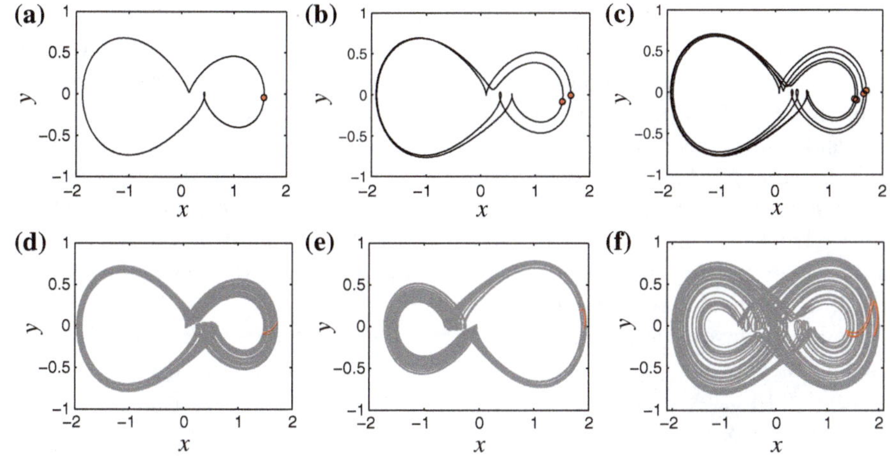

Fig. 14.7 Phase portraits and Poincaré sections of period 1 doubling for $\zeta = 0.027, \lambda = 0, f = 0.40, \omega = 0.26$, **a** $\mu = 0.960$, **b** $\mu = 0.977$, **c** $\mu = 0.980$, **d** $\mu = 0.984$, showing the period 1, 2, 4, and chaotic motion corresponding to the blue bifurcation diagram in Fig. 14.6c; **e** $\mu = 0.984$, **f** $\mu = 1.000$

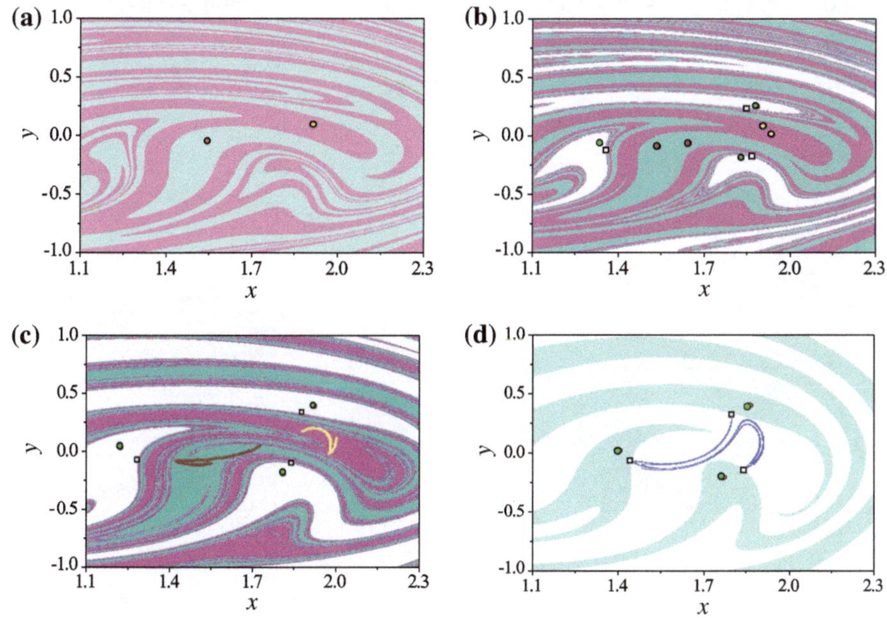

Fig. 14.8 Attraction basins and their corresponding co-existed attractors for $\zeta = 0.027, \lambda = 0, f = 0.40, \omega = 0.26$: **a** $\mu = 0.960$, **b** $\mu = 0.977$, **c** $\mu = 0.984$, **d** $\mu = 1$

3 (green) solutions when $\mu = 0.977$, on the boundaries of which there exists an unstable period 3 solution (the hollow black squares) accompanied with the stable one. In the same way, as shown in Fig. 14.8c for $\mu = 0.984$, an unstable period 3 solution (the hollow black squares) also appears on the boundaries of basins (cyan, pink and white, respectively) of the pair of chaotic attractors (red and yellow) led by period doubling from the periodic solutions and the stable periodic solution (green) above. Figure 14.8d presents the basins (white) of the chaotic attractor (marked blue) merged from the pair of small chaotic attractors above, and the stable period 3 solution (green bullets), where the unstable period 3 attractor (hollow black squares) still remains on the basin boundary.

Numerical investigations have been carried out in order to exhibit the bifurcation phenomena displayed in Fig. 14.6d which is the enlargement of the right rectangular area of Fig. 14.6a. At the critical value $\mu = 1.1798$ a crisis occurs such that the chaotic attractor disappears suddenly jumping to a period 7 orbit (the phase portrait for $\mu = 1.1802$ shown in Fig. 14.9a). The period 7 solution stops suddenly at $\mu = 1.1817$ via a doubling-period bifurcation cascade ($\mu = 1.1814$) [27] with a remaining pair of stable period 5 solutions occurring via a saddle-node bifurcation (SN) at $\mu = 1.1804$ (Fig. 14.9b plots the period 5 trajectories with the Poincaré sections for $\mu = 1.1812$), which leads to a merged chaotic attractor at $\mu = 1.1851$ shown in Fig. 14.9d.

A transient chaos is also found at $\mu = 1.1812$ to be a non-attracting chaotic saddle, which settles down to one of the period 5 motions depending on the initial

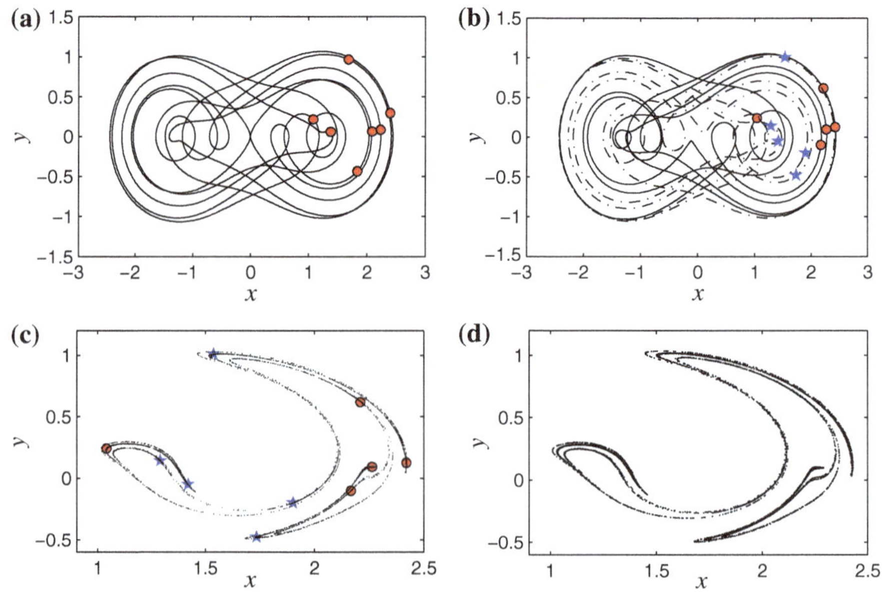

Fig. 14.9 Trajectories and Poincaré sections for $\zeta = 0.027, \lambda = 0, f = 0.40, \omega = 0.26$: **a** $\mu = 1.1802$, **b**, **c** $\mu = 1.1812$, transient chaos and period five solutions, and **d** $\mu = 1.1851$, chaos attractor

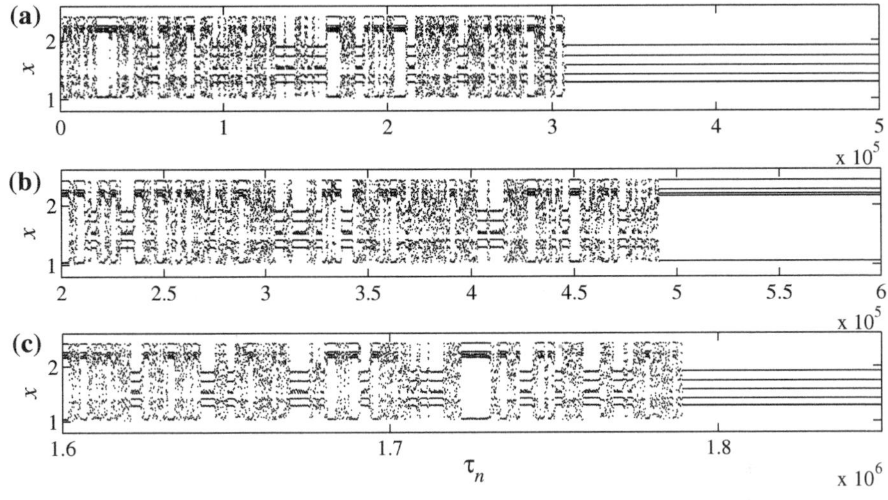

Fig. 14.10 Time-series of stroboscopic map for discrete displacements x versus time τ_n ($\tau_n = 2n\pi/\omega, n = 0, 1, 2, \ldots$) for $\mu = 1.1812$

conditions, as shown in Fig. 14.9c. The life of the transient chaos can be established via time-series, as shown in Fig. 14.10a–c for initial values $(-1.0, 0)$, $(2.1, 0)$ and $(1.0, 0)$ respectively, with different durations of the chaotic saddles.

14.4 Isolator with Stable-Quasi-Zero-Stiffness

This section presents the vibration isolation effects of the system by a parameter optimization to achieve the SQZS system of a maximum interval of low frequency and good isolation performances.

14.4.1 Parameter Optimization for SQZS

The stable quasi-zero-stiffness and the maximum interval of low frequencies of the model will now result from parameter optimization.

The stiffness of system (14.4) is written as

$$K(x) = 1 - \frac{\mu}{(1 + x^2)^{3/2}}, \tag{14.20}$$

which is plotted in Fig. 14.11a for different values of parameter μ. The zero stiffness equilibrium points and the corresponding parameters are obtained when

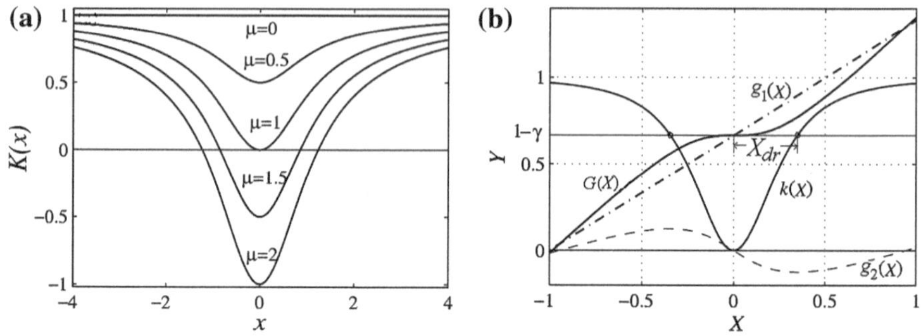

Fig. 14.11 Graphs for the dimensionless stiffness and restoring forces: **a** the stiffness of system (14.4) for parameter $\mu = 0, 0.5, 1.0, 1.5, 2.0$, **b** the total stiffness of system (14.2), marked $k(X)$, and the corresponding restoring force, marked $G(x)$, comprising the components of the vertical and the oblique springs, marked $g_1(X)$ with the stiffness $1-\gamma$ and $g_2(X)$ respectively, for $\alpha = \gamma = 0.5443$

$K(x) = 0$, $F(x) = 0$, that is, $x = 0, \mu = 1, \lambda = 0$; $x = -(\mu^{2/3} - 1)^{1/2}, \mu > 1, \lambda = (\mu^{2/3} - 1)^{3/2}$; $x = (\mu^{2/3} - 1)^{1/2}, \mu > 1, \lambda = -(\mu^{2/3} - 1)^{3/2}$, as shown in Fig. 14.3 ($\mathscr{B}_0$), ($\mathscr{B}_1$) and ($\mathscr{B}_2$), respectively. It can be seen from the stability results shown on Fig. 14.3 ($\mathscr{B}_0$), ($\mathscr{B}_1$) and ($\mathscr{B}_2$) that $x = 0, \mu = 1, \lambda = 0$ is the unique SQZS equilibrium.

To obtain the maximum interval of lower frequency from which the isolation is activated, further optimization is necessary for parameter $\gamma(=\alpha)$. Let us come back to Eq. (14.2) with parameter $\mu = \gamma/\alpha$ the dimensionless stiffness of which is

$$k(X) = 1 - \frac{\alpha^2\gamma}{(\alpha^2 + X^2)^{3/2}}, \tag{14.21}$$

and the stiffness of the vertical spring is denoted as $k_1 = 1 - \gamma$ correspondingly.

Consider the following control inequality

$$k(X) = 1 - \frac{\alpha^2\gamma}{(\alpha^2 + X^2)^{3/2}} \leq \varepsilon k_1, \tag{14.22}$$

where $\varepsilon \in [0, 1]$ is the standard ratio. The excursion from $X = 0$ satisfying (14.22) can be obtained as

$$X_d = \gamma\sqrt{n^{-2/3} - 1}, \tag{14.23}$$

where $n = 1 - \varepsilon(1 - \gamma)$ and the derivative of X_d with respect to γ is

$$X_d'(\gamma) = \sqrt{n^{-2/3} - 1} - \frac{\varepsilon\gamma}{3n^{5/3}\sqrt{n^{-2/3} - 1}}, \tag{14.24}$$

from which $\gamma \in [0, 1]$ can be optimized to maximize the lower frequency interval of X_d by letting $X_d'(\gamma) = 0$.

Figure 14.11b plots the restoring forces, $G(X)$ and $g_1(X)$, the corresponding stiffness, $k(X)$ and k_1, in system (14.2) and the maximum dimensionless excursion X_d at the optimum value $\gamma = 0.5443$ when $\varepsilon = 1.0$; the details of the symbols are given from the above discussions and the corresponding captions.

14.4.2 Vibration Isolation

Herein the performance of vibration isolation of the system in Fig. 14.1 is considered. Suppose the base moves in the form of a harmonic vibration $Y = B\cos(\Omega t)$, which could simulate the ambient disturbance in the ground vibration test and the influence of vibration of vehicles and ships on their equipments [28]. The equation of motion of such a dynamical model can be written as

$$m\ddot{Z} + c\dot{Z} + K_1 Z + 2K_2 Z\left(1 - \frac{L}{\sqrt{a^2 + Z^2}}\right) = mB\Omega^2 \cos(\Omega t), \qquad (14.25)$$

where $Z = X - Y$ is the relative displacement between the base and the supported mass m. Equation (14.25) is written in a dimensionless form

$$z'' + 2\zeta z' + z + rz\left(1 - \frac{1}{\sqrt{\alpha^2 + z^2}}\right) = A\omega^2 \cos(\omega\tau) \qquad (14.26)$$

by letting $z = \frac{Z}{L}, \omega_1^2 = \frac{K_1}{m}, \omega_2^2 = \frac{2K_2}{m}, 2\zeta = \frac{c}{m\omega_1}, r = \frac{\omega_2^2}{\omega_1^2}, \alpha = \frac{a}{L}, \tau = \omega_1 t, A = \frac{B}{L}, \omega = \frac{\Omega}{\omega_1}$. In the dimensionless equation (14.26), the absolute displacement and the excitation are denoted by $U = \frac{X}{L}$, $y = A\cos(\omega\tau)$ instead of the preceding form X, Y. From the foregoing analysis of statics, it has been known that Eq. (14.26) has an equilibrium with quasi-zero stiffness when $\alpha = \frac{r}{1+r}$.

Now consider the non-dimensional dynamical system with Eq. (14.26) and suppose that it has a solution of the first order mode $z = b\cos(\omega t + \theta)$. Using the averaging approach [18, 29, 30] and substituting

$$\begin{cases} z = b\cos(\omega t + \theta), \\ z' = -b\omega\sin(\omega t + \theta), \end{cases} \qquad (14.27)$$

into Eq. (14.26), we can obtain the equations of slowly varying amplitude and phase as that

$$\begin{cases} b' = -\frac{1}{\omega} S(b, \theta)\sin(\omega t + \theta), \\ \theta' = -\frac{1}{b\omega} S(b, \theta)\cos(\omega t + \theta), \end{cases} \qquad (14.28)$$

in which

$$S(b,\theta) = b\omega^2\cos(\omega t+\theta) + 2b\zeta\omega\sin(\omega t+\theta) - b(1+r)\cos(\omega t+\theta) + \frac{br\cos(\omega t+\theta)}{\sqrt{\alpha^2+b^2\cos^2(\omega t+\theta)}} + A\omega^2\cos(\omega t). \tag{14.29}$$

By averaging $\psi = \omega t + \theta$ over a period from 0 to 2π, the differential system (14.28) for b, θ reads

$$\begin{cases} b' = -\dfrac{1}{2\omega}\left(2b\zeta\omega + A\omega^2\sin\theta\right), \\ \theta' = -\dfrac{1}{2b\omega}\left(-b(1+r) + b\omega^2 + A\omega^2\cos\theta + \dfrac{br}{\pi}I(b)\right), \end{cases} \tag{14.30}$$

where

$$\begin{aligned} I(b) &= \int_0^{2\pi} \frac{\cos^2\psi}{\sqrt{\alpha^2+b^2\cos^2\psi}}\mathrm{d}\psi \\ &= \frac{4}{b^2}\left(\sqrt{\alpha^2+b^2}\mathrm{EllipticE}\left(\sqrt{\frac{\mathrm{b}^2}{\alpha^2+\mathrm{b}^2}}\right) - \frac{\alpha^2}{\sqrt{\alpha^2+\mathrm{b}^2}}\mathrm{EllipticK}\left(\sqrt{\frac{\mathrm{b}^2}{\alpha^2+\mathrm{b}^2}}\right)\right), \end{aligned} \tag{14.31}$$

and EllipticK (k), EllipticE (k) denotes the complete elliptic integral of the first and second kind with its elliptic modulus k, respectively.

Let $b' = 0,\ \theta' = 0$ to derive the singular point (b_s, θ_s) in the (b, θ) plane, which corresponds to the periodic solution of Eq. (14.26), that is

$$\begin{cases} 2b\zeta\omega + A\omega^2\sin\theta = 0, \\ -b(1+r) + b\omega^2 + A\omega^2\cos\theta + \frac{br}{\pi}I(b) = 0, \end{cases} \tag{14.32}$$

such that the averaging frequency-response relationship follows

$$(2b\zeta\omega)^2 + \left(-b(1+r) + b\omega^2 + \frac{br}{\pi}I(b)\right)^2 - \left(A\omega^2\right)^2 = 0. \tag{14.33}$$

In order to obtain the stability of the solution (14.27), perturbation variables $\xi = b - b_s,\ \eta = \theta - \theta_s$ can be introduced into the averaging Eq. (14.30), of which (b_s, θ_s) is a singular point in the phase plane (b, θ). For simplicity, we can apply the following designations

$$\begin{cases} P = 2b\zeta\omega, \\ Q = -b(1+r) + b\omega^2 + \frac{br}{\pi}Q(b), \end{cases} \tag{14.34}$$

such that

$$\begin{cases} \xi' + \frac{1}{2\omega}\left(\left(\frac{\partial P}{\partial b}\right)_s \xi - Q_s\eta\right) = 0, \\ \eta' + \frac{1}{2\omega b_s}\left(\left(\frac{\partial Q}{\partial b}\right)_s \xi + P_s\eta\right) = 0, \end{cases} \tag{14.35}$$

where the subscript s represents the substitution $b = b_s, \theta = \theta_s$ into the above formulae.

The stability can be derived by the characteristic equation of the linear perturbation Eq. (14.35)

$$\begin{vmatrix} \lambda + \frac{1}{2\omega}\left(\frac{\partial P}{\partial b}\right)_s & -\frac{1}{2\omega}Q_s \\ \frac{1}{2\omega b_s}\left(\frac{\partial Q}{\partial b}\right)_s & \lambda + \frac{1}{2\omega b_s}P_s \end{vmatrix} = \lambda^2 + b_1\lambda + b_2 = 0, \tag{14.36}$$

where

$$b_1 = \frac{1}{2\omega}\left(\frac{\partial P}{\partial b}\right)_s + \frac{1}{2\omega b_s}P_s, \quad b_2 = \frac{1}{4b_s\omega^2}\left[\left(\frac{\partial P}{\partial b}\right)_s\right]P_s + \frac{1}{4b_s\omega^2}Q_s\left[\left(\frac{\partial Q}{\partial b}\right)_s\right]. \tag{14.37}$$

Because the (b_s, θ_s) is a simple zero of Eq. (14.32), we can exploit the first order approximate Lyapunov stability theory and the Routh–Herwitz criterion [18, 30] to determine the stability of the related solutions: if $b_1 > 0, \quad b_2 > 0$, the singular point is stable asymptotically; as long as $b_2 < 0$, the singular point is unstable, and $b_2 = 0$ is the boundary between the stable and unstable region of solutions.

14.4.3 Transmissibilities

For the SQZS vibration isolation system with Eq. (14.26) under the vibrating base excitation, the relative and absolute displacement transmissibilities [31, 32] can be defined as

$$T_R = \frac{\|z\|_{\max}}{A} = \frac{b}{A} \tag{14.38}$$

and

$$T_D = \frac{\|U\|_{\max}}{A} \tag{14.39}$$

respectively, which have the same form as that of a linear isolation system and where $\|\cdot\|$ denotes the displacement for a corresponding variable.

From the foregoing analysis of the SQZS model, the maximum absolute displacement response can be given as

$$\begin{aligned} \|U\|_{\max} &= \|y + z\|_{\max} \\ &= \sqrt{b^2 + A^2 + 2bA\cos\theta}, \end{aligned} \tag{14.40}$$

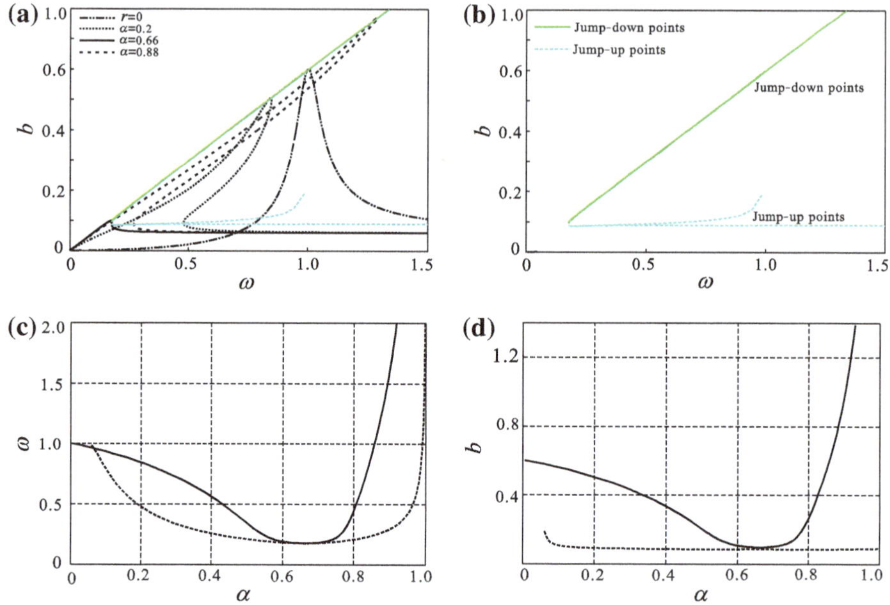

Fig. 14.12 **a** Amplitude-frequency response of Eq. (14.26) for $\zeta = 0.05$, $A = 0.06$ and varying $\alpha = 0.2, 0.66, 0.88$, as well as $r = 0$ (the linear isolator without horizontally auxiliary springs), respectively. **b** Trajectories of the frequency-jump points for varying values of parameter α. **c, d** Frequencies and amplitudes of the jumping-down (*solid line*) and jumping-up (*dashed line*) points with respect to varying α, respectively

such that the estimated formula of transmissibility of the absolute displacement for the analytical solution is

$$T_{\mathrm{Da}} = \sqrt{\left(\frac{b}{A}\right)^2 + 1 + 2\left(\frac{b}{A}\right)^2 \frac{(1+r) - \omega^2 - \frac{r}{\pi} Q(b)}{\omega^2}}, \tag{14.41}$$

which can be applied for evaluating the vibration isolation performance for a single harmonic excitation of the base.

We now describe the influence of the geometrical parameters of the system on the frequency responses of the non-dimensional SQZS Eq. (14.26). Firstly, in Fig. 14.12a, the amplitude-frequency responses with respect to parameter α are shown for varying $\alpha = 0.2$ (plotted in dotted line), 0.66 (solid line), 0.88 (thin-dashed curve), and $r = 0$ (dot-dot-dashed line) and for other parameters $\zeta = 0.05$, $A = 0.06$. Meanwhile, it displays the trajectories of the jumping-down (green solid line) and jumping-up (cyan thick-dashed) points for continuously varying α, namely the two intersection points between the frequency-response curves obtained from Eq. (14.34) and the stability boundary $b_2 = 0$ in Eq. (14.37). In particular, the trajectories of the frequency jump points are given in Fig. 14.12b alone for the purpose of a noticeable clarity. To further

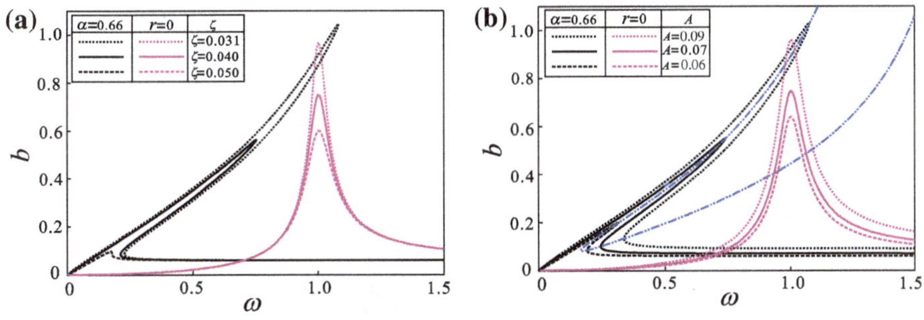

Fig. 14.13 Amplitude-frequency responses of the SQZS system with Eq. (14.26) for parameters $\alpha = 0.66$ and $r = 0$: **a** $\zeta = 0.031, 0.04, 0.05$ with $A = 0.06$; **b** $A = 0.06, 0.07, 0.09$ with $\zeta = 0.047$, and meanwhile the stability boundary of solutions for $\zeta = 0.047$ shown in the *dot-dot-dashed curve* in (**b**)

characterize the relations of frequency and amplitude with respect to the geometrical parameters α, the details are shown in Fig. 14.12c and d, respectively, where the solid curves are plotted for jumping-down points and dashed curves for jumping-up points, respectively. To suppress the frequency jump phenomena, the value of parameter α can be optimized by making the jumping-down points move towards a frequency as low as possible. For instance, $\alpha = 0.66$ is an optimum value approximately for the SQZS system with Eq. (14.26) when parameters $\zeta = 0.05$, $A = 0.06$.

To make the effects of other parameters on the amplitude-frequency responses more clear for the non-dimensional SQZS system with Eq. (14.26), the amplitude-frequency response curves for varying parameters ζ and A are shown in Fig. 14.13a and b respectively. Figure 14.13a shows the amplitude-frequency response curves in black for varying values of $\zeta = 0.031$ (dotted), 0.040 (solid), 0.050 (dashed) for $\alpha = 0.66$, $A = 0.06$. Meanwhile, the responses for the linear isolation system without the horizontally auxiliary springs, that is $r = 0$, are presented for the same damping ratios ζ. In Fig. 14.13b, the curves of frequency responses are plotted by varying $A = 0.06$ (dashed), 0.07 (solid), 0.09 (dotted) for $\alpha = 0.66$, $\zeta = 0.047$. For detailed information about these lines, see the upper-left sub-panel of each sub-figures in Fig. 14.13. It can be seen that both the jumping-down frequency and amplitude of the response increase as the damping ratio decreases or the external base vibration amplitude increases. Especially, in the case when the moving amplitude of the base increases over some value, the advantage of the vibration isolation of the SQZS system can be lost compared to the corresponding linear isolator without horizontally auxiliary springs. Moreover, the stability boundary $b_2 = 0$ in Eq. (14.37) are also shown in Fig. 14.13b in a dot-dot-dashed curve, along which the jumping-down points moves towards a higher frequency and a larger amplitude as the vibration amplitude of the base increases.

In order to facilitate the numerical calculations for the absolute displacement transmissibility, the system with Eq. (14.26) is rewritten according to the formula (14.39). Letting $u = U/A$ leads to

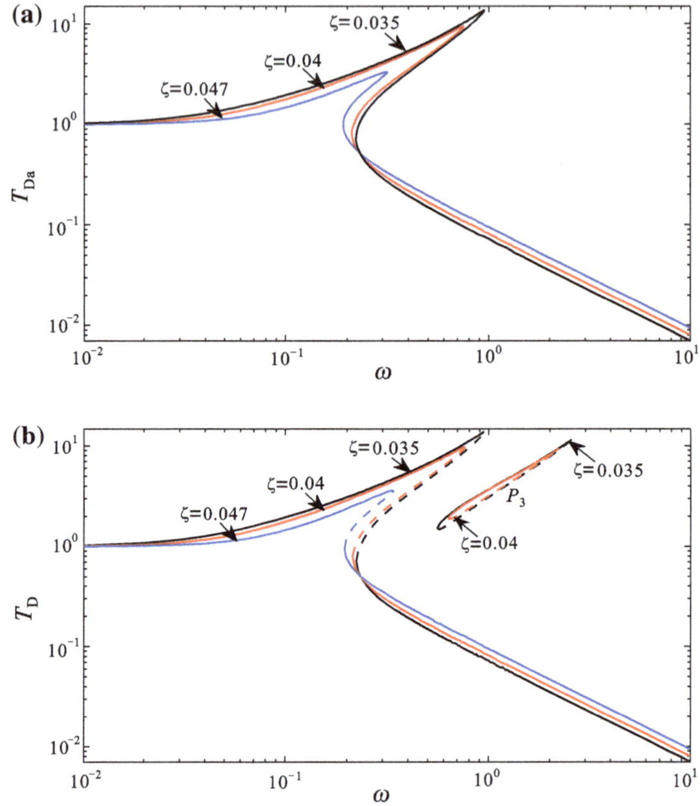

Fig. 14.14 Transmissibility of the absolute displacement for different damping ratios $\zeta = 0.035, 0.04, 0.047$ (indicated by *arrows* associated with separate texts): **a** analytical results ($T_{\mathrm{Da}} - \omega$), **b** results ($T_{\mathrm{D}} - \omega$) of directly numerical simulations from Eq. (14.42)

$$\begin{cases} u' = v, \\ v' = -2\zeta u' - u - r\,(u - \cos(\omega t))\left(1 - \dfrac{1}{A\sqrt{\left(\frac{\alpha}{A}\right)^2 + (u - \cos(\omega t))^2}}\right) - 2\zeta\omega \sin(\omega t) + \cos(\omega t), \end{cases} \tag{14.42}$$

where the absolute displacement transmissibility can be obtained by directly numerical simulation and expressed as T_D-ω ($T_D = \|u\|_{\max}$ according to Eq. (14.39)).

By integrating the SQZS dynamical system with Eq. (14.42) for different excitation frequencies ω and using a sufficient number of initial conditions taken randomly, the amplitude-frequency ($|u|_{\max} - \omega$) response is obtained as shown in Fig. 14.14b for parameters $\alpha = 0.66$, $A = 0.06$ and $\zeta = 0.035$ (black line, where the solid and the dashed lines represent the stable and the unstable solutions respectively), 0.04 (red), 0.047 (blue). The analytical transmissibilities obtained by Eq. (14.41) are also shown for $\zeta = 0.035, 0.04, 0.047$ in Fig. 14.14a (the stabilities of solutions are not determined herein), which are in good agreement with the numerical results in Fig. 14.14b for primary resonance cases. It can be found that the subharmonic solutions will reduce the frequency range of vibration isolation according to the traditional knowledge for vibration isolation, but the subharmonic resonance could be suppressed if the damping ratio ζ increases; however, as the value of ζ increases, the efficiency of vibration isolation will be cut down. Hence, an appropriate damping ratio should be chosen for a desired isolation performance through making synthesis of the transmissibility and the complexity of dynamics.

14.5 Summary

A nonlinear isolator has been proposed in this chapter for vibration isolation due to a stable quasi-zero-stiffness (SQZS). This isolator can be used to a nonlinear supporting system for ground vibration tests (GVT) for large scale machines like aircrafts. The dynamical analysis shows the asymmetrical dynamics due to the consideration of the weight of the mass to the system to be beneficial for the nonlinear system and an isolation. An extra parameter is introduced to the system due to the asymmetrical dynamics of the original system. The SQZS can be achieved for the proposed isolator by optimization of the parameters to get the maximum interval of low frequency with a lower transmissibility in comparison with the linear system shown in both analytical and numerical analysis.

References

1. Molyneux, W. G. (1957). *Supports for vibration isolation*. ARC/C-322. London: Aeronautical Research Counci.
2. Lorrain, P. (1974). Low natural frequency vibration isolator or seismograph. *Review of Science Instrument*, *4*(5), 198–202.
3. Denoyer, K., & Johnson, C. (2001). Recent achievements in vibration isolation systems for space launch and on-orbit applications. In *52nd Int Astronautics Congress*, Toulouse, France.
4. Winterflood, J. (2001). *High performance vibration isolation for gravitational wave detection*. PhD thesis, University of Western Australia, Western Australia.
5. Ibrahim, R. A. (2008). Recent advances in nonlinear passive vibration isolators. *Journal of Sound and Vibration*, *314*(3–5), 371–452.
6. Arafat, R. M. D., Park, S. T., & Sajal, C. B. (2010). Design of a vehicle suspension system with negative stiffness system. *IST Transaction of Mechanical System Theoretic and Application*, *1*(2), 1–7.
7. Green, G. S. (1945). The effect of flexible ground supports on the pitching vibrations of an aircraft. *R&M* (Vol. 2291).
8. Xing, J. T., Xiong, Y. P., & Price, W. G. (2005). Passive-active vibration isolation systems to produce zero or infinite dynamic modulus. *Journal of Sound and Vibration*, *286*(3), 615–636.
9. Blair, D. G., Winterflood, J., & Slagmolen, B. (2002). High performance vibration isolation using springs in euler column buckling mode. *Physics Letters A*, *300*(2–3), 122–130.
10. Platus, D. L. (2014). Vibration isolation systems.
11. Zhang, J. Z., Li, D., & Chen, M. J. (2004). An ultra-low frequency engineering material. *Journal of Engineering Materials*, *57*(258), 231–236.

12. Alabuzhev, P., Gritchin, A., & Kim, L. (1989). *Vibration protecting and measuring systems with quasi-zero stiffness*. New York: Hemisphere Publishing Co, Taylor and Francis Group.
13. Carrella, A., Brennam, M. J., & Waters, T. P. (2007). Static analysis of a passive vibration isolator with quasi-zero-stiffness characteristic. *Journal of Sound and Vibration, 301*(3–5), 678–689.
14. Kovacic, I., Brennam, M. J., & Waters, T. P. (2008). A study of a non-linear vibration isolator with a quasi-zero stiffness characteristic. *Journal of Sound and Vibration, 315*(3), 700–711.
15. Dimminie, D. C., & Haberman, R. (2000). Slow passage through a saddle-center bifurcation. *Journal of Nonlinear Science, 10*(2), 197–221.
16. Lichtenberg, A. J., & Lieberman, M. A. (1992). *Regular and chaotic dynamics*. New York: Springer.
17. DE Freitas, M. S. T., Viana, R. L., & Grebogi, C. (2004). Multistability, basin boundary structure, and chaotic behavior in a suspension bridge model. *International Journal of Bifurcation and Chaos, 14*(3), 927–950.
18. Guckenheimer, J., & Holmes, P. (1983). *Nonlinear oscillation, dynamical system and bifurcation of vector fields*. New York: Springer.
19. Hong, L., & Xu, J. X. (1999). Crisis and chaotic transients studied by the generalized cell mapping digraph method. *Physics Letters A, 262*(4–5), 361–375.
20. Hong, L., & Xu, J. X. (2004). Crisis and chaotic saddle and attractor in forced duffing oscillator. *Communication of Nonlinear Science Numerical Simulation, 9*(3), 313–329.
21. Nusse, H. E., Ott, E., & Yorke, J. A. (1995). Saddle-node bifurcations on fractal basin boundaries. *Physical Review Letters, 75*(13), 2482–2485.
22. Souza, S. L. T., Caldas, I. L., & Viana, R. L. (2004). Sudden changes in chaotic attractors and transient basins in a model for rattling in gearboxes. *Chaos, Solition and Fractals, 21*(3), 763–772.
23. Souza, S. L. T., Caldas, I. L., Viana, R. L., & Balthazar, J. M. (2008). Control and chaos for vibro-impact and non-ideal oscillators. *Journal of theoretical and applied mechanics, 46*(3), 641–664.
24. Wolf, A., Jack, B., Swift, H., Swinny, L., & John, A. (1985). Determining lyapunov exponents from a time series. *Physica D-Nonlinear Phenomena, 16*(3), 285–317.
25. Hsu, C. S. (1992). Global analysis by cell mapping. *International Journal of bifurcation and chaos, 2*(4), 727–771.
26. Nusse, H. E., & Yorke, J. A. (1994). *Dynamics: Numerical explorations* (Vol. 101)., Applied Mathematical Sciences. New York: Springer.
27. Thompson, J. M. T., & Stewart, H. B. (2002). *Nonlinear dynamics and chaos* (2nd ed.). Chichester: Wiley.
28. Hao, Z. F., & Cao, Q. J. (2015). The isolation characteristics of an archetypal dynamical model with stable-quasi-zero-stiffness. *Journal of Sound & Vibration, 340*, 61–79.
29. Li, Z. X., Cao, Q. J., Wiercigroch, M., & Léger, A. (2013). Analysis of the periodic solutions of a smooth and discontinuous oscillator. *Acta Mechanica Sinica, 29*(4), 575–582.
30. Liu, Y. Z., & Chen, L. Q. (2001). *Nonlinear Vibration*. Beijing, China: Higher Education Press.
31. Xing, J. T. (2015). *Energy flow theory of nonlinear dynamical systems with applications*. London: Springer.
32. Xing, J. T., & Price, W. G. (1999). A power flow analysis based on continuum dynamics. *Proceedings of the Royal Society A Mathematical Physical & Engineering Sciences, 455*(1982), 401–436.

Chapter 15
Secure Communication Scheme

The last engineering application presented here deals with secure communication, for which only the theoretical foundations are presented based on non-autonomous compound remodulating scheme. In this chapter, we show that the SD oscillator might be used to satisfy the requirements of security of communications using its chaotic characteristics. In the proposed scheme, the driving signal is generated by the combination of multiple state variables. This scheme not only breaks cryptanalysis from the phase space reconstruction of the transmitted dynamics, but also enlarges the key space of the cryptographic system. The investigation shows that the return maps of the SD oscillator as a cryptosystem are chaotic, which enables the robust security of the secure communication schemes. The synchronization performance of the scheme is also supported by the mathematical conclusions and simulation results of the SD oscillator. An implementation has been done using the SD oscillator to get a more secure scheme for chaotic based communication.[1]

15.1 Introduction

Even a large amount of works on chaos-based cryptosystems have been published since the discovery of the chaotic synchronization in 1993, see [1] for details, schemes like chaotic masking, chaotic shift keying, parameter modulation, non-autonomous chaotic modulation [2–4] and those based on their modification [5–7] are difficult to implement in practice with a reasonable degree of security [8–10]. It is always a real challenge to design robust chaotic secure communication systems against the available attacks [11, 12].

---

[1]This part relies on a common work with Mr. Shengliang Fu, Centre for Nonlinear Dynamics Research, Department of Mathematics and Physics, Shijiazhuang Tiedao University, 050043 China.

Q. Cao and A. Léger, *A Smooth and Discontinuous Oscillator*, Springer Tracts in Mechanical Engineering, DOI 10.1007/978-3-662-53094-8_15

The following types of countermeasures, [13], have been studied to achieve the robustness of the schemes in the literature:

1. More complex dynamical systems, such as hyperchaotic systems or multiple cascaded heterogeneous chaotic systems, see in [14–16] for details;
2. More complicated synchronization modes, seen in [17];
3. Coupled fast-spiking neurons with chemical synapses, seen in [18] for instance;
4. More complicated chaotic cryptosystems into traditional cyphers, see also in [19];
5. Remodulating the driving signal [20];
6. An impulsive driving signal instead of a continuous signal to realize modulation and synchronization [21].

After specific attacks, it was found that the first four systems remained unsecure, and some secure defects of the last have also been reported. We shall prove the applicability of the SD oscillator to a chaotic cryptosystem. Unlike most of the secure communication schemes proposed a signal based on a single state variable, a new scheme called compound non-autonomous remodulating (CNAR) scheme is based on multi-state variables. Using the method of return map analysis used in [22] shows that the return maps of the SD oscillator are chaotic, which can be of a good performance against attack. Generally speaking, among the most known attacks, the return map based attacks [23, 24], spectral attack [25] and short time period attack [26] is the powerful tool to attack the system, which is treated to be a powerful criterion of judging whether a chaotic secure communication scheme is actually secure or not.

The return map which is a partial reconstruction of the dynamics of the transmitted signal, can extractly trail the message by analyzing the evolution of the signal on the attracting set. Suppose that a state variable x of the adopted chaotic system is the transmitted signal, then there are several kinds of return maps used in decoding chaotic systems, see details in [25].

Let $t_n^{\max}$ be the time when $x(t)$ attains its nth local maximum $V_{\max}(n)$ and $t_n^{\min}$ be the time when $x(t)$ attains its nth local minimum $V_{\min}(n)$.

Let $T_{\max}(n) = t_n^{\max} - t_{n-1}^{\max}$ and $T_{\min}(n) = t_n^{\min} - t_{n-1}^{\min}$ be two time intervals, then the following return maps are obtained:

$$
\begin{aligned}
r_{\max}^A &: V_{\max}(n) \mapsto V_{\max}(n+1),\\
r_{\min}^A &: V_{\min}(n) \mapsto V_{\min}(n+1),\\
r_{\max}^T &: T_{\max}(n) \mapsto V_{\max}(n),\\
r_{\max}^T &: T_{\min}(n) \mapsto V_{\max}(n).
\end{aligned}
$$

The methods consist in extracting secret signals which are masked by chaos, regardless of the chaotic transmitters and of the way of reconstructing the dynamics or identifying the parameters of the chaotic transmitters. From the point of view of the return map, a more complex chaotic systems is necessary to provide a higher degree

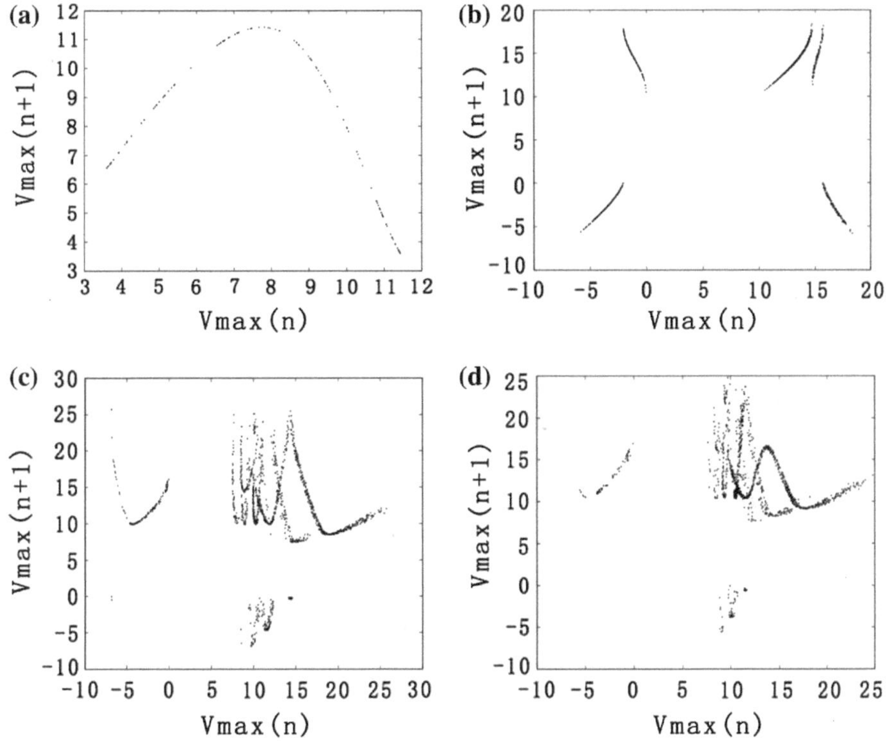

Fig. 15.1 Return maps: **a** for Rössler oscillator, **b** for Lorenz system, **c** for Lü oscillator and **d** for Chen oscillator, respectively

of security if the transmitted signal is generated by the chaotic systems, such as [27–31] and new types of chaotic systems such as [32, 33]. Following, the return maps constructed for Rossler attractor, Lorenz, Lu and Chen attractors in Fig. 15.1, shown in 15.1a–d, respectively, it is clear that those maps are so simple and concentrated, nearly one-dimensional attractors, which cannot guarantee the corresponding secure communication schemes based on such attractors are really secure. It is not difficult to find, to overcome the security problem of most traditional chaos-based secure communication schemes. The most important point is to choose a more complicated chaotic systems.

The following subsections present the potential and feasibility of the SD oscillator and the theory of synchronization using a general compound remodulating chaotic signal and also the synchronization of chaos in the SD oscillator. Finally, the scheme of the compound non-autonomous remodulating communication and secure communication system are shown with an experimental implementation.

15.2 Synchronization Based Upon SD Oscillator

15.2.1 Modelling

We start from the SD oscillator with damping and external harmonic excitation in the dimensionless form

$$\ddot{x} + 2\xi\dot{x} + x\left(1 - \frac{x}{\sqrt{x^2 + \alpha^2}}\right) = f_0 \cos\omega\,\tau \tag{15.1}$$

where x is the displacement from the stable equilibrium position, overdots represent a differentiation with respect to the time τ, ξ, f_0 and ω stand for the damping ratio, the amplitude and the frequency of the excitation. As presented previously, the value of the parameter α determines the transition from smooth to discontinuous. The oscillator (15.1) is re-written in the autonomous form using the transformations $x_1 = x$, $x_2 = \dot{x}$, $x_3 = t$:

$$\begin{cases} \dot{x}_1 = x_2 \\ \dot{x}_2 = -2\xi x_2 - x_1\left(1 - \dfrac{1}{\sqrt{x_1^2 + \alpha^2}}\right) + f_0 \cos(\omega x_3) \\ \dot{x}_3 = 1 \end{cases} \tag{15.2}$$

All of the return maps constructed from the SD oscillator (15.2) displayed in Fig. 15.2 are chaotic attractors. This illustrates that secure communication schemes based on the SD oscillator might be more secure than those based on Lorenz or other attractors and will be difficult to break by the return-map attack methods.

15.2.2 A General Compound Remodulating Chaotic Signal

Consider a first order n-dimensional dynamical system of the form:

$$\begin{cases} \dfrac{dx}{dt} = G(x, y), \\ \dfrac{dy}{dt} = H(x, y), \end{cases} \tag{15.3}$$

where x and y are respectively of dimensions 1 and $n-1$. In the literature, the variable x is commonly taken as representing the transmitted signal. Since 1995, Kocarev and Parlitz chose $s = h(x)$ or $\dot{s} = h(x, s)$ as the transmitted signal. Murali et al. adopted $s = x + u(y)$ as the transmitted signal, where $u(\cdot)$ was either a linear or a nonlinear function. Bu et al. and Wu et al. considered $s(x, t) = A\cos(\omega t + \varphi_0)y(t)x$ and

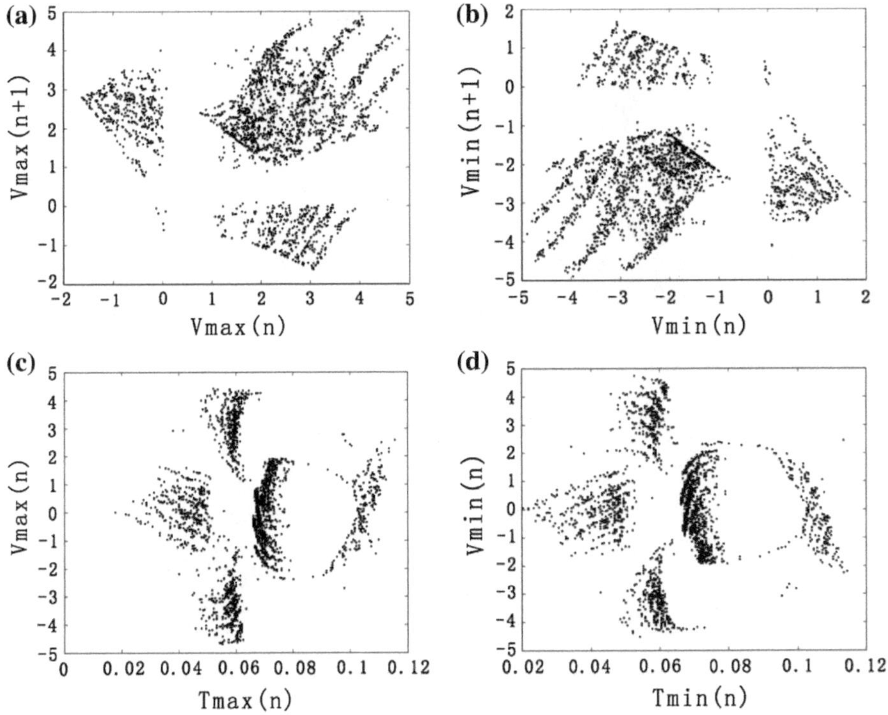

Fig. 15.2 Return maps of the SD oscillator based on the presented four types: **a** the Vmax(n), **b** Vmin(n), **c** Tmax(n) and **d** Tmin(n) respectively

$s(x, t) = A[\cos(\omega t + \varphi_0) + M]y(t)x$ respectively. But it has been checked that the secure communication schemes based on the above compound transmitted signals remained insecure.

Now consider a synchronization using a new compound signal. The set of equations for this new kind of synchronization system is, for the driving part:

$$\begin{cases} \dot{x} = G(x, y), \\ \dot{y} = H(x, y), \\ x_t = \alpha x + \beta u(y), \end{cases} \tag{15.4}$$

where α, β are parameters and $u(\cdot)$ can be either a linear or a nonlinear function. The response system is of the form:

$$\begin{cases} x_d = \dfrac{x_t - \beta u(y')}{\alpha}, \\ \dot{x}' = g(x_d, x', y'), \\ \dot{y}' = h(x_d, x', y'). \end{cases} \tag{15.5}$$

Here x_t is the compound chaotic signal used as the drive and x_d is the signal generated in the response system, so that

$$\begin{cases} x_d = \dfrac{x_t - \beta u(y)}{\alpha} = x \\ \dot{x}' = g(x, x, y) = G(x, y) = \dot{x} \\ \dot{y}' = h(x, x, y) = H(x, y) = \dot{y} \end{cases} \tag{15.6}$$

Let us introduce the differential system for the difference $e = (e_1, e_2)$ with $e_1 = x' - x,\ e_2 = y' - y$. This differential system can be written as,

$$\begin{cases} \dot{e}_1 = g(x_d, x', y') - G(x, y) \\ \\ \dot{e}_2 = h(x_d, x', y') - H(x, y). \end{cases} \tag{15.7}$$

The solution e to the differential system (15.7) is globally asymptotically stable with respect to the influence of signal x_d, then the response variables x' and y' will be synchronized with the driving variables x and y.

15.2.3 Synchronization of Chaos in SD Oscillator

We now apply the idea of complete synchronization through compound chaotic signal to the SD oscillator. Let the driving system be such that:

$$\begin{cases} \dot{x}_1 = x_2, \\ \dot{x}_2 = -2\xi x_2 - x_1 + \dfrac{x_1}{\sqrt{x_1^2 + \alpha^2}} + f_0 \cos \omega\, x_3, \\ \dot{x}_3 = 1, \\ x_t = l x_1 + m x_2, \end{cases} \tag{15.8}$$

where l and m are strictly positive coefficients, and let the response as:

$$\begin{cases} x_d = \dfrac{x_t - m y_2}{l} \\ \dot{y}_1 = y_2 - k(y_1 - x_d) \\ \dot{y}_2 = -2\xi y_2 - y_1 + \dfrac{x_d}{\sqrt{x_d^2 + \alpha^2}} + f_0 \cos \omega\, y_3 \\ \dot{y}_3 = 1 \end{cases} \tag{15.9}$$

where $k(>0)$ is the value of an arbitrary chosen parameter. Let $e = (e_1, e_2)$ be the difference between systems (15.8) and (15.9), with $e_1 = y_1 - x_1, e_2 = y_2 - x, y_3 = x_3$, then the equations for both components of e are:

$$\begin{cases} \dot{e}_1 = -ke_1 + \left(1 - \dfrac{kl_2}{l_1}\right) e_2 \\ \dot{e}_2 = -e_1 - 2\xi e_2 + \left(\dfrac{x_d}{\sqrt{x_d^2 + \alpha^2}} - \dfrac{x_1}{\sqrt{x_1^2 + \alpha^2}}\right). \end{cases} \tag{15.10}$$

Now, let us consider the function $F(x) = \dfrac{x}{\sqrt{x^2 + \alpha^2}}, \dfrac{dF}{dx} = \dfrac{\alpha^2}{\sqrt{(x^2 + \alpha^2)}} > 0.$ Then by intermediate value theorem we have

$$\frac{x_d}{\sqrt{x_d^2 + \alpha^2}} - \frac{x_1}{\sqrt{x_1^2 + \alpha^2}} = -\left[\frac{m}{l} F'(\beta)\right] e_2,$$

and system (15.10) can be written as:

$$\begin{pmatrix} \dot{e}_1 \\ \dot{e}_2 \end{pmatrix} = \begin{pmatrix} -k & 1 - \dfrac{km}{l} \\ -1 & -2\xi - \dfrac{m}{l} F'(\beta) \end{pmatrix} \begin{pmatrix} e_1 \\ e_2 \end{pmatrix}. \tag{15.11}$$

The synchronization then amounts to studying the stability of the origin for system (15.11). On the one hand, when $100 - \frac{km}{l} = 0$, both eigenvalues of the system (15.11) have negative real part which means that the origin (0, 0) is asymptotically stable. When this is satisfied, the synchronization of two SD systems (15.8) and (15.9) is achieved. Let us introduce a Lyapunov function given by $E(t) = \frac{1}{2}(e_1^2 + e_2^2)$, which is a continuously differentiable decreasing positive definite function. It is easily seen that the derivative of E along the trajectories is:

$$\dot{E}(t) = -ke_1^2 - \frac{km}{l} e_1 e_2 - \left(\xi + \frac{m}{l} F'(\beta)\right) e_2^2, \tag{15.12}$$

which leads to the following inequality:

$$\dot{E}(t) \leq -ke_1^2 - \frac{km}{l} e_1 e_2 - 2\xi e_2^2 \tag{15.13}$$

and

$$-ke_1^2 - \frac{km}{l} e_1 e_2 - 2\xi e_2^2 = -k(e_1 - \frac{m}{2l} e_1)^2 - (2\xi - \frac{km^2}{4l^2}) e_2^2.$$

When we take $200\xi - \frac{km^2}{4l^2} < 0$, $E(t)$ is a negative definite function and by Lyapunov direction method, the origin of the error system is asymptotically stable and $e \to 0$ as $t \to 0$, hence the synchronization of two SD systems (15.8) and (15.9) is also achieved.

In the following simulation, the fourth-order Runge–Kutta method was used to solve the systems of differential equations (15.8) and (15.11). In addition, a variable step size was employed. The results of this simulation is given by Fig. 15.4. Figure 15.4a displays the time series x_1 (the dashed line) generated in the driving system (15.8) feedback loop together with the variable y_1 (the solid line) of the response system (15.9). Figure 15.4b displays the behaviour of the trajectories $e_1 = y_1 - x_1$ of the error system which is seen as tending to zero. Accordingly, we may note that the response system variables synchronize their drive counterpart. The asymptotic stability of the stationary point (0, 0) of the difference $(e_1, e_2) \equiv (y_1 - x_1, y_2 - x_2)$ gives the theoretical proof of the synchronization. It is represented on Figs. 15.3 and 15.4.

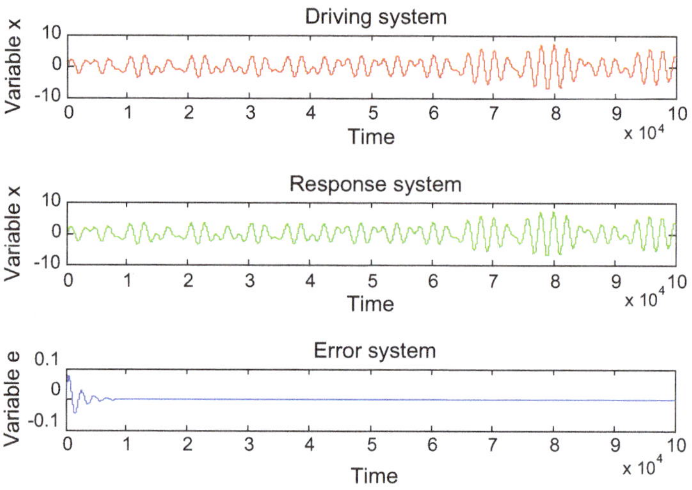

Fig. 15.3 Calculation of the difference

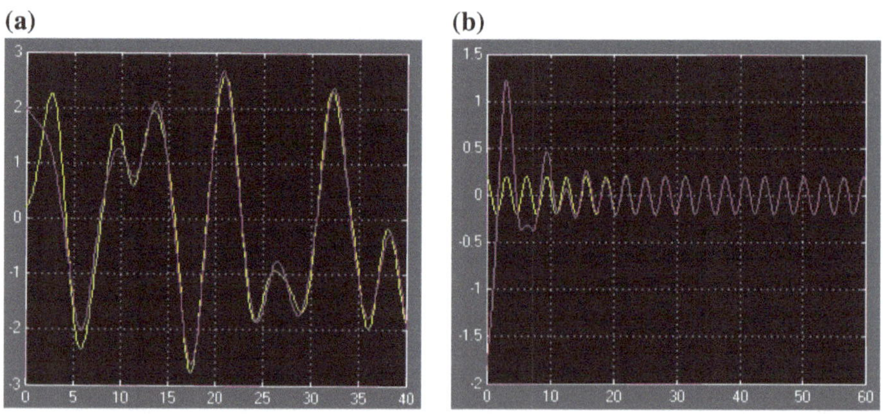

Fig. 15.4 **a** The chaotic signal x_1 generated in the driving system (Eq. (15.8)) feedback loop and y_1 in the response system (Eq. (15.9)). **b** The origin is seen to be an asptotically stable stationary state for the difference $(y_1 - x_1)$

15.3 The Compound Non-autonomous Remodulating Communication

To overcome the security problem of most traditional chaos-based secure communication schemes, the new compound non-autonomous remodulating (CNAR) communication scheme is proposed here combining two countermeasures to get more security. Let us consider the sender and the receiver systems respectively as follows:

The sender

$$\begin{cases} \dot{x} = G(x, y, m(t)), \\ \dot{y} = H(x, y, m(t)), \\ x_t = \alpha(x + m(t)) + \beta u(y), \end{cases} \tag{15.14}$$

and the receiver

$$\begin{cases} x_d = \dfrac{x_t - \beta u(y')}{\alpha}, \\ \dot{x}' = g(x_d, x', y'), \\ \dot{y}' = h(x_d, x', y'), \\ m'(t) = x_d - x'. \end{cases} \tag{15.15}$$

The information signal $m(t)$ is coded with the chaotic signals x and y using a coding function $x_t = \alpha(x + m(t)) + \beta u(y)$, so that the information signal can be decoded uniquely through $m'(t) = x_d - x'$. The main idea of the CNAR proposed method is to modulate the chaotic carrier through the signals generated by other variables of the sender system.

15.4 Secure Communication System Using the SD System

Based upon the SD oscillator, the CNAR scheme can be build up with the sender and receiver written in the explicit form:

$$\begin{cases} \dot{x} = y + km(t), \\ \dot{y} = -x - \xi y + \dfrac{x + m(t)}{\sqrt{(x + m(t))^2 + \alpha^2}} + f_0 \cos(\omega z), \\ \dot{z} = 1, \\ x_t = l_1[x + m(t)] + l_2 y; \end{cases} \tag{15.16}$$

and

$$\begin{cases} x_d = \dfrac{x_t - l_2 y_1}{l_1} \\ \dot{x}_1 = y_1 - k(x_1 - x_d) \\ \dot{y}_1 = -x_1 - 2\xi y_1 + \dfrac{x_d}{\sqrt{x_d^2 + \alpha^2}} + f_0 \cos(\omega z_1), \\ \dot{z}_1 = 1, \\ m'(t) = x_d - x_1. \end{cases} \tag{15.17}$$

Setting $e = (e_1, e_2), e_1 = x_1 - x, e_2 = y_1 - y, z_1 = z$, the difference of Equations of (15.16) and (15.17) satisfies the system

$$\begin{cases} \dot{e}_1 = -ke_1 + \left(1 - \dfrac{kl_2}{l_1}\right) e_2, \\ \dot{e}_2 = -e_1 - 2\xi e_2 + \left(\dfrac{x_d}{\sqrt{x_d^2 + \alpha^2}} - \dfrac{x + m(t)}{\sqrt{x + m(t)^2 + \alpha^2}} \right). \end{cases} \tag{15.18}$$

The proof of system (15.18) is the same as for the above system (15.11). In the two following cases:

1. $\dfrac{kl_2}{l_1} - 1 = 0, l_1 l_2 > 0, k > 0;$
2. $-ke_1^2 + \dfrac{kl_2}{l_1} e_1 e_2 - 2\xi e_2^2 < 0.$

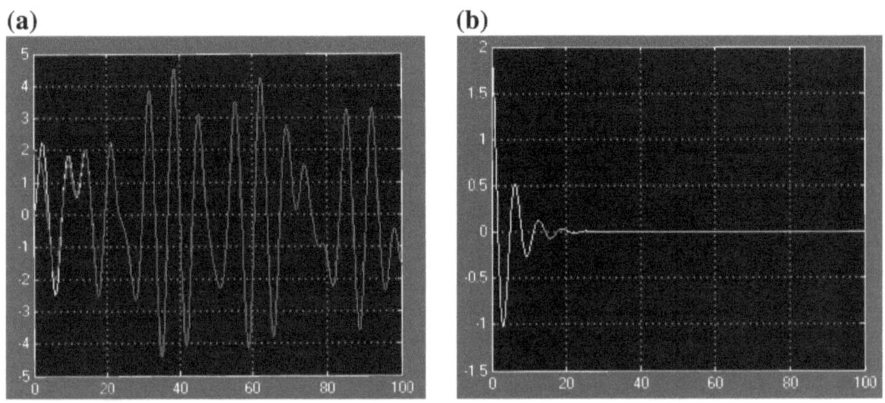

Fig. 15.5 **a** Segments of the transmitted signal $x_t = l_1[x + m(t)] + l_2 y$. Here the parameters are respectively chosen as $\xi = 0.01415$, $k = 2.0$, $l_1 = 1.0$ and $l_2 = 1.0$. **b** The real line denotes the message signal $m(t) = 0.1 sin(10\pi t + \frac{\pi}{2})$ and the signal which is recovered

The origin of the error system is (uniformly) asymptotically stable and $e \rightarrow 0$ as $t \rightarrow \infty$, Thus the trajectories are such that $x_1 \rightarrow x$, $y_1 \rightarrow y$ as $t \rightarrow \infty$.

Figure 15.5a, b show respectively computer simulations of the transmitted signal x_t and the time waveform of the message information $m(t)$, and then the recovered information $m'(t) = x_d - x_1$. We see that, after some transient behaviour, the recovered waveform $m'(t)$ approaches the sine wave $m(t)$.

15.5 Summary

This chapter has proposed a secure communication scheme based upon the chaotic behaviour of SD oscillator. Even only the theoretical foundations are presented based on non-autonomous compound remodulating scheme, the potential for secure communication scheme based SD oscillator is possible. Herein this chapter, we have shown that SD oscillator may satisfy the requirements of security of communications using its chaotic characteristics depending on SD return maps analysis. In the proposed scheme, the driving signal is generated by the combination of multiple state variables. This scheme not only breaks cryptanalysis from the phase space reconstruction of the transmitted dynamics, but also enlarges the key space of the cryptographic system. The investigation showed that the return maps of SD oscillator as a cryptosystem are chaotic, which enables the robust security of the secure communication schemes. The synchronization performance of the scheme is also supported by the mathematical conclusions and simulation results of SD oscillator, which suggests that an implementation is possible for a more secure chaotic communication scheme.

References

1. Lai, Y. C., & Grebogi, C. (1993). Synchronization of chaotic trajectories using control. *Physical Review E Statistical Physics Plasmas Fluids and Related Interdisciplinary Topics*, *47*(4), 2357–2360.
2. Halle, K. S., Wu, C. W., Itoh, M., & Chua, L. O. (1993). Spread spectrum communication through modulation of chaos in Chua's circuit. *International Journal of Bifurcation and Chaos*, *3*(2), 469–477.
3. Pecora, L. M., & Caroll, T. L. (1990). Synchronization in chaotic system. *Physical Review Letters*, *64*(8), 821–824.
4. Yang, T., & Chua, L. O. (1996). Secure communication via chaotic parameter modulation. *IEEE Transactions on Circuits and Systems I: Fundamental Theory and Applications*, *43*(9), 817–819.
5. Ding, M., & Ott, E. (1994). Enhancing synchronization of chaotic systems. *Physical Review E*, *49*(2), 945–948.
6. Murali, K., & Lakshmanan, M. (1997). Synchronization through compound chaotic signal in Chua's circuit and Murall-Lakshmanan-Chua circuit. *International Journal of Bifurcation and Chaos*, *7*(2), 1619–1627.

7. Wu, X., Hu, H., & Zhang, B. (2004). Analyzing and improving a chaotic encryption method. *Chaos, Soliton and Fractals*, *22*(4), 367–373.
8. Alvarez, G., & Li, S. J. (2006). Some basic cryptographic requirements for chaos-based cryptosystems. *Internatioal Journal of Bifurcation and Chaos*, *16*(8), 2129–2151.
9. Alvarez, G., Montoya, F., Romera, M., & Pastor, G. (2004). Breaking parameter modulated chaotic secure communication system. *Chaos Solitons and Fractals*, *21*(4), 783–787.
10. Li, S., Alvarez, G., & Chen, G. (2005). Breaking a chaos-based secure communication scheme designed by a improved modulation method. *Chaos, Soliton and Fractals*, *25*(1), 109–120.
11. Senouci, A., Boukabou, A., Busawon, K., Bouridane, A., & Ouslimani, A. (2014). Robust chaotic communication based on indirect coupling synchronization. *Circuits Systems and Signal Processing*, *34*(2), 393–418.
12. Wang, X. Y., & Zhang, H. (2013). A robust secondary secure communication scheme based on synchronization of spatiotemporal chaotic systems. *Zeitschrift Fur Naturforschung A*, *68*, 573–580.
13. Chandrasekaran, J., Subramanyan, B., & Selvanayagam, R. (2011). *A chaos based approach for improving non linearity in S box design of symmetric key cryptosystems*. Berlin: Springer.
14. Matsumoto, T., Chua, L. O., Kobayashi, K. (1986). Hyperchaos: Laboratory experiment and numerical confirmation. *IEEE Transactions on Circuits and Systems*, *CAS-33*(11), 1143–1147.
15. Murali, K. (2000). Heterogeneous chaotic systems based cryptography. *Physics Letters A*, *272*(3), 184–192.
16. Murali, K. (2001). Digital signal transmission with cascaded heterogeneous chaotic systems. *International Journal of Bifurcation and Chaos*, *63*(10), 2489–2497.
17. Freeman, G. A. (2012). Synchronizing chest compression and ventilation in cardiac resuscitation.
18. Wang, Q. Y., Lu, Q. S., & Chen, G. R. (2008). Synchronization transition induced by synaptic delay in coupled fast-spiking neurons. *International Journal of Bifurcation and Chaos*, *18*(4), 1189–1198.
19. Hsiao, F. H., Hsieh, K. P., Lin, Z. H. (2014). Exponential optimal synchronization of chaotic cryptosystems: Neural-network-based approach. In *2014 International Conference on Cyber-Enabled Distributed Computing and Knowledge Discovery (CyberC)* (pp. 35–42).
20. Okura, J. (1999). Time-base corrector circuit capable of adding color burst signal to output signal based on kind of input video signal.
21. Zhang, R., Hu, M., & Xu, Z. Y. (2007). Impulsive synchronization of rø"sler systems with parameter driven by an external signal. *Physics Letters A*, *364*(3–4), 239–243.
22. Li, S. J., Chen, G. R., & Alvarez, G. (2005). Return-map cryptanalysis revisited. *International Journal of Bifurcation and Chaos*, *16*(5), 1557–1568.
23. Perez, G., & Cerdeira, H. A. (1995). Extracting messages masked by chaos. *Physical Review Letters*, *74*(11), 1970–1973.
24. Guedes de Oliveira, A., & Jones, A. J. (2011). Synchronization of chaotic maps by feedback control and application to secure communications using chaotic neural networks. *International Journal of Bifurcation and Chaos*, *8*(11), 2225–2237.
25. Yang, T., Yang, L. B., & Yang, C. M. (1998). Breaking chaotic secure communications using a spectogram. *Physics Letters A*, *247*(1), 105–111.
26. Alvarez, G., & Li, S. J. (2004). Estimating short-time period to break different types of chaotic modulation based secure communications. *Annales de L'Institut Henri Poincaré*, *22*(5), 597–608.
27. Jovic, B. (2011). *Application of chaotic synchronization to secure communications*. Berlin: Springer.
28. Dedieu, H., Kennedy, M. P., & Hasler, M. (1993). Chaos shift keying: Modulation and demodulation of a chaotic carrier using self-synchronizing. *IEEE Transactions on Circuits and Systems II: Analog and Digital Signal Processing*, *40*(10), 634–641.
29. Skufca, J. D., & Bollt, E. M. (2004). Communication and synchronization in disconnected networks with dynamic topology: Moving neighborhood networks. *Mathematical Biosciences and Engineering MBE*, *1*(2), 347–359.

30. Paul, R. S., Rajasekar, S., & Murali, K. (1999). Coexisting chaotic attractors, their basin of attractions and synchronization of chaos in two coupled duffing ocsilattors. *Physics Letters A*, *264*(4), 283–288.
31. Chen, J. Y., Wong, K. W., Cheng, L. M., & Shuai, J. W. (2003). A secure communication scheme based on the phase synchronization of chaotic systems. *Chaos*, *13*(2), 508–514.
32. Fallahi, K., Raoufi, R., & Khoshbin, H. (2008). An application of Chen system for secure chaotic communication based on extended Kalman filter and multi-shift cipher algorithm. *Communications in Nonlinear Science and Numerical Simulation*, *13*(4), 763–781.
33. Grzybowski, J. M. V., Rafikov, M., & Balthazar, J. M. (2009). Synchronization of the unified chaotic system and application in secure communication. *Communications in Nonlinear Science and Numerical Simulation*, *14*(6), 2793–2806.

Chapter 16
Challenges and Open Problems

This book presents a smooth and discontinuous (SD) oscillator with fundamental theory, methodology and applications.

The fundamental theory followed by the introduction of the SD oscillator builds up the generalized Hooke's law at the origin. A hyperbolic transition from a single stability to a bistability and the transition from smooth to discontinuous dynamics is presented depending on the variation of a geometrical parameter. These properties result in abundant nonlinear phenomena including equilibrium bifurcations, co-dimension two bifurcations, periodic resonances and chaotic attractors.

The methodologies proposed for the SD oscillator aim at obtaining analytical solutions, semi-analytical solutions and computational schemes for both smooth and discontinuous oscillators. The irrational elliptic functions give analytical solutions for the unperturbed system, the extended four dimensional averaging procedure provides periodic behaviours with their stability and the computational schemes based on the triple linear approach or the cell-mapping method offer a better understanding of the oscillator through direct computations.

From the point of view of the applications, the SD oscillator is an abstract form of a nonlinear mechanical system. The applications involve a wide range among physics and engineering disciplines, as suggested by the few examples presented in this book, such as the very basic experimental researches about mechanical and electronic SD oscillators, the rotating SD oscillator and its archeological application, and the SD oscillator with friction and impacts as well.

The interesting application presented in this book about Ground Vibration Test for large scale machineries with low or ultra low required frequency, will be left to the readers' consideration in the list of open problems.

Many open or challenging problems remain to be tackled. Let us now close this book by a provisional list of such subjects.

Q. Cao and A. Léger, *A Smooth and Discontinuous Oscillator*, Springer Tracts in Mechanical Engineering, DOI 10.1007/978-3-662-53094-8_16

16.1 Theoretical Subjects for Dynamical Systems

- Theoretically, the traditional nonlinear dynamics answered successfully the question addressed in the introduction, Chap. 1 about the efficiency of finite truncations of a nonlinear system. If the system satisfies the assumptions of being sufficiently smooth locally with finite codimension, then the local dynamics is qualitatively described by the universal unfolding of the truncated normal form with finite number of unfolding parameters. There is no general theory for the case where one of these assumptions fails among that of smoothness, local assumption and finite codimension. Even when the system satisfies these three assumptions, there are no general tools for calculating the exact dynamics.
- Introducing irrational elliptic functions for both the smooth and the discontinuous cases provides the way to get an analytical approach to the unperturbed system. There still remains an open problem of how to investigate the further properties of these functions including the imaginary periodicity to pave a way for a deep understanding of the theory and properties of irrational nonlinear systems that occur in engineering.
- As it has been seen in a previous chapter, the construction of the isolator with the SD oscillator can provide a negative vertical stiffness and cancel the weight so as to make up a quasi-zero stiffness. This property enables the benefits of lowering the frequency and enlarging the stroke of low stiffness of the system. The open problem is how to enlarge the soft stroke into the maximum interval as required and how to extend the vibration isolation frequency to the ultra low frequency as needed to make the best use of higher order stable quasi-zero stiffness.
- The micro-vibration environment is in particular needed in engineering applications, such as laser beam travelling to the moon, photography with long distance of fine resolution and rocket launching platform, as a small error may rise to a disaster difference. The open problem here is how to construct a stable quasi-zero stiffness environment for the isolated object in six degree of freedom by using the composed SD oscillator with a ultra low frequency and maximal stroke with low stiffness.

16.2 Material Properties and Coupling of Nonlinearities

- Hooke's law gives the linear elastic deformation of a system submitted to applied forces. Using SD oscillator in engineering applications has been shown to produce devices with a negative stiffness or a stable quasi-zero stiffness. The property of having a negative stiffness results from the fact that the full geometrical nonlinearity leads to hyperbolicity broken at the degenerate point at $\alpha = 1$ where the transition happens from a single stability to a bistability. Being able to build special devices with multiple stability might lead to high degree of degeneration. This might have a potential to construct higher order stable quasi-zero stiffness system to reduce the isolation frequency and enlarge the soften stroke.

- The nonlinearity of the SD oscillator originates from the geometrical configuration of the linear springs, which is usually known as the geometrical nonlinear system. A future generalization would be the introduction of nonlinear springs. Nonlinear elasticity, obtained by a polynomial stress-strain relation in place of Hooke's law for example, would probably not bring qualitatively new properties than those investigated here. But changing the elastic springs into irreversible links such as elastic plastic or damageable springs, might be of the utmost importance and might lead to new properties.
- The introduction of the SD oscillator starts from the first buckling mode of an arch. In fact, the SD oscillator got an autonomy as a new oscillator of one degree of freedom with large deformation. Coming back to a continuous body from the SD oscillator with large strains would be next application to simulate multiple modes with large deformation: an arch can be constructed by a line of SD oscillators, a shell can be composed by a two dimensional series of SD oscillators and a body can be formulated as layers of shells of SD oscillators. This might lead to a new material close to those introduced as meta-materials, with a miro-scale negative stiffness or stable quasi-zero stiffness property. This is open to all for the formulation and the properties of the material.

16.3 Problems Requiring Modeling

- The negative stiffness property of the restoring force gives the possibility of the stable quasi zero stiffness, which provides an opportunity to construct a nonlinear isolation system with low isolation frequency. A model involving quasi zero stiffness of higher order would probably lead to more efficient systems, specially in the domain of isolation against excitations or impacts. Suggesting systems with a stable quasi-zero stiffness of order high enough to meet the requirements of the isolation application is addressed to the readers.
- As presented in Chap. 11, many nonlinear phenomena happen when the mass of the SD oscillator is subjected with friction and impact. A challenging subject is whether a model based upon the whole knowledge of the SD oscillator with moving contact may give a good understanding of multiple stick-slip, or related behaviours such as sudden jumps or debonding.
- Dynamics of lattices has widely been studied at small strains, but taking geometrical nonlinearities might give understanding of new phenomena, specially in micro or quantum physics. 1-D, 2D and 3D models could be constructed with SD oscillators to describe the motion of a collection of particles, which could result in new properties of crystals. Laser-solid interaction requires a modeling within the framework of quantum mechanics, and SD oscillators might lead to the necessary understanding of the entanglement mechanism with multiple stability.

Erratum to: A Smooth and Discontinuous Oscillator

Qingjie Cao and Alain Léger

Erratum to:
Q. Cao and A. Léger, *A Smooth and Discontinuous Oscillator*, Springer Tracts in Mechanical Engineering, DOI 10.1007/978-3-662-53094-8

The book was inadvertently published without updating the belated corrections from one of the co-authors of the book. The erratum chapters and the book have been updated with the changes.

The original version of the book was revised: Author belated corrections have been updated throughout the book. The erratum PDF is available at
DOI 10.1007/978-3-662-53094-8

Q. Cao (✉)
School of Astronautics, Harbin Institute of Technology, Harbin, China
e-mail: Qingjiecao@hotmail.com

A. Léger
Laboratoire de Mécanique et d'Acoustique, CNRS, Marseille Cedex 20, France

Q. Cao and A. Léger, *A Smooth and Discontinuous Oscillator*, Springer Tracts in Mechanical Engineering, DOI 10.1007/978-3-662-53094-8_17

Appendix A

The first integral in (7.21) can be written as

$$-2\xi\int_{-\infty}^{\infty}(y_{\pm}(\tau))^2\,d\tau = -4\xi\int_{0}^{\infty}(y_{\pm}(\tau))^2\,d\tau$$

$$= -4\xi\int_{x_0}^{0}(\omega_1 x(\tau))\,dx - 4\xi\int_{x_c}^{x_0}\sqrt{R^2-\omega_2^2\left(x_{\pm}-\sqrt{1-\alpha^2}\right)^2}\,dx$$

$$= 2\xi\left[\omega_1 x_0\sqrt{1-\alpha^2}+\frac{R^2}{\omega_2}\left(\arccos\frac{\omega_2\left(x_0-\sqrt{1-\alpha^2}\right)}{R}-\frac{\pi}{2}\right)\right].$$

and the second integral as

$$\int_{-\infty}^{\infty}\left[y_{\pm}(\tau) f_0\cos\omega(\tau+\tau_1)\right]d\tau = f_0\int_{-\infty}^{\infty}y_{\pm}(\tau)\cos\omega(\tau+\tau_1)d\tau$$

$$= f_0\left(\int_{-\infty}^{-\tau_0}+\int_{-\tau_0}^{0}+\int_{0}^{\tau_0}+\int_{\tau_0}^{\infty}\right)y_{\pm}(\tau)\cos\omega(\tau+\tau_1)d\tau$$

$$= f_0\omega_1 x_0\int_{-\infty}^{-\tau_0}\cos\omega(\tau+\tau_1)e^{\omega_1(\tau+\tau_0)}d\tau - f_0R\int_{-\tau_0}^{0}\cos\omega(\tau+\tau_1)\sin\omega_2 d\tau$$

$$-f_0R\int_{0}^{\tau_0}\cos\omega(\tau+\tau_1)\sin\omega_2\tau d\tau - f_0\omega_1 x_0\int_{\tau_0}^{\infty}\cos\omega(\tau+\tau_1)e^{-\omega_1(\tau-\tau_0)}d\tau.$$

Determining the following integrals from the previous steps

$$\int_{-\tau_0}^{\tau_0}\cos\omega(\tau+\tau_1)\sin\omega_2\tau d\tau = \int_{-\tau_0}^{\tau_0}(\cos\omega\tau_1\cos\omega\tau\sin\omega_2\tau - \sin\omega\tau_1\sin\omega\tau\sin\omega_2\tau)\,d\tau$$

Q. Cao and A. Léger, *A Smooth and Discontinuous Oscillator*, Springer Tracts in Mechanical Engineering, DOI 10.1007/978-3-662-53094-8

$$= \frac{\sin \omega\tau_1}{\omega^2 - \omega_2^2} \left[(\omega + \omega_2)\sin(\omega - \omega_2)\tau_0 - (\omega - \omega_2)\sin(\omega + \omega_2)\tau_0\right],$$

$$x_0 e^{\omega_1\tau_0} \int_{-\infty}^{-\tau_0} \cos\omega(\tau + \tau_1) e^{\omega_1\tau} d\tau = x_0 e^{\omega_1\tau_0} \int_{-\infty}^{-\tau_0} (\cos\omega\tau_1 \cos\omega\tau - \sin\omega\tau_1 \sin\omega\tau)\, e^{\omega_1\tau} d\tau$$

$$= \frac{x_0}{\omega^2 + \omega_1^2} (\cos\omega\tau_1 (-\omega\sin\omega\tau_0 + \omega_1\cos\omega\tau_0) + \sin\omega\tau_1 (\omega\cos\omega\tau_0 + \omega_1 \sin\omega\tau_0)),$$

$$x_0 e^{\omega_1\tau_0} \int_{\tau_0}^{\infty} \cos\omega(\tau + \tau_1) e^{-\omega_1\tau} d\tau = x_0 e^{\omega_1\tau_0} \int_{\tau_0}^{\infty} (\cos\omega\tau_1 \cos\omega\tau - \sin\omega\tau_1 \sin\omega\tau)\, e^{-\omega_1\tau} d\tau$$

$$= \frac{x_0}{\omega^2 + \omega_1^2} (\cos\omega\tau_1 (-\omega\sin\omega\tau_0 + \omega_1\cos\omega\tau_0) - \sin\omega\tau_1 (\omega\cos\omega\tau_0 + \omega_1 \sin\omega\tau_0)),$$

and substituting these terms into (7.21) allows to write finally the Melnikov function as

$$M_\pm(\tau_1) = 2\xi\Omega(\omega_1, \omega_2, x_0, R) \pm f_0 \sin\omega\tau_1 (\Psi_1(\omega, \omega_2, \tau_0, R) + \Psi_2(\omega, \omega_1, x_0, \tau_0))$$
$$= 2\xi\Omega(\omega_1, \omega_2, x_0, R) \pm f_0\Psi(\omega, \omega_1, \omega_2, \tau_0, R)\sin\omega\tau_1, \tag{A.1}$$

where

$$\begin{aligned}
\Omega(\omega_1, \omega_2, x_0, R) &= \omega_1 x_0\sqrt{1-\alpha^2} + \frac{R^2}{\omega_2}\left(\arccos\frac{\omega_2\left(x_0 - \sqrt{1-\alpha^2}\right)}{R} - \frac{\pi}{2}\right),\\
\Psi(\omega, \omega_1, \omega_2, \tau_0, R) &= \Psi_1(\omega, \omega_2, \tau_0, R) + \Psi_2(\omega, \omega_1, x_0, \tau_0),\\
\Psi_1(\omega, \omega_2, \tau_0, R) &= \frac{R}{\omega^2 - \omega_2^2}\left[(\omega - \omega_2)\sin(\omega + \omega_2)\tau_0 - (\omega + \omega_2)\sin(\omega - \omega_2)\tau_0\right],\\
\Psi_2(\omega, \omega_1, x_0, \tau_0) &= \frac{2x_0}{\omega^2 + \omega_1^2}(\omega\cos\omega\tau_0 + \omega_1\sin\omega\tau_0).
\end{aligned} \tag{A.2}$$

It is noticed that, for the limit case, $\alpha = 0$, there follows $\Omega(\omega_1, \omega_2, x_0, R) = \frac{\pi}{2}$, and

$$\Psi_1(\omega, \omega_2, \tau_0, R) = \frac{R}{\omega^2 - 1}\left[(\omega - 1)\sin(\omega + 1)\pi - (\omega + 1)\sin(\omega - 1)\pi\right]. \tag{A.3}$$

Appendix B

This short appendix just aims at getting the pole of the hyperbolic function $tsd(\tau)$ of Eq. (9.21).

Using the half-angle formulae, such as

$$t = \tan\frac{\varphi}{2}, \qquad \frac{d\varphi}{dt} = \frac{2}{1+t^2},$$

$$\cos\varphi = \frac{1-t^2}{1+t^2}, \qquad \sin\varphi = \frac{2t}{1+t^2},$$

integral (9.21) is rewritten as

$$\tau = 2\int_0^t \frac{1+t^2+(1-\alpha)(1-t^2)}{(1+t^2)\sqrt{(1+t^2+(1-\alpha)(1-t^2))^2-(1+t^2)^2\alpha^2}}dt. \tag{B.1}$$

Whittaker [1] proved that the integral of the motion is still real when t is replaced by $\sqrt{-1}t$ and the initial conditions $\beta_1 \cdots \beta_n$ by $\sqrt{-1}\beta_1 \cdots \sqrt{-1}\beta_n$, respectively, if the force is independent of time. The expression thus obtained represents the same motion with the same initial condition of the system. In this way, the variable is extended to the imaginary axis as follows: Setting

$$\tau = iu, \qquad t = is,$$

integral (B.1) is rewritten as

$$u = 2\int_0^s \frac{1-s^2+(1-\alpha)(1+s^2)}{(1-s^2)\sqrt{(1-s^2+(1-\alpha)(1+s^2))^2-(1-s^2)^2\alpha^2}}ds, \tag{B.2}$$

which leads to the following definitions.

Q. Cao and A. Léger, *A Smooth and Discontinuous Oscillator*, Springer Tracts in Mechanical Engineering, DOI 10.1007/978-3-662-53094-8

$$tsd(\tau, \alpha) = tsd(iu, \alpha) = \frac{2t}{1+t^2} = i\frac{2s}{1-s^2} = i\frac{tsd(u, \alpha)}{tad(u, \alpha)}. \tag{B.3}$$

Let us denote

$$H(\alpha) = 2\int_0^1 \frac{1-s^2+(1-\alpha)(1+s^2)}{(1-s^2)\sqrt{(1-s^2+(1-\alpha)(1+s^2))^2-(1-s^2)^2\alpha^2}}ds, \tag{B.4}$$

then $u = H(\alpha)$ when $s = 1$, and $tsd(\tau, \alpha)$ has a pole $\tau = iH(\alpha)$ when $s = 1$.

Reference

1. Whittaker, E. T. (1937). *A Treatise on the analytical dynamics of particles and rigid bodies* (4th ed.). Cambridge: Cambridge University Press.

Index

Q. Cao and A. Léger, *A Smooth and Discontinuous Oscillator*, Springer Tracts in Mechanical Engineering, DOI 10.1007/978-3-662-53094-8

MIX
Papier | Fördert gute Waldnutzung
FSC® C083411

Kolibri 360 GmbH
Ferdinand-Jühlke-Straße 7
99095 Erfurt, Deutschland
produktsicherheit@kolibri360.de